Efficient Sensor Interfaces, Advanced Amplifiers and Low Power RF Systems

Kofi A.A. Makinwa • Andrea Baschirotto
Pieter Harpe
Editors

Efficient Sensor Interfaces, Advanced Amplifiers and Low Power RF Systems

Advances in Analog Circuit Design 2015

Springer

Editors
Kofi A.A. Makinwa
Department of Microelectronics
Delft University of Technology
Delft, The Netherlands

Andrea Baschirotto
Department of Physics "G. Occhialini"
University of Milan
Milano, Italy

Pieter Harpe
Department of Electrical Engineering
Eindhoven University of Technology
Eindhoven, The Netherlands

ISBN 978-3-319-21184-8 ISBN 978-3-319-21185-5 (eBook)
DOI 10.1007/978-3-319-21185-5

Library of Congress Control Number: 2015949137

Springer Cham Heidelberg New York Dordrecht London

Printed on acid-free paper

Springer International Publishing AG Switzerland is part of Springer Science+Business Media (www.springer.com)

Preface

This book is part of the Analog Circuit Design series and contains the contributions from all 18 speakers of the 24th workshop on Advances in Analog Circuit Design (AACD). The local chairs were Christian Enz (Ecole polytechnique fédérale de Lausanne (EPFL), Neuchâtel) and Alain-Serge Porret (Swiss Center for Electronics and Microtechnology (CSEM), Neuchâtel). This year, the sponsors of the workshop were Nano-Tera, Melexis, IEEE Switzerland Section, Republique et Canton de Neuchâtel, and Ville de Neuchâtel. The workshop was held at the Institute of Microengineering (IMT) of EPFL in Neuchâtel, Switzerland, from April 21 to 23, 2015.

The book comprises three parts, covering topics in advanced analog and mixed-signal circuit design that we consider to be of great interest to the circuit design community:

- Efficient Sensor Interfaces
- Advanced Amplifiers
- Low-Power RF Systems

Each part consists of six chapters written by experts in the field.

The aim of the AACD workshop is to bring together a group of expert designers to discuss new developments and future options. Each workshop is then followed by the publication of a book by Springer as part of their successful series on Analog Circuit Design. This book is the 24th in this series (a full list of the previous topics can be found on the following page). The series can be seen as a reference for all people involved in analog and mixed-signal design.

We are confident that this book, like its predecessors, will prove to be a valuable contribution to our analog and mixed-signal circuit design community.

Delft, The Netherlands — Kofi A. A. Makinwa
Milano, Italy — Andrea Baschirotto
Eindhoven, The Netherlands — Pieter Harpe

The Topics Covered Before in This Series

2014	Lisbon (Portugal)	High-Performance AD and DA Converters
		IC Design in Scaled Technologies
		Time-Domain Signal Processing
2013	Grenoble (France)	Frequency References
		Power Management for SoC
		Smart Wireless Interfaces
2012	Valkenburg (The Netherlands)	Nyquist A/D Converters
		Capacitive Sensor Interfaces
		Beyond Analog Circuit Design
2011	Leuven (Belgium)	Low-Voltage Low-Power Data Converters
		Short-Range Wireless Front-Ends
		Power Management and DC-DC
2010	Graz (Austria)	Robust Design
		Sigma Delta Converters
		RFID
2009	Lund (Sweden)	Smart Data Converters
		Filters on Chip
		Multimode Transmitters
2008	Pavia (Italy)	High-Speed Clock and Data Recovery
		High-Performance Amplifiers
		Power Management
2007	Oostende (Belgium)	Sensors, Actuators and Power Drivers for the Automotive and Industrial Environment
		Integrated PAs from Wireline to RF
		Very High Frequency Front Ends
2006	Maastricht (The Netherlands)	High-Speed AD Converters
		Automotive Electronics: EMC Issues
		Ultra Low Power Wireless

2005	Limerick (Ireland)	RF Circuits: Wide Band, Front-Ends, DACs
		Design Methodology and Verification of RF and Mixed-Signal Systems
		Low Power and Low Voltage
2004	Montreux (Swiss)	Sensor and Actuator Interface Electronics
		Integrated High-Voltage Electronics and Power Management
		Low-Power and High-Resolution ADCs
2003	Graz (Austria)	Fractional-N Synthesizers
		Design for Robustness
		Line and Bus Drivers
2002	Spa (Belgium)	Structured Mixed-Mode Design
		Multi-bit Sigma-Delta Converters
		Short-Range RF Circuits
2001	Noordwijk (The Netherlands)	Scalable Analog Circuits
		High-Speed D/A Converters
		RF Power Amplifiers
2000	Munich (Germany)	High-Speed A/D Converters
		Mixed-Signal Design
		PLLs and Synthesizers
1999	Nice (France)	XDSL and Other Communication Systems
		RF-MOST Models and Behavioural Modelling
		Integrated Filters and Oscillators
1998	Copenhagen (Denmark)	1-V Electronics
		Mixed-Mode Systems
		LNAs and RF Power Amps for Telecom
1997	Como (Italy)	RF A/D Converters
		Sensor and Actuator Interfaces
		Low-Noise Oscillators, PLLs and Synthesizers
1996	Lausanne (Swiss)	RF CMOS Circuit Design
		Bandpass Sigma Delta and Other Data Converters
		Translinear Circuits
1995	Villach (Austria)	Low-Noise/Power/Voltage
		Mixed-Mode with CAD Tools
		Voltage, Current and Time References
1994	Eindhoven (The Netherlands)	Low-Power Low-Voltage
		Integrated Filters
		Smart Power
1993	Leuven (Belgium)	Mixed-Mode A/D Design
		Sensor Interfaces
		Communication Circuits
1992	Scheveningen (The Netherlands)	OpAmps
		ADCs
		Analog CAD

Contents

Part I
Efficient Sensor Interfaces

The first part of the book discusses recent developments in the design and implementation of efficient sensor interfaces. Driven by the gradual emergence of an internet of things, and the accompanying proliferation of battery-powered sensor systems, e.g. in mobile phones and wearable electronics, the energy/power efficiency of sensor interfaces has become an important performance metric. The papers in this section describe low-power incremental ADCs, low-power capacitor and temperature-to-digital interfaces, as well as low-power systems for the acquisition of biomedical signals.

The first paper, by Edoardo Bonizzoni et al., describes the design of a second-order 3-bit incremental ADC. A key feature is the use of a Smart-DEM algorithm, which shuffles the elements of the feedback DAC such that their mismatch is averaged in a manner that takes into account their time-dependent contribution to the ADC's output. Behavioral simulations show that, compared to traditional DEM schemes, this approach can lead to significantly better linearity. For a 1.42Vpp maximum input signal, a prototype converter achieves an SNDR (and SNR) of 102.4 dB into a 5 kHz bandwidth, which translates into a Schreier FoM of 175 dB.

The second paper, by Chia-Hung Chen, Gabor Temes, et al., describes the design of a two-step incremental ADC, which first performs a normal second order conversion, and is then reconfigured to perform a first order conversion of the resulting residue, present at the output of the second integrator. This approach results in performance close to that of a third order incremental ADC, while using less hardware. For a 2.2Vpp maximum input signal, the converter achieves a measured dynamic range of 99.8 dB and an SNDR of 91 dB in a bandwidth of 250 Hz. This translates into a Schreier FOM of 173.5 dB.

The third paper, by David Blaauw, Sechang Oh, et al., describes several novel capacitance-to-digital converters (CDCs). By combining techniques such as correlated-double-sampling, base-line cancellation and zooming with incremental delta-sigma, successive approximation and single/dual-slope ADC architectures, these CDCs are able to achieve 7–14 bit resolution while dissipating hundreds of nWs to tens of μWs, respectively. The corresponding energy-efficiencies range from 0.2 to 5 pJ/conversion-step, respectively, which begins to approach the efficiency of conventional ADCs.

The fourth paper, by Samira Zaliasl, Aaron Partridge, et al., describes the design of a low-power temperature-to-digital converter (TDC) which is used for the temperature compensation of an ultra-low-power MEMS-based 32 kHz oscillator. Based on NPN sensing elements, a 15-bit incremental ADC and a novel DEM scheme, the TDC (and its digital backend) draw 4.5 μA from a 1.8 V supply, while achieving a resolution of 25 mK in a conversion time of 6 ms. This corresponds to a resolution FoM of 24 pJK^2, one of the lowest ever reported for a commercial design.

The fifth paper by, Jiawei Xu, Firat Yazicioglu, et al., provides an overview of recent advances in the design of energy-efficient instrumentation amplifiers intended for wearable and implantable biomedical systems. Such amplifiers must have low-noise, high CMRR and must be able to cope with low-frequency aggressors such as electrode offset, motion artefacts, flicker noise and power-line hum. Recent designs can do all this while drawing only a few μA and achieving noise-efficiency factors well below ten.

The last paper, by Mahsa Shoaran and Alexander Schmid, describes the first fully-integrated system capable of realizing multi-channel compressed-domain feature extraction. Intended for use in cortical implants for the detection of the onset of epileptic seizures, the system consists of 16 channels, which each operate at sub-μW levels and occupy an active area of 250 μm by 250 μm.

Smart-DEM for Energy-Efficient Incremental ADCs

Edoardo Bonizzoni, Yao Liu, and Franco Maloberti

Abstract This paper describes a dynamic element matching (DEM) algorithm, the so-called Smart-DEM (SDEM) algorithm, suitable for multi-bit incremental converters. The effectiveness of the algorithm is studied and compared with conventional DEM methods at the behavioural level with the help of two case studies. The paper, moreover, presents the design and the experimental verification of a second-order 3-bit incremental converter which employs the SDEM algorithm to compensate for the mismatch among unity elements of its multi-level digital-to-analog converter (DAC). The circuit, fabricated in a mixed 0.18–0.5 μm CMOS technology, achieves a resolution of 16.7 bit over a 5-kHz bandwidth by using 256 clocks periods per sample.

1 Introduction

Instrumentation and measurement applications, such as the readout of bridge transducers and biomedical acquisition systems [1, 2], require monotonic analog-to-digital converters (ADCs) with high resolution, good linearity, low offset and, typically, low power. Incremental converters, directly derived from $\Sigma\Delta$ schemes, are particularly suitable for those needs. Although they share the same structure as a $\Sigma\Delta$ ADC, in an incremental ADC the integrators are reset at the start of each conversion.

As a result, no information about past samples can be used, and so there is no need to describe the quantization error as noise and even to talk about noise shaping.

E. Bonizzoni (✉) • F. Maloberti
Department of Electrical, Computer, and Biomedical Engineering, University of Pavia, Via Ferrata 5, 27100 Pavia, Italy
e-mail: edoardo.bonizzoni@unipv.it

Y. Liu
Department of Electrical, Computer, and Biomedical Engineering, University of Pavia, Via Ferrata 5, 27100 Pavia, Italy

Institute of Microelectronics and Microprocessor, National University of Defense Technology, Changsha, China

K.A.A. Makinwa et al. (eds.), *Efficient Sensor Interfaces, Advanced Amplifiers and Low Power RF Systems*, DOI 10.1007/978-3-319-21185-5_1

Thus an incremental ADC can be regarded as a Nyquist-rate converter. The input should be constant during the entire conversion, otherwise the digital result becomes a weighed average of the analog input, in a manner similar to that of a FIR filter.

The equivalent number of bits (ENOB) of an incremental converter depends on the order of the scheme, the number of clock cycles per sample, the resolution of the quantizer and the digital post processing. Generally, to obtain a given resolution, higher order modulators are more efficient since the required number of clock periods is reduced. However, when the order of the scheme is higher than two, stability requirements can limit the effectiveness of the architecture.

An incremental ADC can use single or multi-bit quantizers. Most designs adopt single-bit quantization [3–7]. In this case, since the DAC uses only 2 levels, it is inherently linear. Nevertheless, the relatively large output swing of integrators may result in operational amplifiers (op-amps) working in slewing mode. In addition, when the order of the scheme is larger than two, as discussed shortly, stability of the feedback loop demands the use of fractional coefficients in the signal path, which strongly degrades the conversion efficiency.

This paper focuses on the use of multi-bit quantization in incremental converters and on the use of the so-called Smart-DEM (SDEM) algorithm. The main advantages of this approach are: reduced op-amps output swing, higher conversion efficiency due to the possible use of larger (maximally equal to 1) coefficients in the signal path, and greater resolution to an extent equal to the number of bits used in the quantizer. On the other hand, however, the non-linearity introduced by the multi-level DAC severely degrades the converter performance, if not properly compensated for with a suitable technique.

In this paper the working principle of the SDEM algorithm is explained and its effectiveness is first tested and compared with conventional solutions at the behavioural level with the help of case studies. In the second part of the paper, the design of a second-order incremental converter that includes a 3-bit quantizer is described [8, 9]. The circuit, fabricated in a mixed 0.18–0.5-μm CMOS technology, uses 256 clock periods to achieve a resolution of about 17 bit. The mismatch of the multi-level DAC unity elements is compensated for by means of the SDEM technique. The chosen second-order architecture enables low swing of the operational amplifier outputs and a nearly rail-to-rail input range. The measured power consumption is 280 μW and the obtained figure of merit (FoM), applying Schreier's definition [10], is 174.95 dB.

2 Incremental Converter Schemes

An incremental ADC scheme may be regarded as the combination of a ΣΔ modulator and a dual-slope ADC [11]. Figure 1 shows the block diagram of a first-order incremental ADC. It consists of a delayed-integrator, a comparator and a 2-level DAC. The operation principle is as follows: when a new conversion cycle starts, the output of the integrator, V_{res}, is reset. Since the frequency of the

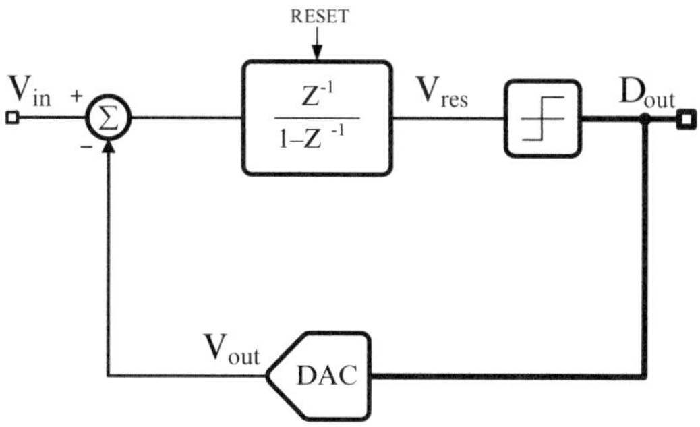

Fig. 1 First-order incremental ADC block diagram

input signal of incremental ADCs is usually low, V_{in} can be regarded as a constant signal. In each clock period, V_{in} subtracts V_{out} (the analog version of D_{out}) and the difference is accumulated by the delayed integrator. At the end of N clock cycles, the residue voltage at the output of the integrator is given by

$$V_{res} = \sum_{i=1}^{N-1} V_{in}(i) - \sum_{i=1}^{N-1} D_{out}(i) V_{ref} \tag{1}$$

Due to the stability of the feedback loop, the voltage of V_{res} is limited, namely $-V_{ref} < V_{res} < V_{ref}$, where $\pm V_{ref}$ are the reference voltages. The input signal V_{in} can be, hence, represented as

$$V_{in} = \frac{\sum_{i=1}^{N-1} D_{out}(i) V_{ref}}{N-1} + \frac{V_{res}}{N-1} \tag{2}$$

and the resolution of a first-order incremental ADC can be expressed as

$$R_{1-ord} = log_2(N-1) \tag{3}$$

Unfortunately, the conversion efficiency of a first-order incremental ADC is low. Methods for increasing the resolution are by increasing the number of clock periods N and using more effective schemes with cascaded integrators. High-order incremental ADCs, therefore, contain multiple integrators with reset at the beginning of the conversion cycle. The key points are to increase the accumulation efficiency, while maintaining the stability of the structure and keeping $V_{res}(N)$ minimized.

A conventional second-order incremental ADC is discussed here to illustrate how the conversion efficiency changes compared with the first-order architecture. As seen in Fig. 2, this structure contains two integrators with delay. In the signal path there are two coefficients c_1 and c_2. In order to keep the loop stable, two feed-forward paths are included with coefficients f_1 and f_2. This scheme also employs a comparator and a 2-level DAC. Using the same mathematical approach employed for the first order scheme, $V_{res}(N)$ for this second-order structure can be expressed as

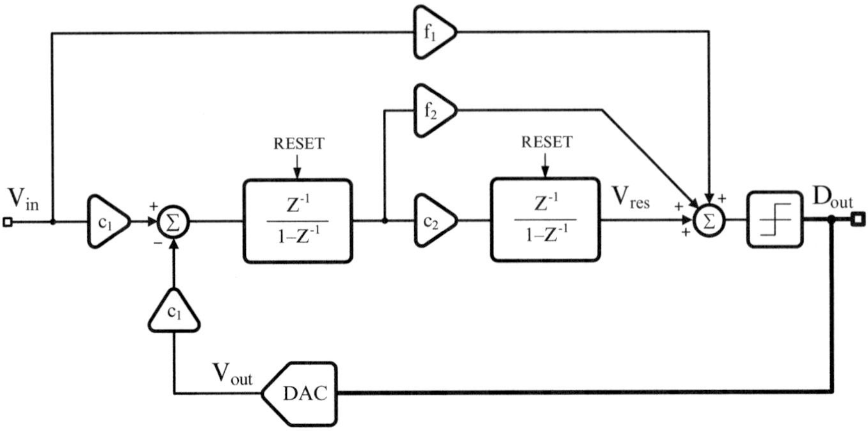

Fig. 2 Second-order incremental ADC block diagram

$$V_{res}(N) = c_1 c_2 \sum_{i=1}^{N-1} \sum_{j=1}^{i-1} V_{in}(i) - c_1 c_2 \sum_{i=1}^{N-1} \sum_{j=1}^{i-1} D_{out}(j) V_{ref} \tag{4}$$

Similarly, the resolution of the second-order structure is described as

$$R_{2-ord} = log_2 \frac{c_1 c_2 (N-1)(N-2)}{2!} \tag{5}$$

To compare the conversion efficiency of first-order and second-order incremental ADCs, parameters $N = 1024$, $c_{1,2} = 1, f_1 = 1$ and $f_2 = 2$ are chosen. According to (3) and (5), the first-order scheme can provide 10-bit resolution while the second-order structure achieves 19 bit.

A generalization of the scheme of Fig. 2 to an Lth-order architecture is illustrated in Fig. 3, which includes L delayed-integrators and feed-forward paths with coefficients $f_{1,2,\ldots,L}$. There is only one feedback path in the scheme and L coefficients along the signal path $c_{1,2,\ldots,L}$. The resolution of the Lth-order incremental ADC can be estimated as

$$R_{L-ord} = log_2 \frac{c_1 c_2 \ldots c_L (N-1)(N-2)\ldots(N-L)}{L!} \tag{6}$$

In order to ensure stability, coefficients $c_{1,2,\ldots,L}$ are generally lower than 1.

As mentioned in the Introduction, the use of a 2-level DAC ensures linearity, but the output swing of integrators is large and may result in the op-amps slewing. Moreover, when the order of the scheme L is larger than 2, stability of the feedback loop demands for the use of fractional coefficients along the signal path, which degrades the conversion efficiency. For example, in the third-order modulator described in [12], $c_1 = 0.5674$, $c_2 = 0.5126$, and $c_3 = 0.3171$. Using (6)

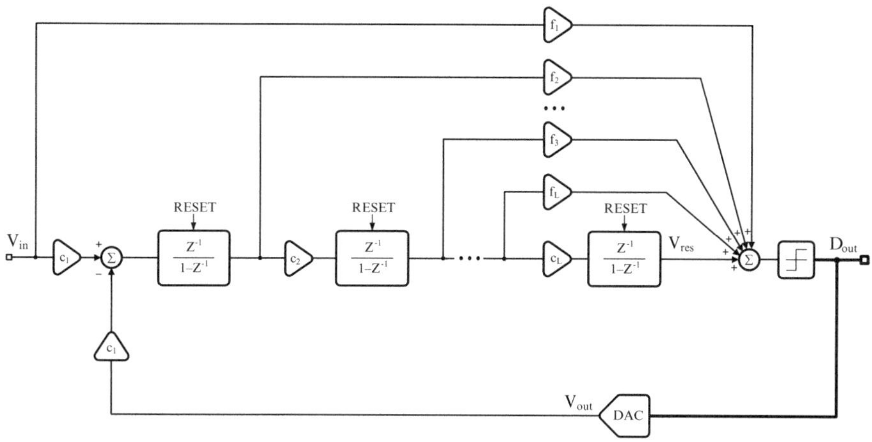

Fig. 3 *L*th-order incremental ADC block diagram

with $N = 128$, the resolution is 14.9-bit, which is 3.4-bit less than the maximum achievable (with $c_{1,2,3} = 1$). Another case is the fourth-order modulator reported in [13]. Coefficients c_1, c_2, c_3 and c_4 are 0.25, 0.4, 0.22, and 0.11, respectively. With $N = 128$, the resolution is 14.6 bit while the maximum theoretical resolution is 23.3 bit. The resolution lost in this case is, hence, more than 8.7 bit. Therefore, the benefit of high order schemes is limited with single-bit quantizers.

On the other hand, the use of multi-bit DACs relaxes the stability requirements, thus allowing the use of larger values for coefficients $c_{1,2,\ldots,L}$, in the limit equal to 1. Nonetheless, the non-linearity of a multi-bit DAC needs to be properly compensated for. For $\Sigma\Delta$ modulators, dynamic calibration, such as dynamic-element-matching (DEM) [14–16], is a well known and effective way to compensate for the non-linearity of the DAC. Those methods work well with $\Sigma\Delta$ schemes, but they are not truly efficient for incremental ADCs.

For high resolution incremental ADCs, the kT/C noise is a key limitation and has to be carefully investigated. In a single-ended switched capacitor (SC) implementation of the second order incremental scheme of Fig. 2, in each clock cycle, the noise injected from the input of the modulator is $2kT/C_S$, where C_S is the sampling capacitance. At the end of the N-th clock period, the total noise power accumulated at node V_{res} is

$$v_{n,tot}^2 = \frac{2kT}{C_s} \sum_{i=1}^{N-2} i^2 \tag{7}$$

Therefore, the input referred noise is

$$v_{n,in}^2 = \frac{v_{n,tot}^2}{G_{2-ord}^2} \tag{8}$$

where G_{2-ord} is the gain of the input signal which is equal to $(N-1)(N-2)/2$. In order to achieve R_{2-ord} resolution, $v_{n,in}$ should be less than half of V_{LSB}, which gives rise to

$$C_S > \frac{8kT\sum_{i=1}^{N-2} i^2}{G_{2-ord}^2 V_{LSB}^2}; \; V_{LSB} = \frac{V_{FS}}{2^{R_{2-ord}}} \tag{9}$$

being V_{FS} the full scale voltage. The above approach and calculations can be easily extended to the L-th order case.

3 The Smart-DEM Algorithm

This section describes the working principle of the Smart-DEM algorithm. Supposing an incremental converter with only one feedback path (like the ones described in the previous Section) and an m-bit DAC, the number of the unity elements is $M = 2^m$ and the mismatch for each element can be represented by $\epsilon_i (i = 1, 2, \ldots, M)$. Observing that the sum of the mismatch goes to zero (otherwise it could be considered as a gain factor), the following equation holds:

$$\epsilon_1 + \epsilon_2 + \ldots + \epsilon_M = 0 \tag{10}$$

The key point behind the Smart-DEM algorithm is that, for incremental converters, the error caused by the mismatch among the M unity elements of the DAC depends on the injection time. Indeed, the weight of the signal injected at the beginning of the conversion is larger than the weight of the ones injected close to the end of the conversion. For an incremental ADC that uses N clock periods to converter one sample, the total error, ϵ_{tot}, is then the summation of the weighted mismatches, which can be represented as

$$\epsilon_{tot} = W_1\epsilon_1 + W_2\epsilon_2 + \ldots + W_M\epsilon_M \tag{11}$$

where W_i are the weights for each mismatch $\epsilon_i (i = 1, 2, \ldots, M)$.

Notice that ϵ_{tot} is predictable and if it can be further controlled with a certain algorithm, the multi-bit DAC can be used directly. The Smart-DEM algorithm during the data conversion balances the W_i, thus minimizing ϵ_{tot}. The ideal case is that, if all the weights W_i are equal, according to (8), ϵ_{tot} is equal to 0.

The estimation of the weights of the error introduced by the multi-level DAC along the conversion depends on the scheme and on the order of the incremental converter. For the multi-bit version of a second-order scheme as the one depicted in Fig. 2, if an error from the DAC enters at the k-th period ($1 \le k \le N$) after the reset, it is stored by the first integrator for one clock period and then it is accumulated as a linear function on the second integrator. The weights of the error caused by the mismatch in the k-th period can thus be calculated as:

$$W_{2-ord}(k) = N - k - 1 \tag{12}$$

For the case of a multi-bit third-order scheme (see Fig. 3 with $L = 3$), if an error from the DAC enters at the k-th period ($1 \leq k \leq N$) after the reset, it is stored at the output of the first integrator, linearly accumulated by the second one, and quadratically accumulated at the output of the third integrator. In this case, the weights of the error can be estimated as:

$$W_{3-ord}(k) = \frac{(N-k-1)(N-k-2)}{2!} \tag{13}$$

In general, for the multi-bit version of the L-th order scheme of Fig. 3, the weights can be expressed as:

$$W_{L-ord}(k) = \frac{(N-k-1)(N-k-2)\ldots(N-L)}{L!} \tag{14}$$

The flow of the Smart-DEM algorithm is as follows:

Step 1: before starting a new conversion cycle, reset the total weights of all the elements W_i ($i = 1, 2, \ldots, M$) to zero.
Step 2: in each clock cycle, select the unity elements with the minimum weight. If it is the last clock period, jump to Step 4.
Step 3: calculate the weight of the current clock period $W(k)$ ($k = 1, 2, \ldots, N$) and update the weights of the elements selected, then go back to Step 2.
Step 4: finish the conversion cycle.

In order to better understand how the SDEM algorithm works, let us consider the scheme of Fig. 4. It is a 3-bit second-order scheme with both integrators with delay. The achievable resolution for this scheme is:

$$R_{2ord-multibit} = log_2\frac{c_1c_2(N-1)(N-2)}{2!} + b_q \tag{15}$$

where b_q is the resolution of the quantizer.

With $N = 256$ and $b_q = 3$, $R_{2ord-multibit} = 18$ bit. Behavioural simulation results show that the swing of the DAC is $1.25V_{ref}$. It means that the quantizer needs two extra comparators to accommodate the two extra levels required in the DAC. In this way, the required number of unity elements is $M = 10$. Figure 5 shows an example on how the weights of the error change with the SDEM algorithm along the conversion for an input signal equal to $0.305V_{ref}$. During the reset phase, the weight of each of the 10 unity elements $C1, \ldots, C10$ is reset to 0. In the first clock period of the conversion, the converter digital output is $D_{out}[10:1] = 0000111111$ which corresponds to a decimal value equal to 6 (the decimal value of $D_{out}[10:1]$

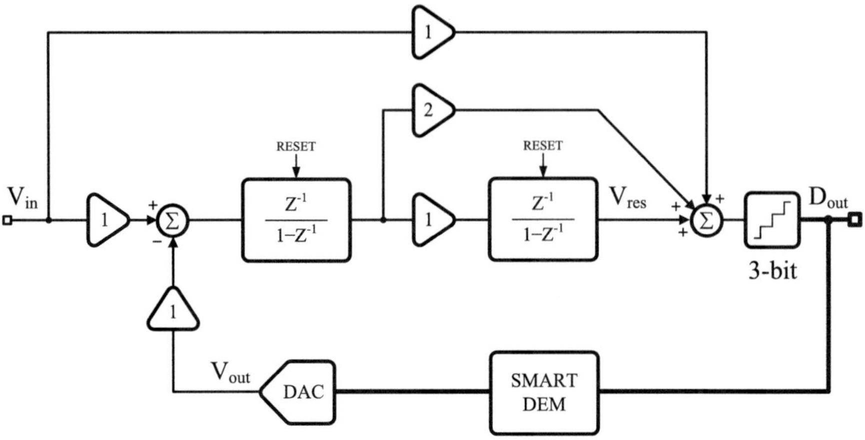

Fig. 4 Block diagram of a 3-bit second-order incremental scheme

RESET		k=1		k=1		k=2		k=2		k=2		End of k=256	
C10	0	C10	0	C6	254	C6	254	C3	507	C3	253	C7	1
C9	0	C9	0	C5	254	C5	254	C2	507	C2	253	C6	1
C8	0	C8	0	C4	254	C4	254	C1	507	C1	253	C5	1
C7	0	C7	0	C3	254	C3	507	C6	254	C6	1	C4	1
C6	0	C6	254	C2	254	C2	507	C5	254	C5	1	C1	1
C5	0	C5	254	C1	254	C1	507	C4	254	C4	1	C10	0
C4	0	C4	254	C10	0	C10	253	C10	253	C10	0	C9	0
C3	0	C3	254	C9	0	C9	253	C9	253	C9	0	C8	0
C2	0	C2	254	C8	0	C8	253	C8	253	C8	0	C3	0
C1	0	C1	254	C7	0	C7	253	C7	253	C7	0	C2	0

$D_{out}[10:1]=0000111111$
W(1)=254

$D_{out}[10:1]=0001111111$
W(2)=253

Fig. 5 Status of the weights for $V_{in} = 0.305V_{ref}$

is from 0 to 10, with a common mode value 5). According to the Smart-DEM algorithm flow, 6 unity elements with minimum weights should be chosen and the corresponding weights need to be updated. Therefore, unity elements C1–C6 are selected. Using (9), the associated weight is calculated as 254. This value is added to the existing weights of C1–C6. After that, the weight array is sorted and the largest values are moved at the top of the stack. In the second clock period, $D_{out}[10:1] = 0001111111$: the corresponding decimal value is 7. 7 elements with the minimum weight are, hence, selected: they are C4–C6 and C7–C10. However, the corresponding weight in this clock period changes to 253. This value is again added to the current weights of these elements. After sorting, the minimum weight in the array is subtracted from all the weights in order to avoid hardware overflow. Finally, after 256 clock periods, $W(i)(i = 1, 2, \ldots, 10)$ is no more than 1 and the total effect of the mismatch is negligible.

4 Design Examples

This Section describes two design examples to demonstrate the effectiveness of the SDEM algorithm when compared with a conventional technique. The algorithm is applied to a second and to a third order multi-bit incremental scheme and converters performance are evaluated at the behavioural level.

4.1 Second-Order Incremental Example

As mentioned in the previous Section, for the scheme of Fig. 4, the swing of DAC is $1.25V_{ref}$ and thus 10 unity capacitors are used. The output swing of the first integrator is below $0.25V_{ref}$ while, for the second one, it is within $0.125V_{ref}$.

To demonstrate the effectiveness of the SDEM algorithm, two groups of simulations have been performed: they use $N = 256, b_q = 3$ and $V_{FS} = 3.3$. In the first group, the mismatch among the unity elements of the DAC obeys a normal distribution with $\mu = 0$ and $\sigma = 0.08\,\%$. Figure 6 illustrates the performance comparison of the second-order scheme of Fig. 4 in three different cases: mismatch of unity elements not compensated for, assisted with conventional DWA method, and compensated for with the SDEM algorithm. The input is a

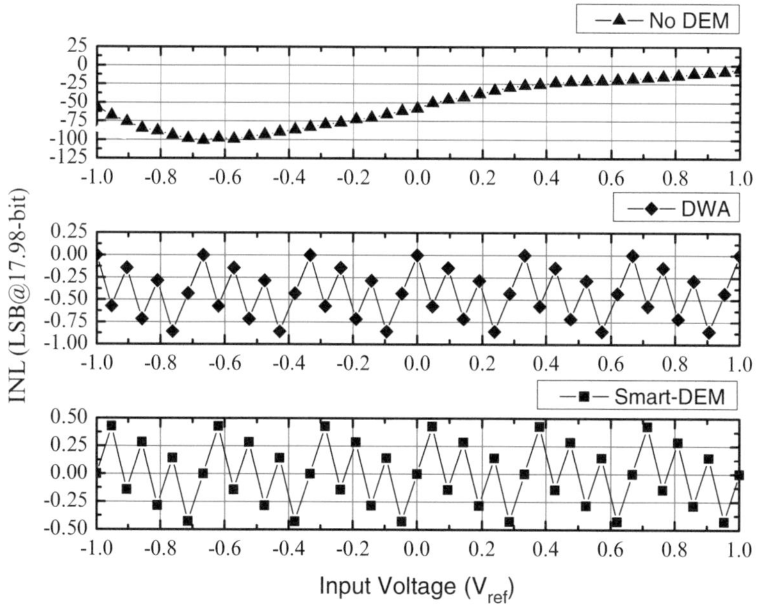

Fig. 6 INL comparison: without DEM (*top*), with DWA (*middle*) and with Smart-DEM (*bottom*). The mismatch of unity elements obeys a normal distribution with $\mu = 0$ and $\sigma = 0.08\,\%$

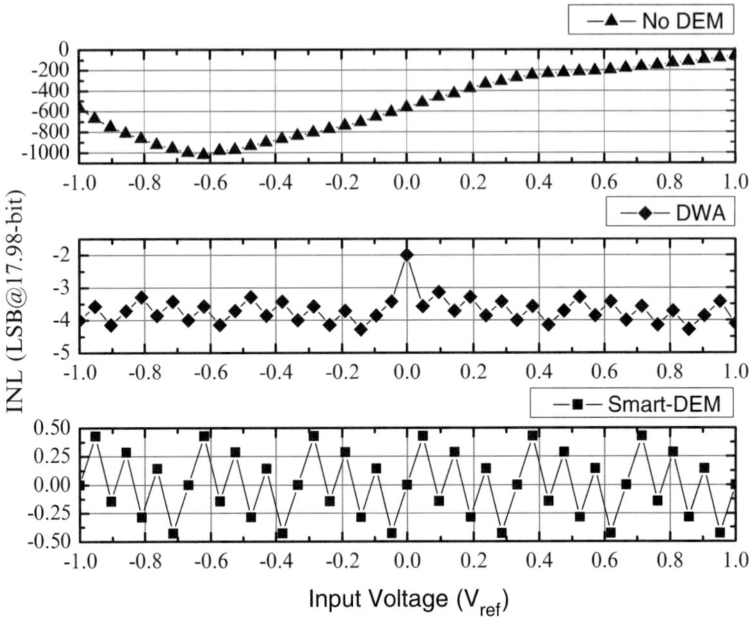

Fig. 7 INL comparison: without DEM (*top*), with DWA (*middle*) and with Smart-DEM (*bottom*). The mismatch of unity elements obeys a normal distribution with $\mu = 0$ and $\sigma = 0.8\,\%$

constant voltage ranging from $-V_{ref}$ to V_{ref}. As seen in Fig. 6, the maximum INL with no compensation is about 101 LSB. When using DWA, the maximum INL is reduced to 0.9 LSB. The Smart-DEM is able to keep the error within 0.5 LSB for the entire range.

Moreover, it is useful to compare the performance of different DEM techniques when a larger mismatch is considered. Figure 7 illustrates the performance comparison when the mismatch obeys a normal distribution with $\mu = 0$ and $\sigma = 0.8\,\%$. The maximum INL without DEM algorithm is 1019.6 LSB and for DWA is 4.3 LSB. In the SDEM case, the maximum INL is still limited within 0.5 LSB over the entire input range.

Figure 8 shows the weight accumulation of the 10 unity elements of the multi-level DAC. Since 10 curves overlap with each other and the difference is difficult to be observed, only the last 50 clock periods are plotted. The input signal is $0.123V_{ref}$. At the end of the conversion, the maximum weight is 17,786 while the minimum one is 17,784. The weights for all the elements converge to the same level with a maximum difference of 2, which results in a negligible residual error.

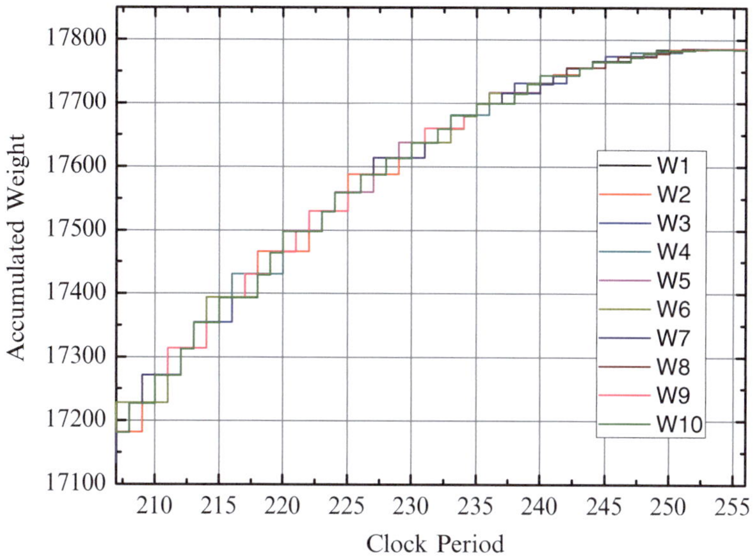

Fig. 8 Accumulated weights of the 10 unity elements for $V_{in} = 0.123V_{ref}$

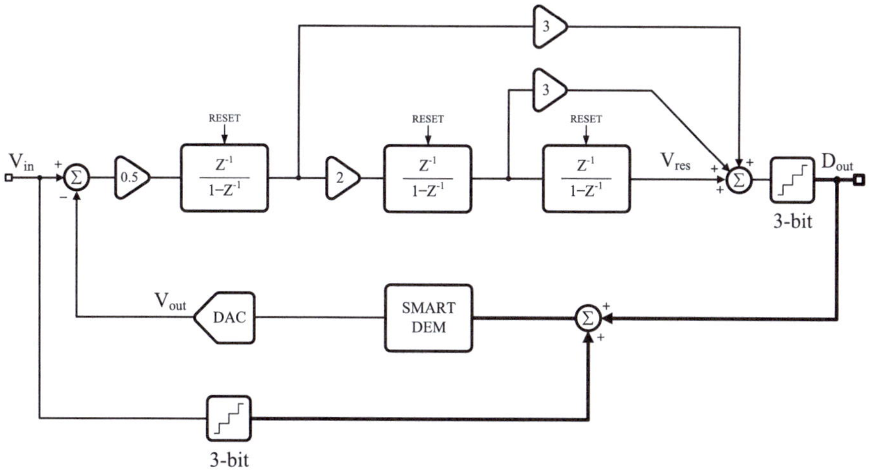

Fig. 9 Block diagram of a 3-bit third-order incremental scheme

4.2 *Third-Order Incremental Example*

Figure 9 shows the block diagram of a multi-bit third-order incremental scheme. The architecture is similar to the architectures published in [7, 12]. Several features, however, are included to convert it into a high-resolution multi-bit incremental converter. The use of multi-bit quantizer and DAC benefits the converter extra bit of resolution and reduces the output swing of op-amps. Feed-forward paths are added to maintain the stability of structure while the coefficients along the integration path do not degrade the overall conversion efficiency. In order to reduce the output swing of the first op-amp, coefficients 0.5 and 2 are introduced at the input of first and second integrators. Finally, a quantized version of V_{in} is added to D_{out}, which means that the amplitude of the real input signal is reduced to less than half quantization interval of the quantizer. The cost is the increased swing of the multi-bit DAC. In order to achieve 18-bit resolution, design parameters $N = 61$ and $b_q = 3$ are selected. The output swing of all the three op-amps is below $0.25V_{ref}$. Nevertheless, the output of the DAC is 75 % more than the full scale. The DAC needs, hence, 6 extra levels. The number of unity elements is 14.

To demonstrate the effectiveness of the SDEM algorithm also for this structure two groups of simulations have been performed. Figure 10 shows the performance comparison of the third-order scheme in three different cases. The mismatch for the 14 unity elements obeys a normal distribution with zero mean value and $\sigma = 0.035\,\%$. The input is a constant voltage ranging from $-V_{ref}$ to V_{ref}. The maximum INL without compensation is about 127.8 LSB. When using the DWA method, the error is not linear with a maximum of 3.9 LSB. The SDEM is able to keep the error within 0.5 LSB for the entire range (Fig. 11).

For the second group of simulations, the mismatch for the 14 unity elements obeys a normal distribution with zero mean value and $\sigma = 0.35\,\%$. The input is

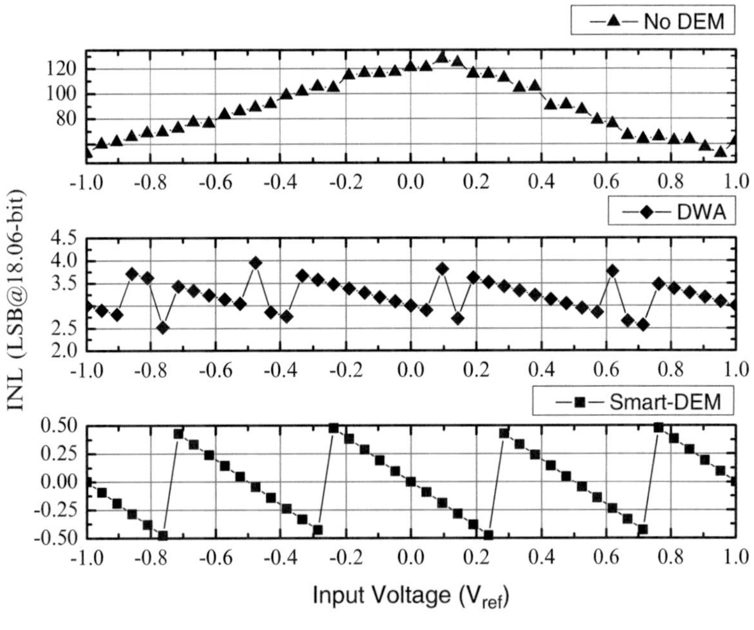

Fig. 10 INL comparison: without DEM (*top*), with DWA (*middle*) and with Smart-DEM (*bottom*). The mismatch of unity elements obeys a normal distribution with $\mu = 0$ and $\sigma = 0.035\,\%$

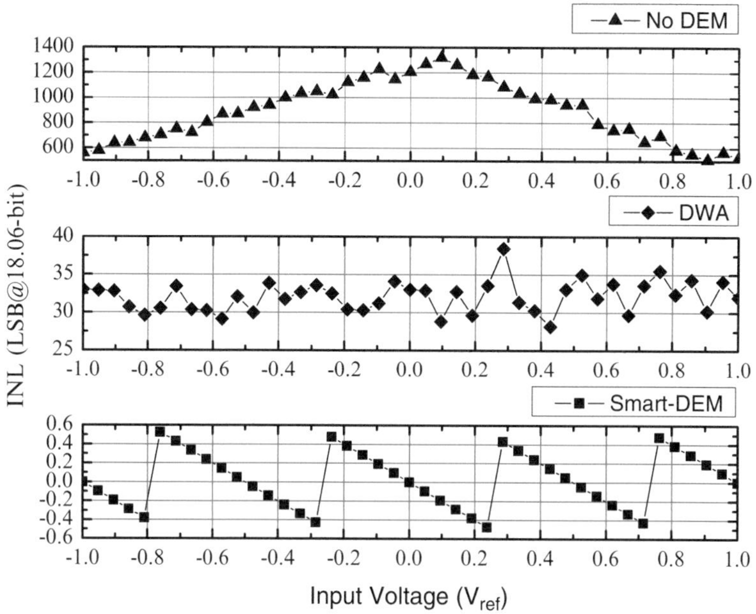

Fig. 11 INL comparison: without DEM (*top*), with DWA (*middle*) and with Smart-DEM (*bottom*). The mismatch of unity elements obeys a normal distribution with $\mu = 0$ and $\sigma = 0.35\,\%$

again a constant voltage ranging from $-V_{ref}$ to V_{ref}. The maximum INL without compensation, however, is 1319.8 LSB. With the DWA method, the maximum error is 38.4 LSB. The Smart-DEM is able to keep the error within 0.53 LSB for the entire range. This value is slightly larger than the expected 0.5 LSB because of the unbalance of the weights at the end of N-th clock period.

5 Test Chip Design

The SDEM algorithm has been applied and implemented in a 3-bit second order incremental converter [8, 9]. Figure 12 shows the architecture. It replaces the analog feed-forward paths used in the $\Sigma\Delta$ counterpart [17], with digital feed-forwards. It is the cascade of two sampled-data integrators (one without delay, the other with delay) with three ADCs, which digitize the input signal and the outputs of the two integrators. The digital output is the addition of the three A/D conversions. The quantization step of each ADC is $V_{FS}/8$, where V_{FS} is the full scale voltage whose value is twice of the analog reference voltage V_{ref}. With $N = 256$ and $b_q = 3$, the achievable resolution is equal to 17.99-bit. The use of digital feed-forward operates similarly to the analog counterpart. It reduces the swing of op-amps which becomes lower than $0.4V_{ref}$. This allows using for ADC2 and ADC3 only 4 comparators.

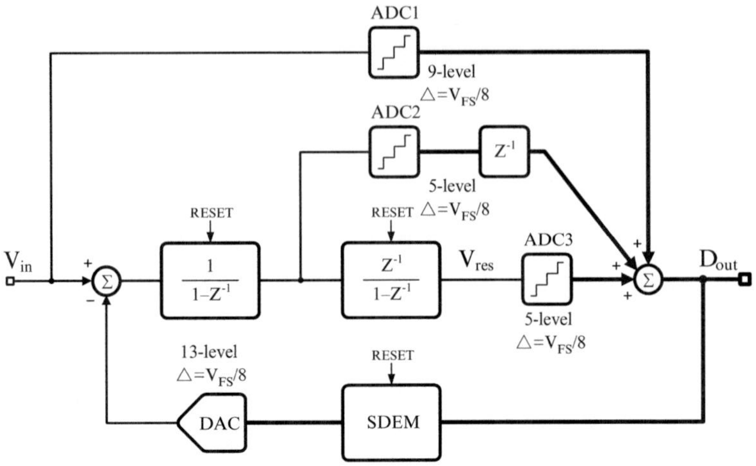

Fig. 12 Designed second-order 3-bit incremental ADC block diagram

The use of digital feed-forward paths allows avoiding active elements or complicated passive networks in front of the quantizer at the cost of a number of comparators negligible in terms of power consumption or area overhead. The SDEM block processes the modulator output to properly select the unity elements to be used in the multi-level DAC.

5.1 *Circuit Description*

The designed circuit is the fully differential version of the schematic of Fig. 13. The first SC integration includes chopper at input and output to cancel the offset. At the beginning of the conversion, switches reset the operational amplifiers. Six switched unity capacitors used in a bipolar fashion implement the DAC. The permitted overrange enables a full dynamic of the input signal. The use of a bipolar DAC allows us to use only 6 unity elements instead of 12. The left terminal of each DAC capacitors, C_u, can be connected to V_{ref} or $-V_{ref}$ either during Φ_q or Φ_2. As specified by the Table in Fig. 13, controls A_1 and A_2 provide positive, negative or null injection. The same capacitors and same references make exactly symmetric positive and negative injections. A unity capacitance of 450 fF makes the kT/C noise negligible.

The first op-amp scheme is a fully differential recycling folded cascode amplifier [18], with discrete time (DT) common mode control. The scheme, derived from conventional folded cascode implementation, boosts the gain, bandwidth and slew-rate without affecting noise performance or introducing additional offset. The simulated DC gain and GBW are 95 dB and 19 MHz, respectively. The second op-amp is a conventional fully differential folded cascode amplifier with 83-dB gain and 21-MHz bandwidth. The offset cancellation in incremental converter is not a

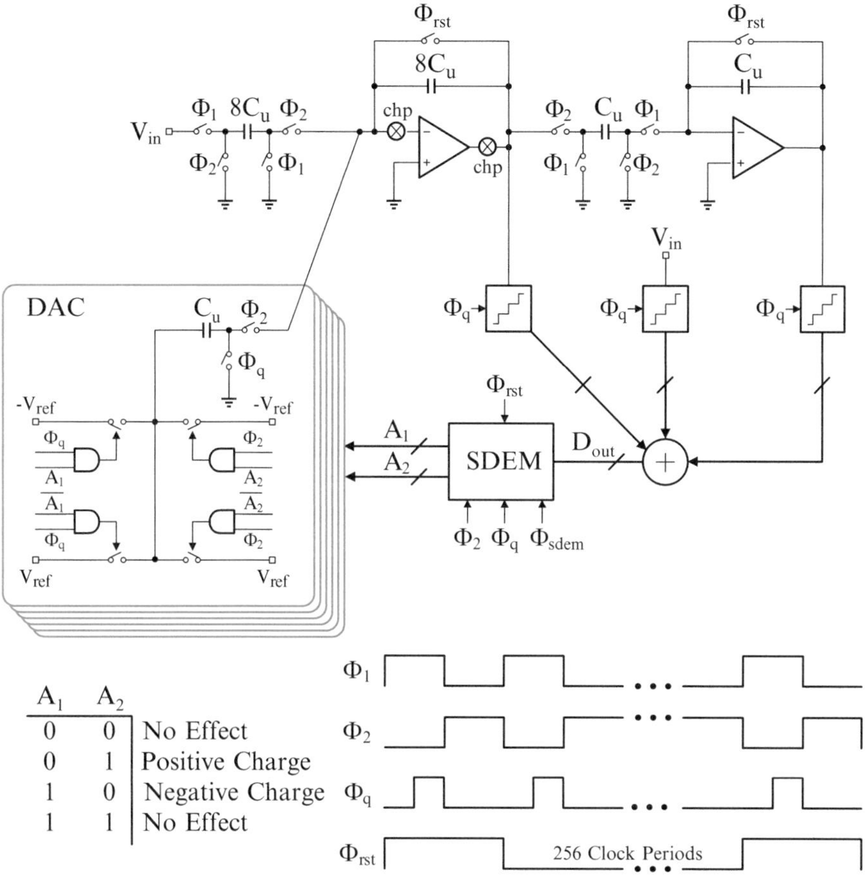

Fig. 13 Second-order 3-bit incremental ADC schematic diagram

plain task as it is for an amplifier. It is necessary to take the offset effect into account throughout the conversion cycle. For this reason, this design use the so-called one-step chopping technique, described in [19].

Figure 14 shows the block diagram of the Smart-DEM algorithm implementation of this design. The summation of the outputs of the three quantizers D_{out} is an integer number ranging from 0 to 12. The SDEM encoder transforms D_{out} to two temporary control signals $A_{1t}[6:1]$ and $A_{2t}[6:1]$. Under the control of the SDEM finite state machine (FSM), the weights corresponding to the current clock period are calculated. The insert-sorting algorithm orders the results, temporarily stored in the auxiliary memory. The cross network uses the new order of elements in the next clock period to give rise to the $A_1[6:1]$ and $A_2[6:1]$ output control signals.

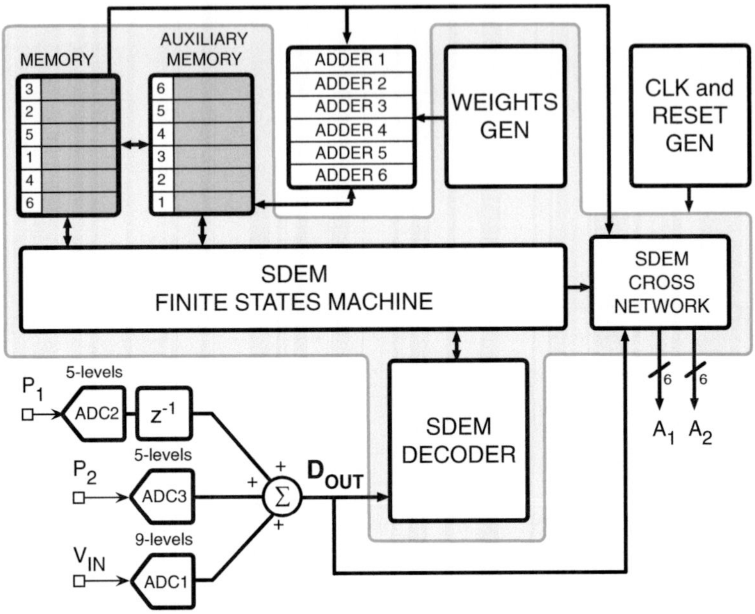

Fig. 14 The block diagram of Smart-DEM implementation

5.2 *Experimental Results*

An experimental prototype, fabricated in a 0.18–0.5 μm CMOS technology with dual poly and six metal layers, verifies the performances. The analog supply voltage is 3.3 V; the digital section operates with 1.5 V supply. The analog sampling frequency is 10 kHz for an analog clock of 2.56 MHz. The digital section (about 1200 gates) implementing the Smart-DEM algorithm runs at 82 MHz. The cascade of two accumulators with reset processes the digital output of the incremental part for 256 clock periods to generate the expected full scale value, 261,120. The digital output is the addition of the three ADCs. The measured converter power consumption is 200 μW for the analog and 80 μW for the digital part.

Figure 15 gives histograms of 2048 repeated measurements on the same part with zero input. The three histograms are for SDEM disabled, SDEM enabled, and SDEM enabled plus single-step chopping at the clock period number 180. The standard deviation of the histograms measures the input referred noise voltage. It is 1.58 LSB (18 μV) and 0.87 LSB (10 μV) without and with SDEM. Indeed, mismatch increases the inaccuracy. The mean of the histogram measures the input offset (160.8 LSB = 1.84 mV and 155.7 LSB = 1.78 mV, respectively). Again, the mismatch causes the difference. The single-step chopping leads to a residual offset equal to 0.85 LSB (9.7 μV).

The linearity of the converter mainly depends on the linearity of the input–output response of the first op-amp. The used scheme shows a gain better than 70 dB

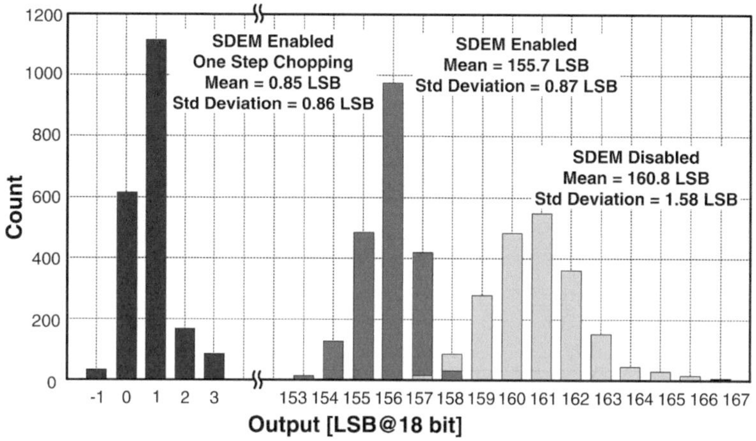

Fig. 15 Histograms of repeated measures with shorted inputs

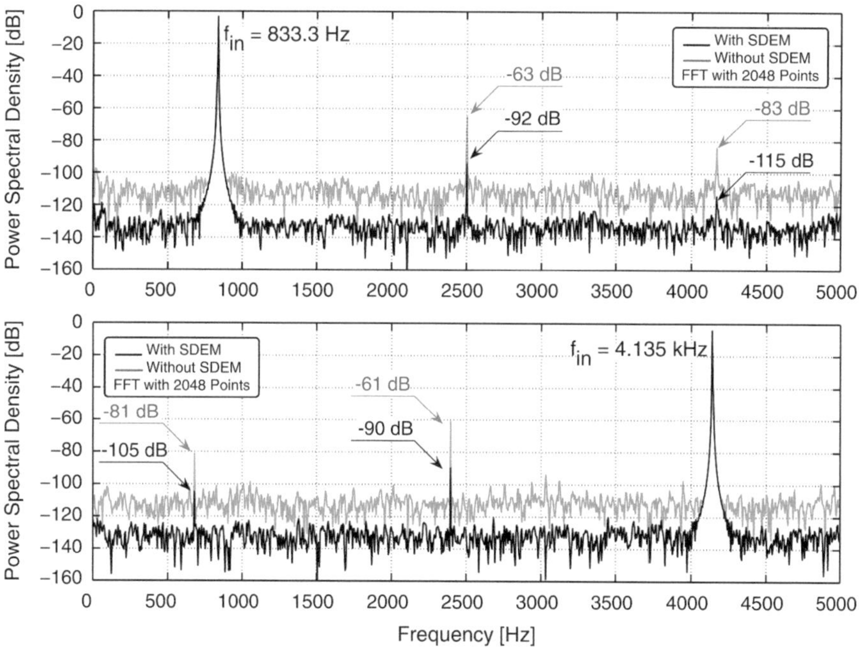

Fig. 16 Measured output spectra

when the op-amp output voltage ranges from 1.05 to 2.25 V with $V_{DD} = 3.3$ V. The use of reference voltages (provided externally) equal to 3.15 and 0.15 V brings the operation in the low gain region, thus giving rise to harmonic distortion, as shown in Fig. 16. It shows the converter output spectra (FFT with 2048 points) with $-2\,\text{dB}_{FS}$ sine waves at 833.3 Hz and 4.135 kHz, respectively, with and

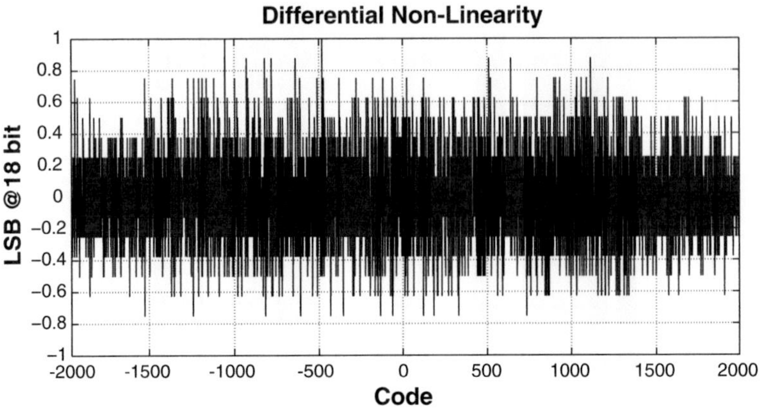

Fig. 17 Measured DNL

without SDEM. The harmonic distortion depends on both capacitors mismatch and first op-amp limitation. Without SDEM, the measured SNR is 82 dB and harmonics are significant. Using SDEM eliminates the capacitors mismatch contribution, showing the harmonic distortion caused by the op-amp. With low frequency input signal, the measured SNR is 105 dB, equivalent to 17.15 bit. The SNR at Nyquist drops by 1.3 dB with a loss of 0.22 bit. Third harmonic distortion dominates the SFDR: −92 dB with low frequency signal and −90 dB when the signal is close to the Nyquist frequency. The use of references equal to 2.775 and 0.525 V limits the swing of the first op-amp output and totally eliminates the distortion tones at the expense of a 2.6-dB loss in the measured SNR, equal to 102.4 dB. The achieved FoM is 174.95 dB, following the Schreier's formula [10], while it is equal to 260 fJ/conversion-step when using the Walden's expression.

Figure 17 illustrates the DNL obtained with the sine wave histogram method and an input sine wave of 50 mV_{pp}. The limited explored interval comes from the instrument's memory (8Mb). The measured DNL within the [−0.8, 1] LSB range confirms the value of the measured SNR.

The chip microphotograph with main circuital blocks highlighted is given in Fig. 18. The active area is $1270 \times 900\,\mu m^2$ (the chip area is $1600 \times 1300\,\mu m^2$). To avoid interferences, a shield of top metal completely covers the DAC capacitive array. The area of the SDEM is only 6 % of the active area.

6 Conclusions

Multi-bit quantization in incremental converters is only effective if the mismatch of the unit elements in their multi-level DACs is carefully compensated for. A Smart-DEM algorithm that effectively reduces the mismatch in the multi-level DACs of

Fig. 18 Chip microphotograph

incremental converters has been described in this chapter. The algorithm has been evaluated through extensive behavioural simulations of a couple of case studies and compared to conventional methods. The technique has been also applied to a 3-bit second-order incremental converter. Experimental results show that the circuit is able to achieve a resolution of about 17 bit over a 5 kHz bandwidth with 256 clock periods while consuming 280 μW.

References

1. Wu R, Chae Y, Huijsing JH, Makinwa KAA (2012) A 20-b ±40 mV range read-out IC with 50-nV offset and 0.04 % gain error for bridge transducers. IEEE J Solid-State Circuits 47(9):2152–2163
2. Garcia J, Rodriguez S, Rusu A (2013) A low-power CT incremental 3rd order ΣΔ ADC for biosensor applications. IEEE Trans Circuits Syst I 60(1):25–36
3. Robert J, Deval P (1988) A second-order high-resolution incremental A/D converter with offset and charge injection compensation. IEEE J Solid-State Circuits 23(3):736–741
4. Lyden C, Ryan J, Ugarte CA, Kornblum J, Yung FM (1995) A single shot sigma delta analog to digital converter for multiplexed applications. In: Proceedings of IEEE custom integrated circuits conference (CICC), May 1995, pp 203–206
5. Caldwell TC, Johns DA (2010) Incremental data converters at low oversampling ratios. IEEE Trans Circuits Syst I 57(7):1525–1537
6. Agah A, Vleugels K, Griffin PB, Ronaghi M, Plummer JD, Wooley BA (2010) A high-resolution low-power incremental ΣΔ ADC with extended range for biosensor arrays. IEEE J Solid-State Circuits 45(6):1099–1110
7. Quiquempoix V, Deval P, Barreto A, Bellini G, Markus J, Silva J, Temes GC (2005) A low-power 22-bit incremental ADC. IEEE J Solid-State Circuits 41(7):1562–1571

8. Liu Y, Bonizzoni E, D'Amato A, Maloberti F (2013) A 105-dB SNDR, 10 kSps multi-level second-order incremental converter with Smart-DEM consuming 280 μW and 3.3-V supply. In: Proceedings of IEEE European solid-state circuits conference (ESSCIRC), Sept. 2013, pp 371–374
9. Liu Y, Bonizzoni E, D'Amato A, Maloberti F (2015) A high-resolution low-power and multi-bit incremental converter with Smart-DEM. In: Analog integrated circuits and signal processing. Springer, New York. (82):663–674. doi:10.1007/s10470-015-0492-4
10. Schreier R, Temes GC (2005)Understanding delta-sigma data converters. Wiley, New York
11. van de Plassche RJ (1978) A sigma-delta modulator as an A/D converter. IEEE Trans Circuits Syst CAS-25(7):510–514
12. Markus J, Silva J, Temes GC (2004) Theory and applications of incremental $\Delta\Sigma$ converters. IEEE Trans Circuits Syst I 51(4):678–690
13. Lyden C, Ryan J, Ugarte CA, Kornblum J, Yung F-M (1995) A single shot sigma delta analog to digital converter for multiplexed applications. In: Proceedings of IEEE custom integrated circuits conference (CICC), pp 203–206
14. Carley LR (1989) A noise-shaping coder topology for 15+ bit converters. IEEE J Solid-State Circuits 28(2):267–273
15. Galton I (1997) Spectral shaping of circuits errors in digital-to-analog converters. IEEE Trans Circuits Syst II 44(10):808–817
16. Baird RT, Fiez TS (1995) Linearity enhancement of multibit A/D and D/A converters using data weighted averaging. IEEE Trans Circuits Syst II 42:753–762
17. Silva J, Moon U, Steensgaard J, Temes GC (2001) Wideband low-distortion delta-sigma ADC topology. IET Electron Lett 37(12):737–738
18. Assaad RS, Silva-Martinez J (2009) The recycling folded cascode: a general enhancement of the folded cascode amplifier. IEEE J Solid-State Circuits 44(9):2535–2542
19. Agnes A, Bonizzoni E, Maloberti F (2012) High-resolution multi-bit second-order incremental converter with 1.5-μV residual offset and 94-dB SFDR. In: Analog integrated circuits and signal processing, vol 72. Springer, New York, pp 531–539

Micropower Incremental Analog-to-Digital Converters

Chia-Hung Chen, Yi Zhang, Tao He, and Gabor C. Temes

Abstract Integrated sensor interfaces require energy-efficient high-resolution data converters. In many applications, the best choice is to use incremental analog-to-digital converters (IADCs) incorporating variants of extended counting. In this chapter, we discuss the design of a micropower IADC. By using a feed-forward architecture, the IADC accumulates the residue voltage, so various hybrid variants of extended counting can be implemented. Several such schemes are reviewed and discussed, as well as the trade-off between higher order modulators, higher oversampling ratio and energy efficiency. A two-step IADC is proposed, which extends the performance of an *Nth*-order IADC close to that of a $(2N-1)$th-order IADC. A design example uses the circuitry of a second-order IADC to achieve a performance nearly equal to that of a third-order IADC. The implemented IADC achieves a measured dynamic range of 99.8 dB, and a SNDR of 91 dB for a maximum input 2.2 V_{PP} and a bandwidth of 250 Hz. Fabricated in 65 nm CMOS and operated from a 1.2 V power supply, the IADC's core area is 0.2 mm^2, and it consumes only 10.7 μW. The measured FoMs are 0.76 pJ/conv.step and 173.5 dB, both among the best reported results for IADCs.

1 Introduction

As semiconductor technology evolves, more sensory functions can be integrated on a system-on-chip (SoC). Such SoCs are found in temperature, magnetic, pressure and image sensors, as well as in weight scales and bio-potential acquisition systems. An energy- and area-efficient high resolution analog-to-digital converter (ADC) is especially critical for battery-operated sensor SoCs. Sensor applications often involve narrow-band signals with frequencies from DC [1–3] up to several hundred Hz [4–6], and so the ADC should achieve high accuracy even in the presence of DC offset voltage and flicker noise. In addition, the integrated ADC must often be

C.-H. Chen (✉) • Y. Zhang • T. He • G.C. Temes
School of EECS, Oregon State University, Corvallis, OR 97330, USA
e-mail: cc2052@gmail.com

K.A.A. Makinwa et al. (eds.), *Efficient Sensor Interfaces, Advanced Amplifiers and Low Power RF Systems*, DOI 10.1007/978-3-319-21185-5_2

multiplexed among many channels. In applications requiring hundreds of channels, such as in image sensors [7, 8] or for bio-potential acquisition [4–6], the ADCs must also be highly efficient in terms of power and chip area.

Incremental analog-to-digital converters (IADCs) are often the best choice for low-frequency high-resolution sensor interfaces [1, 3, 5, 6, 9, 10]. Their advantages [9–11] include simpler decimation filtering, easy multiplexing, and sufficiently low latency. IADCs are also less subject to idle tones [11]. Moreover, the finite-impulse-response (FIR) filtering of the input signal in an IADC reduces aliasing [12]. However, a first-order IADC (IADC1) needs 2^N oversampling clock periods for N-bit accuracy, requiring a high sampling frequency and so it is not energy-efficient. To enhance efficiency, higher-order modulators can be used to increase the accuracy within the same conversion time. However, high-order modulators are more prone to instability, and have a reduced non-overloading input range. As an alternative to single-loop modulators, multi-stage noise-shaping (MASH) IADCs [13, 14], and also hybrid schemes which incorporate an added Nyquist-rate ADC to perform extended counting [6–8, 14, 15] have been proposed. However, MASH modulators and hybrid extended-counting schemes increase circuit complexity, and circuit non-idealities may then cause severe performance degradation. To retain the advantages without too much overhead circuitry, we propose a second-order IADC that uses a two-step architecture.

In this chapter, the design and operation of a conventional single-loop IADC with a feedforward modulator will first be reviewed, followed by a discussion of the operation and advantages of a MASH IADC in Sect. 2. Hybrid IADCs which recycle the hardware to perform extended counting, and thus achieve excellent energy efficiency, are discussed in Sect. 3. The detailed design and theoretical analysis of an IADC that employs a two-step architecture [16, 17] is described in Sect. 4. The novel scheme is applied to a second-order IADC (IADC2). The circuit's design and measured performance are discussed in Sect. 5. By recycling the hardware of the IADC2, the performance of the proposed two-step IADC is nearly as good as that of a conventional third-order IADC (IADC3), but require much less energy for the same conversion time. Section 6 summarizes the chapter and ends with conclusions.

2 Incremental Analog-to-Digital Converters

2.1 Operation and Design of a Second-Order IADC

IADCs are *Nyquist-rate* ADCs which use oversampling and noise shaping to convert a finite number of analog samples into a single digital word. Thus, they are a hybrid of Nyquist-rate and $\Delta\Sigma$ ADCs [9, 18]. Figure 1a depicts the z-domain model of an IADC2 with a low-distortion feed-forward modulator [9, 10]. The simplified timing diagram, including the two-phase non-overlapping clocks and reset pulse, is shown in Fig. 1b. Here, M is the oversampling ratio (OSR), defined as the number

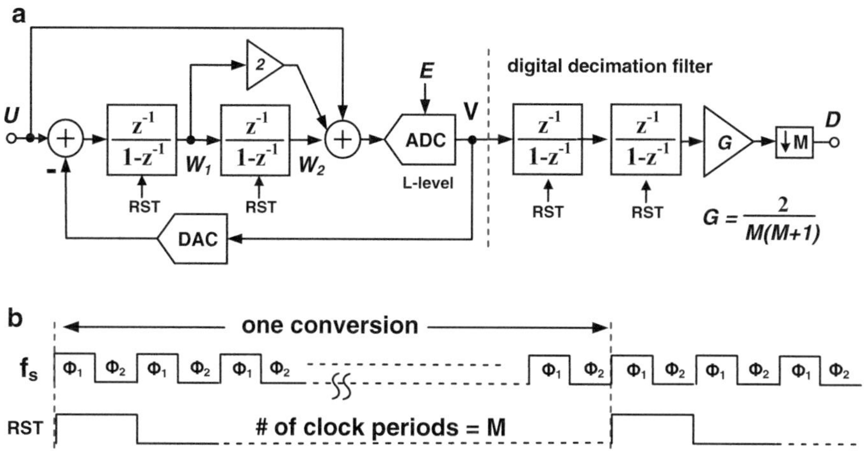

Fig. 1 (**a**) The z-domain model of a single-loop IADC2 with a low-distortion feed-forward modulator. (**b**) The simplified timing diagram

of oversampling clock periods within one conversion period. The operation of the IADC begins with a global reset pulse to clear the memories of all analog and digital blocks. After this reset, the $\Delta\Sigma$ modulator loop quantizes the analog input voltage U, and the digital filter concurrently processes the output bit stream V. After M clock periods, the next reset pulse reads the output word, and clears all memories. The circuit converts analog data *sample-by-sample*, and hence functions as a Nyquist-rate ADC.

The operation of an IADC is best understood by using time-domain analysis [9, 12]. At the end of the conversion (time index $i = M$) in Fig. 1, the variables satisfy the equation

$$U[M] + 2\sum_{i=1}^{M-1} U[i] + \sum_{K=1}^{M-1}\sum_{i=1}^{K-1} U[i] + E[M] = V[M] + 2\sum_{i=1}^{M-1} V[i] + \sum_{K=1}^{M-1}\sum_{i=1}^{K-1} V[i] \tag{1}$$

From (1),

$$\sum_{K=1}^{M}\sum_{i=1}^{K} U[i] + E[M] = \sum_{K=1}^{M}\sum_{i=1}^{K} V[i] \tag{2}$$

The least-significant-bit (LSB) quantization error E of the internal L-level quantizer is $V_{FS}/(L-1)$, where V_{FS} is the full-scale voltage. We may define the *average input voltage* $\tilde{U}$ by the relation

$$\tilde{U} \cong \frac{2}{M\left(M+1\right)}\sum_{j=1}^{M}\sum_{i=1}^{j}U\left[i\right] \tag{3}$$

Note that $\tilde{U}$ represents the input accurately only if U does not vary significantly during the conversion. From (2),

$$\tilde{U} + \frac{2}{M\left(M+1\right)}\frac{V_{FS}}{L-1} = \frac{2}{M\left(M+1\right)}\sum_{K=1}^{M}\sum_{i=1}^{K}V\left[i\right] \tag{4}$$

To reconstruct $\tilde{U}$ from the output bit stream, the digital decimation filter should perform the operation on the right-hand-side of (4). For an IADC2, the decimation filter can thus be simply two counters in cascade (Fig. 1). Alternatively, a multiply and accumulate (MAC) operation may be used. As (4) shows, the loop filter samples the input signal M times in one conversion, and performs finite-impulse-response (FIR) filtering [12] on the input signal.

The equivalent LSB quantization error E_{IADC2} of the IADC2 is

$$E_{IADC2} = \frac{2}{M\left(M+1\right)}\frac{V_{FS}}{L-1} \tag{5}$$

The effective number of bits (ENOB) and the signal-to-quantization-noise-ratio (SQNR) at full-scale input amplitude are given by

$$ENOB_2 = \log_2\left(\frac{V_{FS}}{E_{IADC2}}\right) \tag{6}$$

and

$$\begin{aligned} SQNR_2 &= 10\ \log\left(\frac{V_{FS}^2/8}{E_{IADC2}^2/12}\right) \\ &\approx 20\ \log\left(\frac{V_{FS}}{E_{IADC2}}\right) \approx 2\cdot 20\ \log(M) + 20\ \log\left(L-1\right) - 6 \end{aligned} \tag{7}$$

The analysis can be extended to an *Nth*-order IADC (IADCN). The equivalent quantization error and the maximum SQNR of an IADCN are

$$E_{IADCN} \approx \frac{N!}{M^N}\frac{V_{FS}}{L-1} \tag{8}$$

$$SQNR_N \approx N\cdot 20\ \log(M) + 20\ \log\left(L-1\right) - 20\ \log\left(N!\right) \tag{9}$$

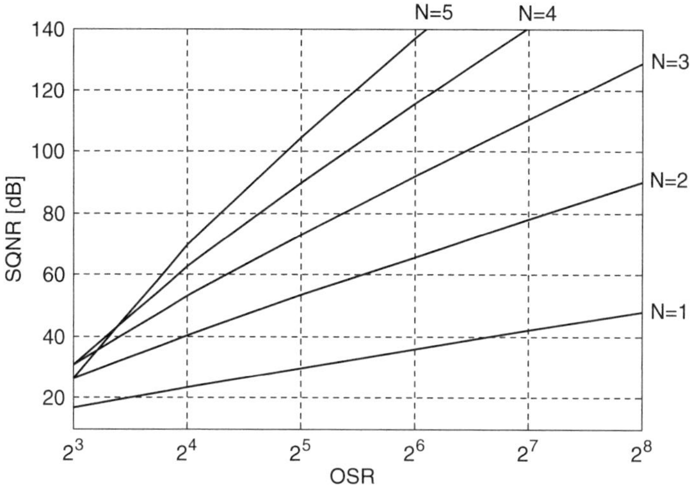

Fig. 2 SQNR versus OSR for one-bit modulator from first-order (N = 1) to fifth-order (N = 5)

The *Nth*-order loop filter scales down the internal quantization error by a factor $N!/M^N$. Figure 2 shows the calculated SQNR versus OSR for two-level ($L = 2$) modulators with orders $N = 1 \sim 5$.

2.2 *MASH IADCs*

To mitigate the stability problem of a higher-order single-loop IADC, the multi-stage noise-shaping (MASH) technique of a conventional $\Delta\Sigma$ ADC can be applied to an IADC [12, 13, 18]. Figure 3 shows an example of 1-1 MASH IADC2 obtained by cascading two IADC1s. A higher-order IADC can be achieved by cascading lower-order modulators and thus the energy efficiency is improved because less peripheral circuitry (quantizer and DAC circuits) is required to sustain a wide non-overloaded range.

Conventional MASH $\Delta\Sigma$ modulators [19, 20] need to use error cancellation logic (ECL) circuits before adding the bit streams of the individual loops, to cancel the quantization error of the MSB loop. The opamp DC gains in the first loop need to be very high, to avoid SQNR degradation caused by mismatch between the analog and digital realizations of the noise transfer function $\left(1 - z^{-1}\right)$. Thus, a MASH $\Delta\Sigma$ ADC for a 16-bit SNR usually requires opamp DC gains of at least 90 dB [20], which is difficult to achieve in a low-voltage design. In MASH IADCs [12, 13], the oversampled bit streams of the $\Delta\Sigma$ loops are accumulated in one or more cascaded counters. The Nyquist-rate data from each loop are accumulated separately, and the

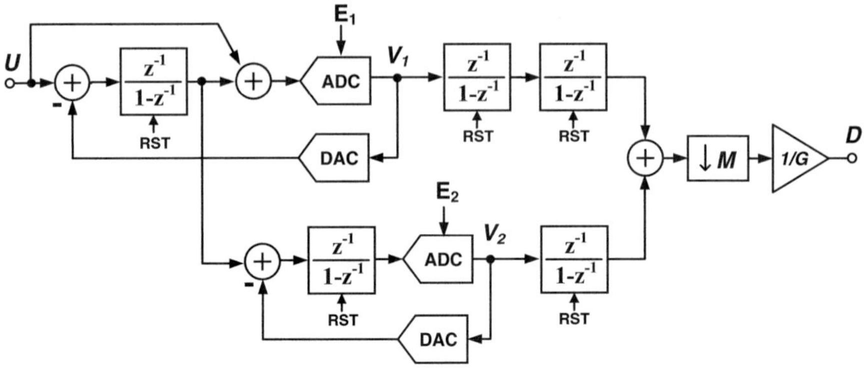

Fig. 3 A 1-1 MASH IADC

counters dump the data before the next reset pulse. The opamp gain requirements in a MASH IADC are hence much more relaxed than in their conventional MASH $\Delta\Sigma$ counterparts [12, 13].

2.3 *IADC Versus* $\Delta\Sigma$ *ADCs*

As discussed above, an IADC functions like a Nyquist-rate ADC, and is more suitable for use in a sensor interface than a conventional $\Delta\Sigma$ ADC. Its advantages are as follows:

- An IADC can be easily multiplexed among many channels. This saves area and the resulting sensor SoC can be cost effective.
- The latency from the analog input to the decimated digital output is only one Nyquist conversion period T_N. For an *Lth* order $\Delta\Sigma$ ADC, it is $(L+1)T_N$ [18–20].
- The idle tone is much less likely than in a conventional $\Delta\Sigma$ counterpart [11]. However, there may be "deadbands" around the quantizer's thresholds (around zero for two-level quantizer). They can be eliminated by injecting dither [9, 13].
- The decimation filter is simpler. It may be a cascade of a few accumulators, or a single MAC stage. Thus, the energy efficiency is improved.

3 Hybrid Schemes Using an IADC and a Nyquist-Rate ADC

In Fig. 1, the feed-forward loop filter processes only the shaped quantization noise. At the end of the conversion, the voltage stored at the last integrator is the residue voltage of the $\Delta\Sigma$ data conversion [6–8, 14, 15], and it is available for fine

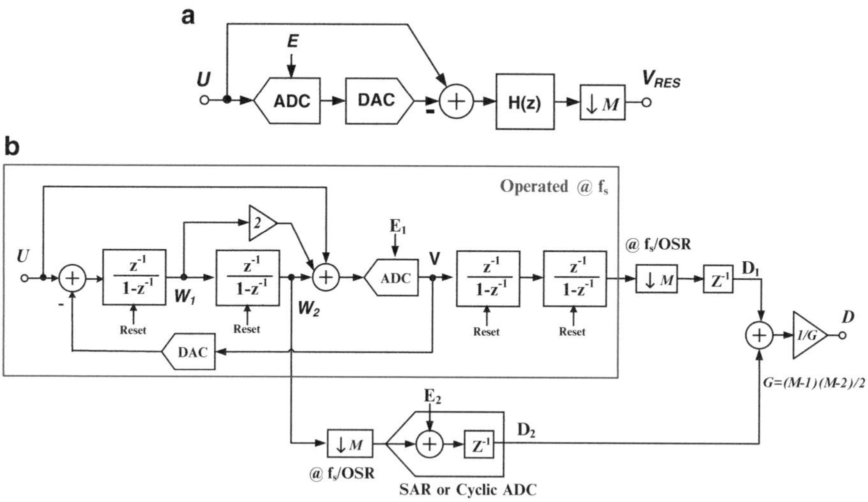

Fig. 4 (**a**) Residue voltage acquisition using a feedforward modulator. (**b**) An IADC using a second Nyquist-rate ADC for extended counting [6]

quantization. The residue voltage acquisition is illustrated in Fig. 4a. Hence, an energy-efficient SAR or cyclic ADC operated at Nyquist rate can sample the residue voltage right before the reset pulse, and perform the fine quantization [6, 21]. An example of such an extended-counting scheme is shown in Fig. 4b [6]. With proper design of the digital summation logic and decimation filter, the two cascaded loops can achieve a very high resolution with good energy efficiency. However, the last integrator needs to drive the large input capacitance of the 11-bit SAR [6], and therefore needs additional power.

For high resolution, the time required for one data conversion is usually quite long. Hence, instead of cascading two loops, the conversion can be performed in two steps, and the hardware can be shared to improve the energy efficiency [7, 8, 14, 15]. An example of a hardware-sharing extended-counting scheme is shown in Fig. 5 [7]. A discrete-time IADC1 performs the coarse quantization (Fig. 5b), and the integrator stores the residue voltage at the end of the first quantization step. In the second step, the hardware is reused and reconfigured as a 10-bit cyclic ADC to continue the fine quantization (Fig. 5c). By sharing the hardware, the energy efficiency is improved significantly.

In [22], a two-step incremental zoom ADC with a 182.7 dB figure-of-merit (FoM) was reported for DC measurements. A coarse ADC finds the six MSBs, without storing the residue, and the MSBs adjust the reference of the second-stage IADC, so as to zoom into a small range around the input signal. Then, the IADC samples the input signals, and performs the fine quantization for 1024 clock periods. Due to its smaller range, however, the input signal must be held very constant during

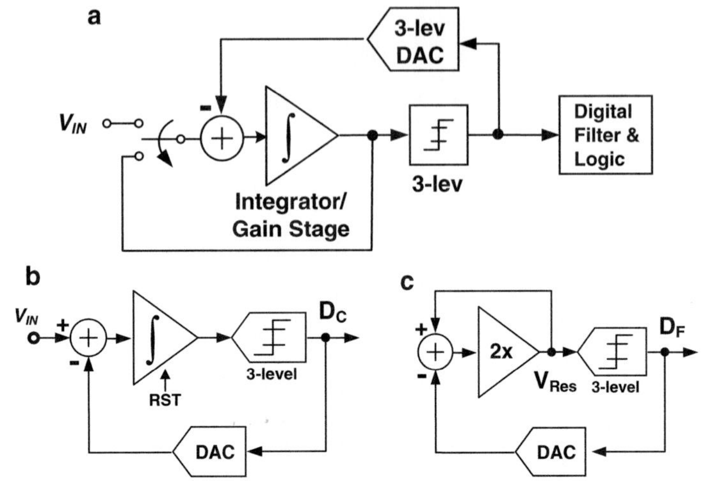

Fig. 5 (**a**) An example of IADC with extended counting using hardware sharing [7]. (**b**) A discrete-time IADC1 acts as the coarse quantization ADC. (**c**) Re-configured as a 10-bit cyclic ADC to perform the fine quantization

the second step. Thus, even though the zoom ADC can measure DC signals with extraordinary energy efficiency, it is not well-suited to wide-band signals, such as occur in bio-potential acquisition systems.

4 Two-Step Incremental ADCs

As shown in Fig. 2, we can improve the SQNR of an IADC by using a higher OSR or a higher-order modulator. For example, for an IADC2 with OSR = 64, doubling the OSR improves the SQNR by 15 dB, while increasing the order by 1 enhances the SQNR by 27 dB. Thus, it is more effective to increase the order of modulation than to raise the OSR. Unfortunately, increasing the order requires extra opamps. Besides, a higher-resolution internal quantizer is usually needed to make a higher-order modulator stable, and the complexity of the peripheral circuitry also increases. The power required increases accordingly, and the ADC becomes less efficient.

Next, a two-step architecture [23] will be described which avoids the excess power dissipation for high-resolution data conversion. Figure 6 shows the z-domain model of the proposed two-step IADC2. During the first step, lasting for M_1 clock periods, the circuit is operated as a conventional IADC2 (Fig. 6a). The residue voltage V_{RES} stored in the second integrator (INT2) after clock period M_1 is given by

$$V_{RES} = W_2\left[M_1\right] = \sum_{K=1}^{M_1-1}\sum_{i=1}^{K-1} U\left[i\right] - \sum_{K=1}^{M_1-1}\sum_{i=1}^{K-1} D_1\left[i\right] \tag{10}$$

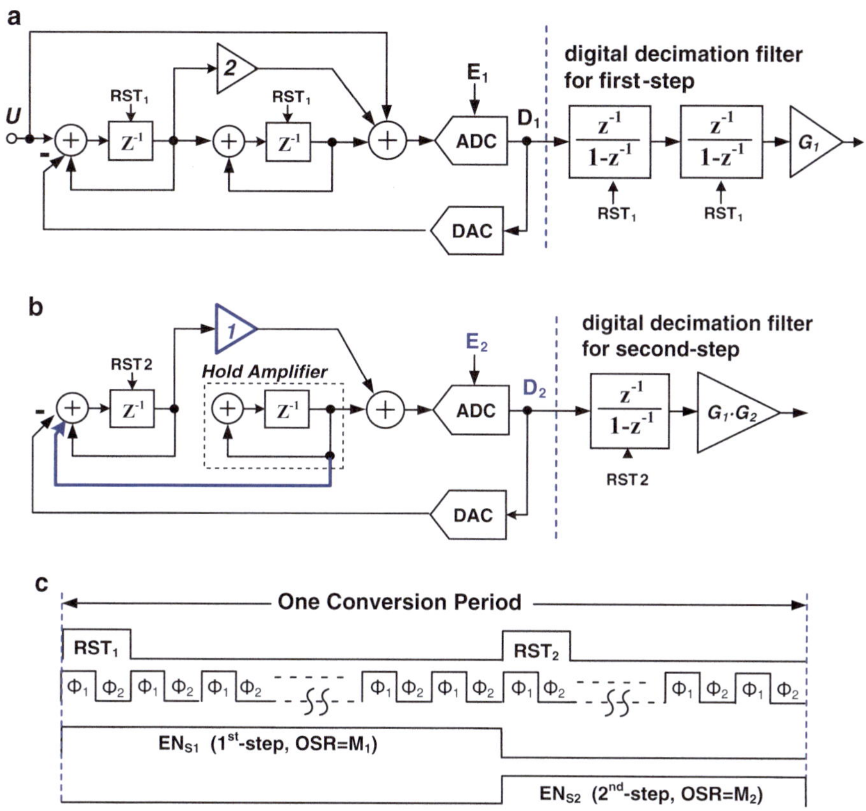

Fig. 6 The proposed IADC2 in two-step operation. (**a**) First step. (**b**) Second step. (**c**) The simplified timing diagram

The direct-input feed-forward modulator generates the residue voltage at the end of first conversion step for fine quantization.

To perform the second step (fine quantization), the analog modulator and the digital filter are reconfigured, as shown in Fig. 6b. The INT2 now stops sampling, and acts as a hold amplifier that feeds the residue voltage V_{RES} into the L-level quantizer and the first integrator (INT1). INT1 is reset again, and then samples the residue voltage V_{RES} from INT2. The reconfigured circuits act as an IADC1 for the remaining M_2 clock periods. Analysis gives

$$\sum_{i=1}^{M_2-1} V_{RES} + E_2 = \sum_{i=1}^{M_2-1} D_2\,[i] \tag{11}$$

Since V_{RES} remains constant during the second step, it can be represented as

$$\sum_{i=1}^{M_2-1} V_{RES} = (M_2 - 1) \cdot V_{RES} \tag{12}$$

The quantization error E_1 of the first step IADC2 is cancelled as in a MASH $\Delta\Sigma$ ADC [12, 13], and only the final error E_2 remains after the two-step operation. While the input voltage U is sampled only during the first step, the average of the input signal $\tilde{U}$ can be re-defined with M replaced by M_1 in (3). After the two steps of conversion, the signals satisfy

$$\tilde{U} + \frac{2}{(M_1-1)(M_1-2)(M_2-1)}E_2 = \frac{2}{(M_1-1)(M_1-2)}\left(\sum_{j=1}^{M_1-1}\sum_{i=1}^{j-1} D_1[i] + \frac{1}{(M_2-1)}\sum_{i=1}^{M_2-1} D_2[i]\right) \quad (13)$$

The decimation filter needed to reconstruct the bit streams of each step can be designed from the right-hand-side of (13). For the first step, the decimation filter can be realized by two cascaded counters. For the second step, one of the counters can be reused. Thus, the IADC2's analog and digital hardware can be used in both steps with a simple reconfiguration. The equivalent quantization error of the two step conversion can be estimated from

$$E_{21} = \frac{2}{(M_1-1)(M_1-2)(M_2-1)}\frac{V_{FS}}{L-1} \quad (14)$$

The SQNR at full-scale input amplitude is given by

$$SQNR_{21}{=}20\ \log(V_{FS}/E_1) \approx 2\cdot 20\ \log(M_1)+20\ \log(M_2)+20\ \log(L-1)-6 \quad (15)$$

With a total OSR $= M = M_1 + M_2$, the optimal selection of the OSR values M_1 and M_2 in a two-step IADC is easily found. Defining the ratio $k = M_1/M_2$, we obtain $M_1 = kM/(k+1)$ and $M_2 = M/(k+1)$. The quantization error of the two-step IADC can then be written in the form

$$E_{21} \approx \frac{2}{M_1^2\cdot M_2}\frac{V_{FS}}{L-1} = \frac{2}{\frac{k^2\cdot M^3}{(k+1)^3}}\frac{V_{FS}}{L-1} \quad (16)$$

The minimum of the quantization error results by setting $k = 2$. The optimum OSRs of the two steps are then $M_1 = 2M_2 = 2M/3$. The maximum $SQNR_{21}$ is

$$SQNR_{21,OPT} \approx 3\cdot 20\ \log(M) + 20\ \log(L-1) - 20\ \log(6) - 20\ \log(9/4) \quad (17)$$

For a total OSR of 192, the SQNR of the two-step IADC versus M_1 is plotted in Fig. 7a, which verifies that the maximum SQNR for $M_1 = 2M_2 = 128$. It can be seen that the optimal ratio is not very sensitive to the exact value of k.

From (9), the SQNR of an IADC3 with the same conversion time is

$$SQNR_3 \approx 3\cdot 20\ \log(M) + 20\ \log(L-1) - 20\ \log(6) \quad (18)$$

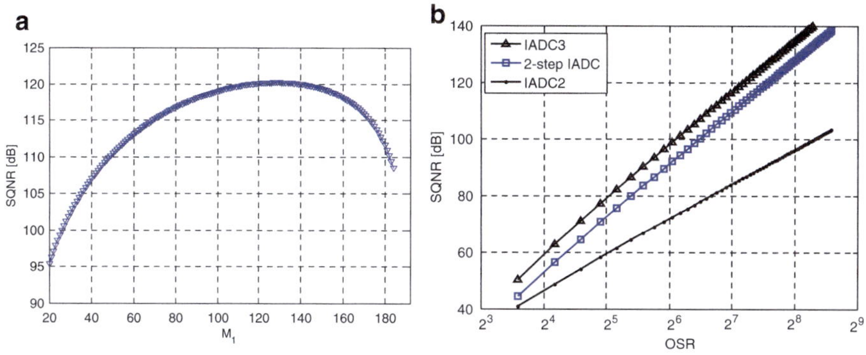

Fig. 7 (**a**) Simulated SQNR versus OSR of the first step (M_1). The input amplitude is −6 dBFS. (**b**) Simulated SQNR versus OSR for a single-loop IADC3, IADC2 and the proposed two-step IADC2. All the IADCs are assumed to have a five-level quantizer, and are tested at −6 dBFS input signal amplitude

Comparison of $SQNR_{21}$ and $SQNR_3$ shows that the two-step IADC2's SQNR is 7 dB lower than that of an IADC3. However, its noise shaping is nearly one order higher than that of a single-loop IADC2. Figure 7b compares the simulated SQNR versus OSR curves for a single-loop IADC2, a single-loop IADC3 and the two-step IADC2. For OSR = 128, an IADC2 can achieve 85 dB, and an IADC3 can achieve 117 dB with a −6 dBFS input signal. Reusing the hardware of an IADC2 in a two-step operation, results in $SQNR_{21}$ ~110 dB, 27 dB higher than $SQNR_2$. Note also that the IADC3 will be overloaded by a −6 dBFS input signal, unless a high-resolution internal quantizer is used. A conventional 2-1 MASH modulator can mitigate the stability issue, but it requires three opamps to achieve third-order noise-shaping performance.

In the proposed two-step operation, the first-step IADC2 is operated only for 2/3 of the total conversion time. The SQNR loss is compensated by the second-step IADC1 operation. The energy efficiency is thereby improved significantly. The higher the OSR, the more significant is the resulting SQNR improvement. The extra circuit cost is low: only an additional timing control is needed to switch the hardware between the two steps. The circuit configuration is much less complex than previously reported hardware-sharing extended-counting schemes [7, 8, 15].

Generally, if the two-step architecture is applied to an *Nth*-order IADC, its performance will be boosted up to nearly that of a *(2N − 1)th*-order one. By using an IADC3 in a two-step operation, as shown in Fig. 8a, an IADC3 performs the first step, and then it is re-configured as an IADC2 for the second step, as shown in Fig. 8b. Figure 8c plots the simulated SQNR of the two-step IADC3 versus the total OSR. Compared to a single-loop IADC3 and IADC2, the two-step IADC3's performance is indeed nearly equal to that of an IADC5. The optimal SQNR can be achieved for an OSR ratio 3:2, which can be derived as in (16). However, the improvement is more significant when the OSR is higher.

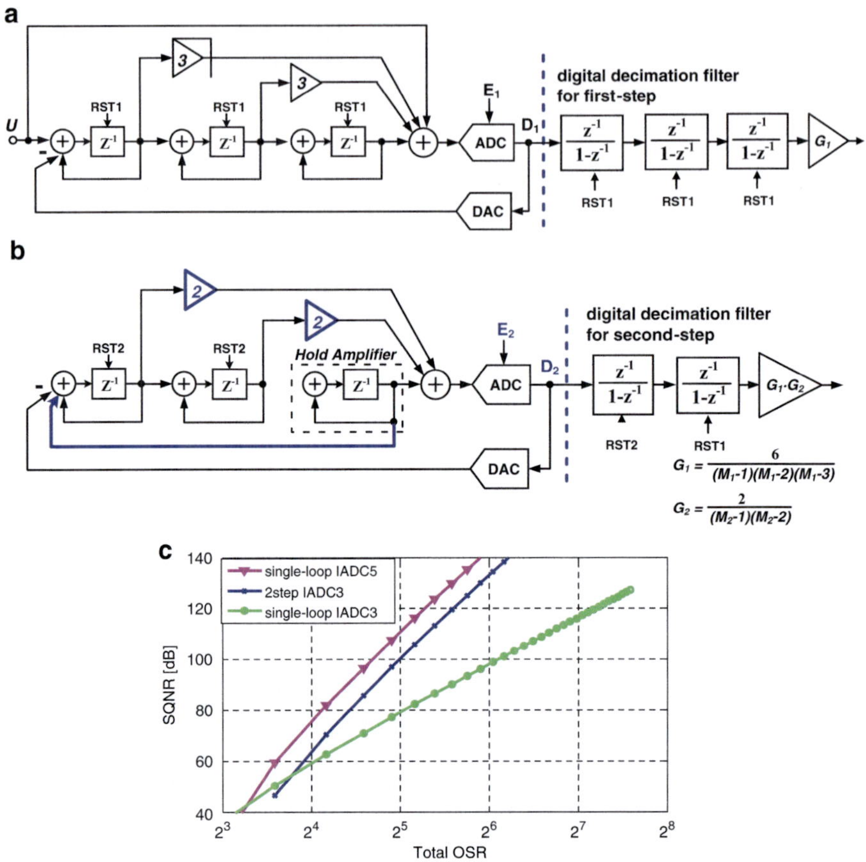

Fig. 8 The IADC3 in two step operation. (**a**) First step. (**b**) Second step. (**c**) Simulated SQNR versus single-loop IADC5 and IADC3

In [24], an algorithmic IADC was proposed with similar two-step operation. However, it requires an extra sample-and-hold stage, and also an additional clock phase to feed back the residue voltage. This complicates the circuit implementation. In our proposed two-step IADC, neither an additional phase, nor an extra active component is needed. All components are reused to accomplish higher SQNR performance, and the power consumption remains the same.

4.1 Multi-Step Operation by an IADC2

The two-step operation can be extended to multiple steps. An example using an IADC2 is shown in Fig. 9 [25]. One more integrator is added to store the residue

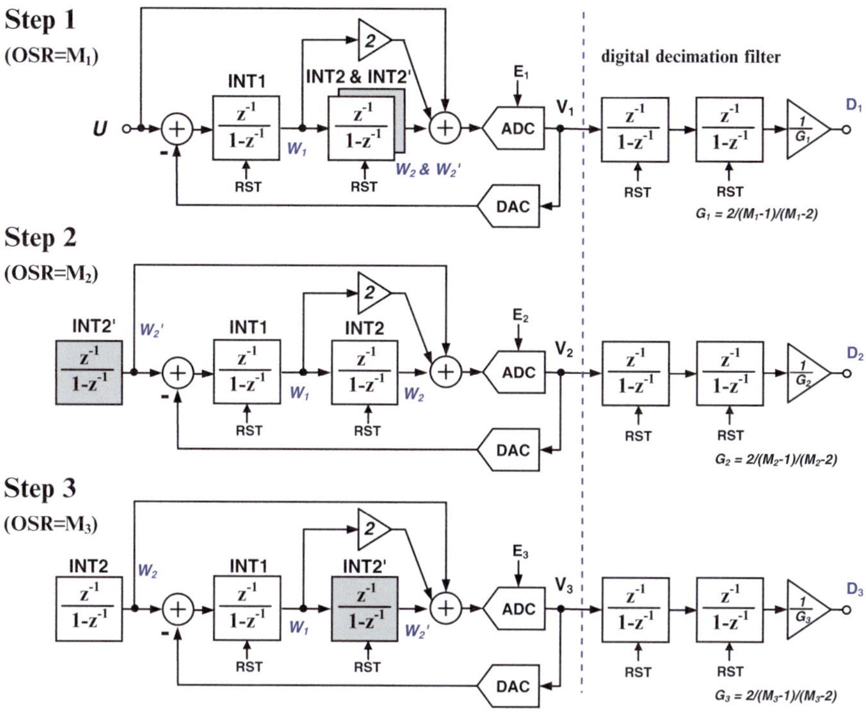

Fig. 9 An example of a multi-step IADC2 [25]

voltage during the first-step IADC operation lasting M_1 clock periods. In the second step of M_2 clock periods, the circuits is re-configured as an IADC2, and INT2′ holds the residue voltage of step one. In the third step, INT2 and INT2′ exchange roles, and the circuit is re-configured again as an IADC2. In each step, a second-order noise shaping is performed. Hence, after an OSR $= M_1 + M_2 + M_3$ clock periods, the order of noise shaping could be nearly 6. Thus, the conversion time can be reduced significantly.

5 Circuit Design Example of the Two-Step IADC

5.1 Switched-Capacitor Circuitry

When the ADC is implemented in a 65 nm technology, the leakage current of the 1-V core MOS devices degrades the performance of a switched-capacitor circuit operated at a low sampling frequency. However, the leakage current of the 2.5 V I/O devices is only 2 pA/μm, which is low enough even for a high-resolution ADC. Hence, 2.5 V I/O devices were used here to implement the prototype IADC.

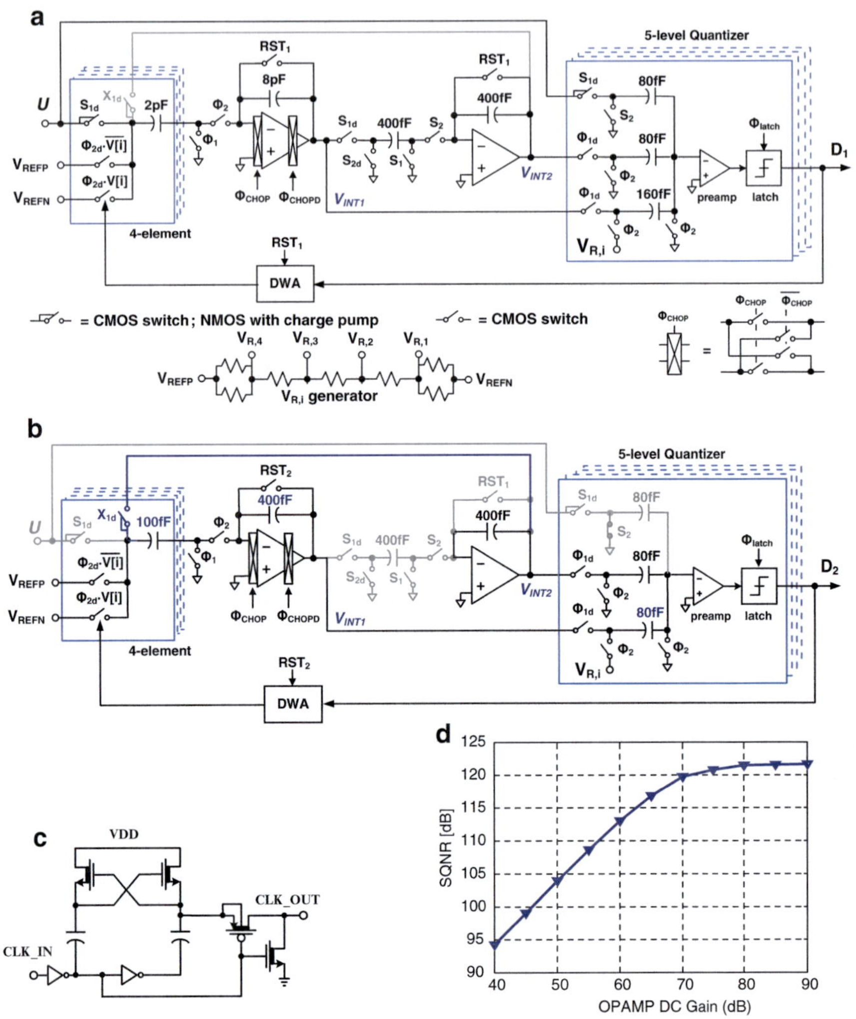

Fig. 10 The equivalent single-ended switched-capacitor circuits implementation of the two-step IADC's modulator. (**a**) First step. (**b**) Second step. (**c**) Voltage doubler. (**d**) Simulated SQNR vs. opamp gain

The switched-capacitor circuit implementation of the proposed two-step IADC's modulator is shown in Fig. 10. Single-ended equivalent circuits are shown for simplicity, but the actual implementation is fully differential. A conventional resistor string was used to generate the five-level reference voltages $V_{R,i}$ for all comparators. To operate the 2.5-V MOS devices with a 1.2-V power supply, the charge pump circuits shown in Fig. 10c were used to double the NMOS gate voltages of the sampling CMOS switches. The I/O devices do not suffer from gate and junction overdrive when operated at 2.5 V, and no extra transistors were needed to improve their reliability.

During the first step lasting $M_I = 128$ clock periods, as shown in Fig. 10a, the gray-scaled paths are not enabled, and the circuit is working as a conventional IADC2. To achieve a 100 dB SNR, the input sampling capacitor of the first integrator is designed to be 8 pF from kT/C thermal noise consideration [6, 9]. During the second step, for $M_2 = 64$ clock periods, as shown in Fig. 10b the two-phase clocks S_1, S_2 are disabled, and X_1 and X_2 establish different input paths reconfiguring the circuit as a first-order modulator. (In the switched-capacitor circuitry used, it is simple to multiplex the different paths and to perform reconfiguration.) The second integrator (INT2), which is now acting as a hold amplifier, drives the INT1's sampling capacitors. The input sampling capacitors of INT1 can therefore be reduced from 8 to 0.4 pF, to ease the loading of INT2.

Since in our circuit the signal bandwidth is 1–250 Hz, it is sensitive to flicker noise. The first opamp's in-band flicker noise is hence mitigated by chopping, at half of the 96 kHz sampling frequency. The signal is chopped during the middle of integrator sampling phase. The input chopping switches are turned off slightly before the output chopping switches, in order to reduce the signal-dependent charge injection from the output chopping switches. Careful layout techniques were employed to make sure that the in-band residual noise caused by chopper non-idealities is low.

In the proposed two-step IADC (Fig. 6), the bit streams of each step are also separately accumulated and decimated. It has the same advantage as the MASH IADC (Fig. 3): the digital circuitry providing $\left(1 - z^{-1}\right)$ is no longer needed, and the opamp gain is much relaxed. Figure 10d shows the simulated SQNR versus the opamp DC gain. The SQNR begins to degrade only when the opamp DC gain falls below 70 dB. The required opamp gain for the proposed two-step IADC is therefore quite low, even for very high-resolution conversion. Although the loop gain in a third-order system can further relax the opamp gain, a second-order system with moderate relaxation can also save cost, and thus improve efficiency.

The detailed timing diagram for the switched-capacitor circuitry is shown in Fig. 11a. An external 96 kHz clock is used to generate the reset signals RST_1, RST_2 and the control signals EN_{S1}, EN_{S2}. The two-phase non-overlapping clock phases Φ_1 and Φ_2 at 96 kHz are used during both steps, while the S_1, S_2 and X_1, X_2 two-phase clock signals are specifically for the first and second step, respectively. Bottom-plate sampling is used to mitigate the switches' non-idealities. The delayed versions of the two-phase clock signals Φ_{1d}, Φ_{2d}, S_{1d}, S_{2d}, X_{1d}, X_{2d} and Φ_{CHOPD} are omitted for simplicity. The simplified circuit used to generate the control timing and two-phase clocks is shown in Fig. 11b. It uses only frequency dividers and simple logic circuits, and hence it is simple and does not need a complicated state machine to generate the timing controls.

For a total OSR of 192, the two-step IADC2 can ideally achieve 120 dB SQNR for a −6 dBFS input amplitude, which is adequate for a 100 dB SNR ADC. For comparison, a single-loop IADC2 with OSR = 128 and OSR = 192 can achieve only 84 dB and 91 dB SQNR, respectively. Increasing the OSR of an IADC2 from 128 to 192 can give only a 7 dB SNQR improvement, while increasing the order of

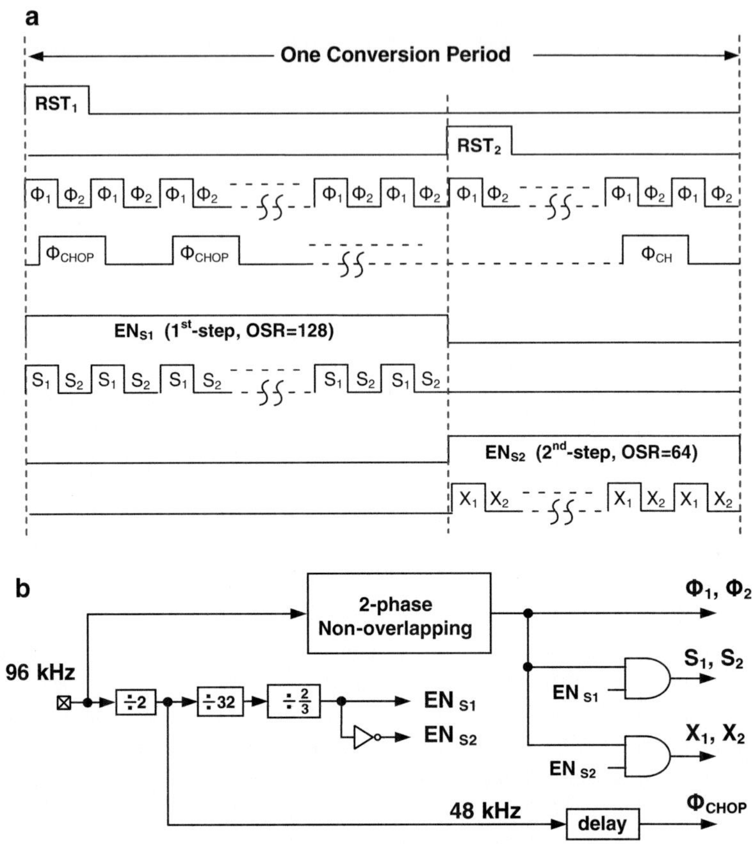

Fig. 11 (**a**) The detailed timing diagrams for the two-step IADC2. (**b**) The circuit for timing control and two-phase non-overlapping clocks

the noise-shaping by 1 can improve the SQNR significantly, by 30 dB. The power penalty for the additional conversion time of 64 clock periods and for the extra control circuitry is small. By just enabling and disabling the control clocks of the switched capacitor circuit, a simple and low-cost operation results.

The digital decimation filter shown in Fig. 6 is not implemented on the chip; the modulator's bit streams were post-processed using MATLAB. The detailed circuit design of the other building blocks can be found in [17].

5.2 Measured Performance

Defining the differential reference 2.4 V_{PP} as 0 dBFS, the measured spectra for a differential 100 μV_{PP} (−87.6 dBFS), 170 Hz, sine-wave input signal are shown in

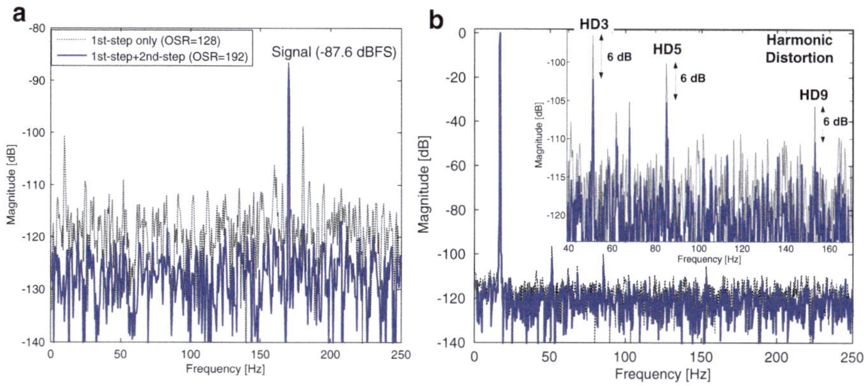

Fig. 12 Measured spectra. The *dotted-line* is for the first step only (IADC2, OSR = 128). The *solid-line* is for the two-step IADC with OSR = 192. (**a**) 100 μV_{PP} (−87.6 dBFS) input amplitude. (**b**) 2.2 V_{PP} (−0.76 dBFS) input amplitude

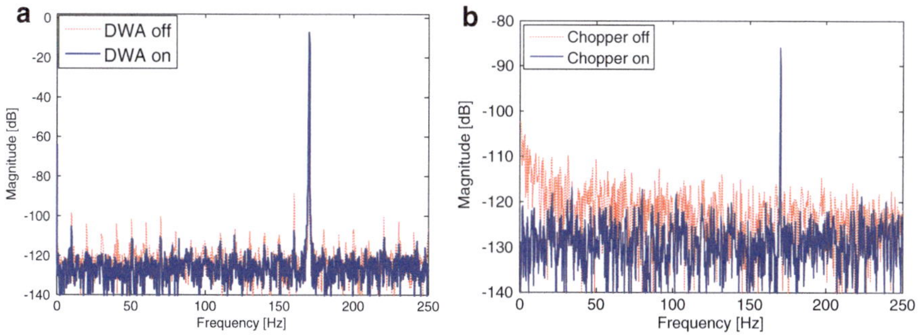

Fig. 13 Measured spectra. (**a**) DWA turned on vs. off with 1 V_{PP} (−6.8 dBFS) input amplitude. (**b**) Chopper on vs. off with 100 μV_{PP} input amplitude

Fig. 12a. The spectra obtained after the first step (OSR = 128) and the second step (OSR = 64) are plotted, showing that the second step enhances the signal-to-noise-and-distortion-ratio (SNDR) by 10.3 dB. Figure 12b shows the measured spectra for a 2.2 V_{PP} (−0.76 dBFS) 17-Hz sine-wave input. The measured SNDR is 84.7 dB for the first step, and 91 dB for two-step operation. Harmonic distortion limits the SNDR for such large signals, and the two-step operation enhances the SNDR by only 7 dB. Figure 13a shows the spectra with the DAC DWA turned on and off. The measured SNDR with a 1 V_{PP} (−7.6 dBFS) input amplitude is 84.6 dB (with the DWA on) and 75.8 dB (DWA off). Nevertheless, the in-band flicker noise degrades the SNR performance significantly. Figure 13b shows the measured spectra with the chopper turned on and off. The chopper stabilization reduces the in-band flicker noise by 11 dB.

The ADC achieves a dynamic range of 99.8 dB and a peak SNDR of 91 dB with a bandwidth from 1 to 250 Hz, consuming only 10.7 μW. Table 1shows a performance

Table 1 Performance summary and comparison

Parameters	[17] **This work**	[27] ESSCIRC '13	[21] TCAS-I '10	[6] JSSC '10	[26] ISSCC '13	[10] JSSC '06
Architecture	**IADC2 + IADC1**	10b SAR + IADC1	IADC2 + 10b cyclic	IADC2 + 11b SAR	Single-loop IADC2	Single-loop IADC3
Process	**65 nm (2.5 V MOS)**	0.6 μm	0.18 μm	0.18 μm	0.16 μm	0.6 μm
Area (mm^2)	**0.20**	1.64	0.50	3.5	0.45	2.08
VDD (V)	**1.2**	3.3	2	1.8	1	3
Sampling freq.	**96 kHz**	5 MHz	115 MHz	45.2 MHz	750 kHz	30.7 kHz
OSR	**192**	256	5	45	80	512
Input range	**2.2 V_{pp}**	2 V_{pp}	3.6 V_{pp}	2 V_{pp}	0.7 V_{pp}	6 V_{pp}
Dyn. range (dB)	**99.8**	84.6	73	90.1	81.9	120
Peak SNDR (dB)	**90.8**	70.7	72	86.3	81.9	120
Bandwidth (Hz)	**250 Hz**	9.75 kHz	11.5 MHz	500 kHz	667 Hz	7.5 Hz
Power	**10.7 μW**	64 μW	48 mW	38.1 mW	20 μW	300 μW
FoM_W (pJ/conv.)	**0.76**	1.17	1.02	1.46	1.48	24.46
FoM_S (dB)	**173.5**	166.4	156.8	161.3	157.1	164.0

summary and comparison with recent state-of-the-art single-loop IADCs [10, 26], as well as with hybrid IADCs using extended-counting schemes [6, 21, 27]. The Walden (FoM_W) and Schreier (FoM_S) figure-of-merits were also calculated, using the formulas

$$FoM_W = \frac{power}{2^{ENOB} \cdot 2BW} \tag{19}$$

$$FoM_S = DR + 10 \cdot \log\left(BW/power\right) \tag{20}$$

For this device, $FoM_W = 0.76$ pJ/conv.-step and $FoM_S = 173.5$ dB were found, both among the best reported results.

6 Conclusions

In this chapter, we first reviewed the design and operation of a conventional single-loop IADC2 using time-domain analysis. The advantages and design considerations of MASH IADCs were also discussed. Using a feedforward modulator, the loop filter accumulates the residue voltage, and stores it at the last integrator's output node. The residue voltage can then be used for fine conversion through an extended counting scheme, which significantly raises the energy efficiency. Several such schemes were reviewed, and their advantages and drawbacks discussed.

To further improve the energy efficiency, we proposed multi-step operation for high-resolution ADCs for use in integrated sensor interface circuits. For example, the components of an *Nth*-order IADC can be re-used to quantize the residue voltage in a second-step operation, resulting in noise-shaping performance close to that of an IADC of order $(2N-1)$. The extra cost is only simple added timing control circuits. Moreover, the required opamp gains can be as low as 60 dB even for 100 dB SNR. The principle can be extended to three- and higher-step operation.

A design example of a two-step IADC2 was demonstrated. The ADC was fabricated using 2.5 V I/O MOS devices in a 65-nm technology, and operated with a 1.2 V power supply. The measured performance showed a 100 dB dynamic range and 91 dB maximum SNDR for a signal bandwidth from 1 to 250 Hz. The device consumed only 10.7 μW. The measured Walden and Schreier FoMs were 0.76 pJ/conversion-step and 173.5 dB, respectively, among the best published IADC FoMs. The active area is 0.2 mm^2, which is the smallest among published designs. The results verify that the proposed two-step IADC is a very area- and energy-efficient solution for integrated sensor systems.

References

1. Wu R, Chae Y, Huijsing JH, Makinwa KAA (2012) A 20-bit ±40mV range read-out IC with 50-nV offset and 0.04% gain error for bridge transducers. IEEE J Solid-State Circuits 47(9):2152–2163
2. Tan Z, Shalmany SH, Meijer GCM, Pertijs MAP (2012) An energy-efficient 15-bit capacitive-sensor interface based on period modulation. IEEE J Solid-State Circuits 47(7):1703–1711
3. Tan Z, Deval P, Daamen R, Humbert A, Ponomarev YV, Chae Y, Pertijs MAP (2013) A 1.2-V 8.3-nJ CMOS humidity sensor for RFID applications. IEEE J Solid-State Circuits 48(10):2469–2477
4. Van Helleputte N et al (2011) A 345 μW multi-sensor biomedical SoC with bio-impedance, 3-channel ECG, motion artifact reduction and integrated DSP. IEEE J Solid-State Circuits 50(1):230–244
5. Chen C-H, Crop J, Chae J, Chiang P, Temes GC (2012) A 12-bit 7 μW/channel 1 kHz/channel incremental ADC for biosensor interface circuits. In: Proceedings of the IEEE international symposium on circuits and systems (ISCAS), pp 2969–2972
6. Agah A, Vleugels K, Griffin PB, Ronaghi M, Plummer JD, Wooley BA (2010) A high-resolution low-power oversampling ADC with extended-range for bio-sensor arrays. IEEE J Solid-State Circuits 45(6):1099–1110
7. Kim J-H et al (2012) A 14b extended counting ADC implemented in a 24Mpixel APS-C CMOS image sensor. In: IEEE ISSCC digest of technical papers, pp 390–392
8. Oike Y, El Gamal A (2013) CMOS image sensor with per-column sigma delta ADC and programmable compressed sensing. IEEE J Solid-State Circuits 48(1):318–328
9. Markus J, Silva J, Temes GC (2004) Theory and applications of incremental delta sigma converters. IEEE Trans Circuits Syst I 51(4):678–690
10. Quiquempoix V, Deval P, Barreto A, Bellini G, Markus J, Silva J, Temes GC (2006) A low-power 22-bit incremental ADC. IEEE J Solid-State Circuits 41(7):1562–1571
11. Kavusi S, Kakavand H, El Gamal A (2006) On incremental sigma-delta modulation with optimal filtering. IEEE Trans Circuits Syst I 53(5):1004–1015
12. Caldwell TC, Johns DA (2010) Incremental data converters at low oversampling ratios. IEEE Trans Circuits Syst I 57(7):1525–1537
13. Robert J, Deval P (1988) A second-order high-resolution incremental A/D converter with offset and charge injection compensation. IEEE J Solid-State Circuits 23(3):736–741
14. Harjani R, Lee TA (1998) FRC: a method for extending the resolution of Nyquist rate converters using oversampling. IEEE Trans Circuits Syst II 45(4):482–494
15. Maeyer JD, Rombouts P, Weyten L (2004) A double-sampling extended-counting ADC. IEEE J Solid-State Circuits 39(3):411–418
16. Chen C-H, Zhang Y, He T, Chiang P, Temes GC (2014) A 11 μW 250 Hz BW two-step incremental ADC with 100 dB DR and 91 dB SNDR for integrated sensor interfaces. In: IEEE custom integrated circuits conference (CICC)
17. Chen C-H, Zhang Y, He T, Chiang P, Temes GC (2015) A micro-power two-step incremental analog-to-digital converter. IEEE J Solid-State Circuits 50(8)
18. Carbone P, Xu F, Kiaei S, Temes GC (2014) Incremental and extended-range data converters. In: Design, modeling and testing of data converters. Springer, Berlin, pp 143–159
19. Schreier R, Temes GC (2005) Understanding delta-sigma data converters. IEEE Press/Wiley, Pascataway
20. Fujimori I et al (2000) A 90-dB SNR 2.5-MHz output-rate ADC using cascaded multibit delta-sigma modulation at 8× oversampling ratio. IEEE J Solid-State Circuits 35(12):1820–1828
21. Lee CC, Flynn MP (2011) A 14b 23 MS/s 48 mW resetting ΔΣ ADC. IEEE Trans Circuits Syst I 58(6):1167–1177
22. Chae Y, Souri K, Makinwa KA (2013) A 6.3 μW 20 bit incremental zoom-ADC with 6 ppm INL and 1 μV offset. IEEE J Solid-State Circuits 48(12):3019–3027

23. Chen C-H, Zhang Y, Jung Y, He T, Ceballos JL, Temes GC (2013) Two-step incremental analogue-to-digital converter. Electron Lett 49(4):250–251
24. Mulliken G, Adil F, Cauwenberghs G, Genov R (2002) Delta-sigma algorithmic analog-to-digital conversion. In: Proceedings of the IEEE international symposium on circuits and system (ISCAS), pp 687–690
25. He T, Zhang Y, Meng X, Chen C-H, Temes GC (2015) Micro-power multi-step incremental ADCs for multi-channel sensor interfaces. In: Proceedings of the IEEE international symposium on circuits and systems (ISCAS), 2015, to appear
26. Chen C, Tan Z, Pertijs MAP (2013) A 1V 14b self-timed zero-crossing-based incremental $\Delta\Sigma$ ADC. In IEEE ISSCC digest of technical papers, pp 274–275
27. Ha S et al (2013) 85 dB dynamic range 1.2 mW 156 kS/s biopotential recording IC for high-density ECoG flexible active electrode array. In: Proceedings of the European solid-state circuits conference (ESSCIRC)

Energy-Efficient CDCs for Millimeter Sensor Nodes

Sechang Oh, Wanyeong Jung, Hyunsoo Ha, Jae-Yoon Sim, and David Blaauw

Abstract Multiple energy-efficient CDCs are proposed for millimeter sensor nodes. Compared to the state-of-the-art, these CDCs achieves excellent energy efficiency, high SNR, and wide input range with a variety of techniques. These include correlated double sampling in front of a SAR ADC, incremental delta-sigma conversion with a zoom-in SAR converter, energy-efficient dual-slope conversion, and fully-digital iterative delay-chain discharge conversion.

1 Introduction

Continuous advances in low-power integrated circuit design and fabrication technology have led to smaller volume computing systems. The system volume can be few cubic millimeter [1, 2] (Fig. 1), and a key challenge is the stringent power budget due to limited battery capacity. Capacitive sensors are well suited to these systems because they do not consume static power. They are widely used in diverse applications to measure environmental signals, such as pressure [2], displacement [3], humidity [4], and acceleration [5]. However, their interfacing circuits can easily dominate system power, which can be as low as few nanowatts [6], and hence energy-efficient capacitance-to-digital converters (CDCs) are required. Therefore, CDC researchers aim at obtaining high resolution with low power consumption. CDCs can be categorized by their conversion schemes including successive approximation [7, 8], delta-sigma modulation [3, 4, 9], and period modulation [2, 10–13].

In the following sections, energy-efficient design techniques are introduced with various state-of-the-art CDC design examples. Section 2 introduces a SAR CDC that

S. Oh (✉) • W. Jung • D. Blaauw
University of Michigan, Ann Arbor, MI, USA
e-mail: chaseoh@umich.edu

H. Ha
Holst Centre and imec, Eindhoven, The Netherlands

J.-Y. Sim
Pohang University of Science and Technology, Pohang, South Korea

K.A.A. Makinwa et al. (eds.), *Efficient Sensor Interfaces, Advanced Amplifiers and Low Power RF Systems*, DOI 10.1007/978-3-319-21185-5_3

Fig. 1 Pressure sensing millimeter sensor node on the edge of US Nickel

uses a correlated double sampling front-end [7]. Section 3 presents an incremental CDC with a 9-bit zoom-in SAR conversion [9]. Section 4 describes a dual slope CDC with an energy-efficient charge subtraction scheme [2]. Section 5 presents a fully-digital CDC using iterative delay-chain discharge [10]. Finally, the paper is concluded in Sect. 6.

2 Energy-Efficient SAR CDC with Correlated Double Sampling

The first design example shows an energy-efficient SAR CDC combining a correlated double sampling (CDS) front-end with a differential asynchronous SAR ADC [7]. It decouples the sensor capacitor from the SAR ADC's capacitive DAC (CDAC), thus preserving the full differential input voltage range for comparisons. It achieves wide conversion range (2.5–75.3 pF) and 1.3 pJ/c-s FoM while consuming 160 nW.

2.1 Correlated Double Sampling

Figure 2 shows the proposed readout scheme with correlated double sampling (CDS). The circuit consists of a sensor capacitor (C_{SENS}), a reference capacitor (C_{REF}), a sampling capacitor (C_{SAMPLE}), and an amplifier. The bottom nodes of

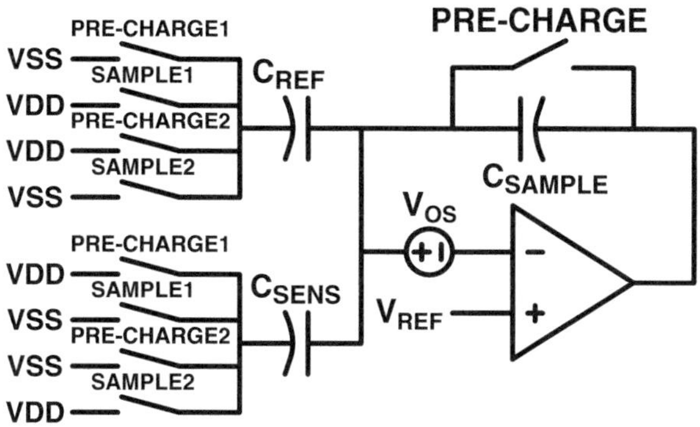

Fig. 2 Correlated double sampling in the proposed CDC

C_{SENS} and C_{REF} are selectively switchable to VDD or VSS. The top node is set to the virtual ground (V_{REF} or VDD/2) through the negative feedback configuration of the amplifier. The charge proportional to the difference between C_{SENS} and C_{REF} are transferred to C_{SAMPLE}, which is performed in four steps: Pre-charge1, Sample1, Pre-charge2, and Sample2. The sequence of Pre-charge1 and Sample1 generates a voltage output proportional to C_{SENS} and C_{REF}. The same operation is repeated in the sequence of Pre-charge2 and Sample2 but with the role of C_{SENS} and C_{REF} switched. Therefore, the two output voltages after Sample1 and Sample2 are equal but with opposite polarity, providing a differential output with double the amount of signal ($Q_{SAMPLE} = 2 \cdot (C_{SENS} - C_{REF}) \cdot VDD$). A key advantage of this approach is that the effect of variations on V_{REF} and offset voltage (V_{OS}) are all canceled in this differential output. Furthermore, parasitic capacitance to ground of off-chip C_{SENS}, which can be several tens of pF, does not affect the sampled result because the top node voltage does not change during the CDS process.

2.2 Detailed CDC Operation

Figure 3 shows the circuit diagram of the CDC, consisting of the sampling front-end, a differential capacitive DAC (CDAC), asynchronous SAR logic, a comparator, and C_{REF} selection logic. Each conversion is performed with CDS followed by an A-D conversion. During CDS, the two capacitor banks of CDAC play the role of C_{SAMPLE} in *Sample1* phase and *Sample2* phase, respectively. Therefore, the differential voltage output is sampled by the CDAC. In the A-D conversion phase, the amplifier is turned off to reduce power consumption, and asynchronous SAR logic generates a 13b output. A transition of the external clock (250 Hz) initiates the conversion process. The timing controller generates all control signals with predefined timing. Though the SAR ADC has a differential input range of −VDD to VDD, the sampling front-end cannot provide such a rail-to-rail output swing due to the amplifier's limited linear output range. In this work, the valid output range of

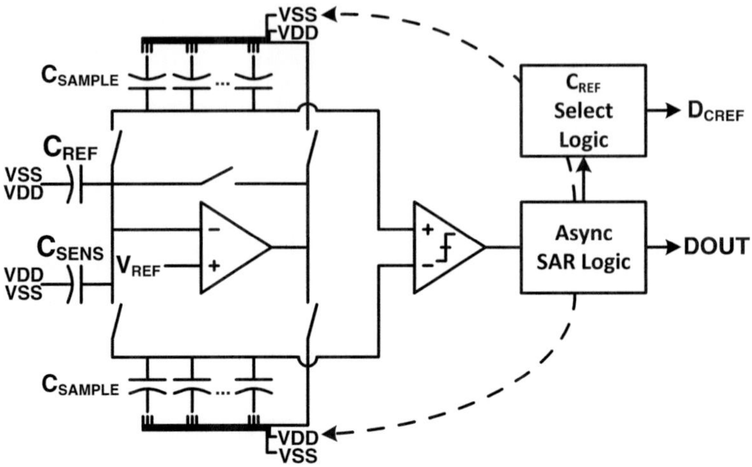

Fig. 3 Circuit diagram of the SAR CDC

the sampling front-end is taken to be from −VDD/2 to VDD/2, corresponding to a C_{SENS} range from $C_{REF} - C_{SAMPLE}/4$ to $C_{REF} + C_{SAMPLE}/4$.

To increase the range of conversion, the C_{REF} select logic chooses one of the eight cases of C_{REF} (from 7.5 to 70.5 pF in 9 pF steps) with a 3b code. The C_{REF} select logic monitors the CDC output code and checks if C_{SENS} is in the valid range of the current C_{REF} value. If not, C_{REF} is incremented or decremented. This procedure is repeated until the CDC output code falls in the valid range, or C_{REF} reaches its maximum or minimum value. The input ranges of neighboring C_{REF}'s overlap by 1 pF to avoid bang-bang updating at the boundaries. The overlap range is also used to calibrate any discontinuities caused by process variations on C_{REF} [14].

2.3 Measurement Results

The CDC is implemented in 0.18 μm standard CMOS and has an active area of 0.49 mm^2. The CDC consumes 119–160 nW as C_{SENS} varies from minimum to maximum. Analog and digital parts use supply voltages of 1.2 V and 0.9 V, respectively. To characterize the CDC's linearity over the entire range of C_{SENS}, the input voltage was swept with fixed C_{SENS} to generate a varying equivalent capacitance instead of a varying physical capacitance. The error in C_{REF} was extracted by measuring the C_{SENS} in the overlap range. There were −0.1 pF and −0.2 pF errors at the two largest cases of C_{REF}, 61.5 pF and 70.5 pF, respectively. Figure 4 shows the measured SNR and FoM in each subrange. The CDC shows SNR of more than 55.4 dB with a resolution of 6.0 fF. The calculated FoM varies from 0.54 to 1.3 pJ/c-s as C_{REF} changes from 7.5 to 70.5 pF.

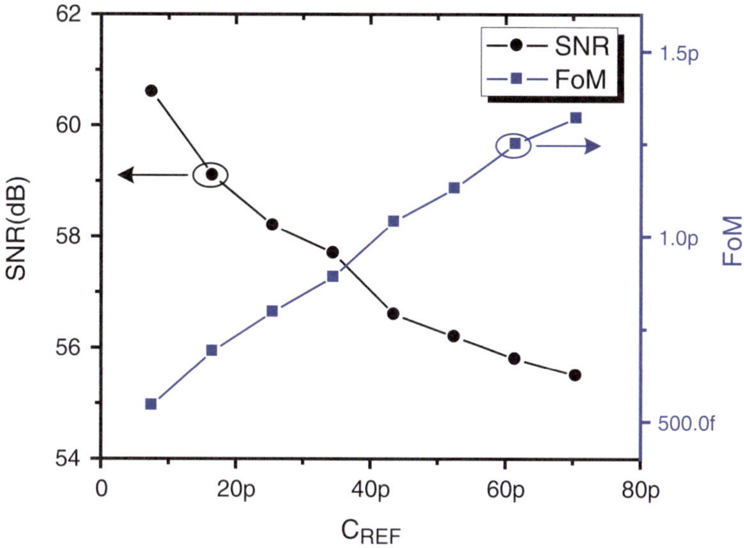

Fig. 4 Measured SNR and FoM across different C_{REF}

3 Incremental CDC with Zoom-In SAR Converter

The second design example describes an energy-efficient incremental CDC with a zoom-in 9b asynchronous SAR converter [9]. By first performing a 9b SAR conversion, the modulator's OSR during the second conversion can be reduced to 32, significantly reducing conversion energy. Unused OTAs in the SAR phase are bypassed, thus further reducing energy. The CDC achieves 94.7 dB SNR and 33.7 μW power consumption with 175 fJ/c-s at 1.4 V supply.

3.1 Zoom-In SAR Converter

Figure 5 shows the block diagram of the proposed CDC. The integer output component, N, is generated by the SAR logic. This is followed by a high-resolution second order incremental $\Delta\Sigma$ conversion that produces the fractional output component, F. During the initial SAR phase, the integration path is bypassed. Although the zoom-in nature restricts the converter to near-DC inputs, it is appropriate for sensor nodes, since environmental signals usually change very slowly.

Figure 6 shows the detailed circuit implementation of the proposed CDC, and the associated waveforms are shown in Fig. 7. The sensed capacitor (C_{sensor}) is an off-chip component, and an on-chip CDAC is used as a reference. In the sampling phase, the nodes n_{cs+} and n_{cs-} are set to the common node voltage (VCM) and VDD,

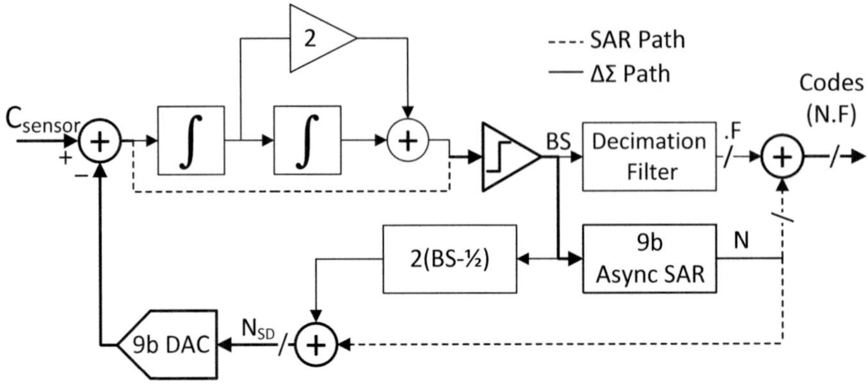

Fig. 5 Block diagram of the proposed incremental CDC

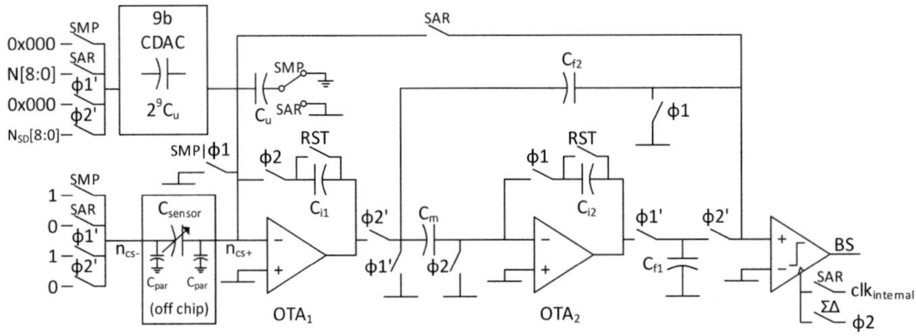

Fig. 6 Detailed circuit diagram showing the integration of the 9b SAR converter and ΔΣ modulator

respectively, and all bottom plates of the CDAC are set to GND. At the beginning of the SAR phase, n_{cs-} becomes GND, and a half of the CDAC bottom plates are switched to VDD. The bottom plate connections are determined using successive approximation, which results in a near VCM final value for n_{cs+}.

Asynchronous logic gates are used for a fast conversion, which allows the SAR conversion to finish within a cycle of the global clock and reduces the static power consumption during the SAR conversion by 90 %. Fifty percent operating margin ($-1 < F < 1$) is given for the ΔΣ phase. The input is shifted by 0.5-bit, and the ΔΣ modulator operates with the elements ($N - 1$) and ($N + 1$) of the CDAC [15]. The 0.5-bit shift is implemented with an additional unit-size capacitor (C_u) of the CDAC. The bottom plate of C_u is set to GND during the sampling phase and to half VDD during the SAR phase. Since the OTAs are bypassed during initial SAR operation, the SAR bits are obtained with negligible energy compared to the bits from the subsequent ΔΣ conversion. The comparator is a two-stage sense amplifier [16] with $\sim$100 μV resolution for the 9b SAR conversion. The maximum SAR resolution is constrained by CDAC mismatch and comparator noise.

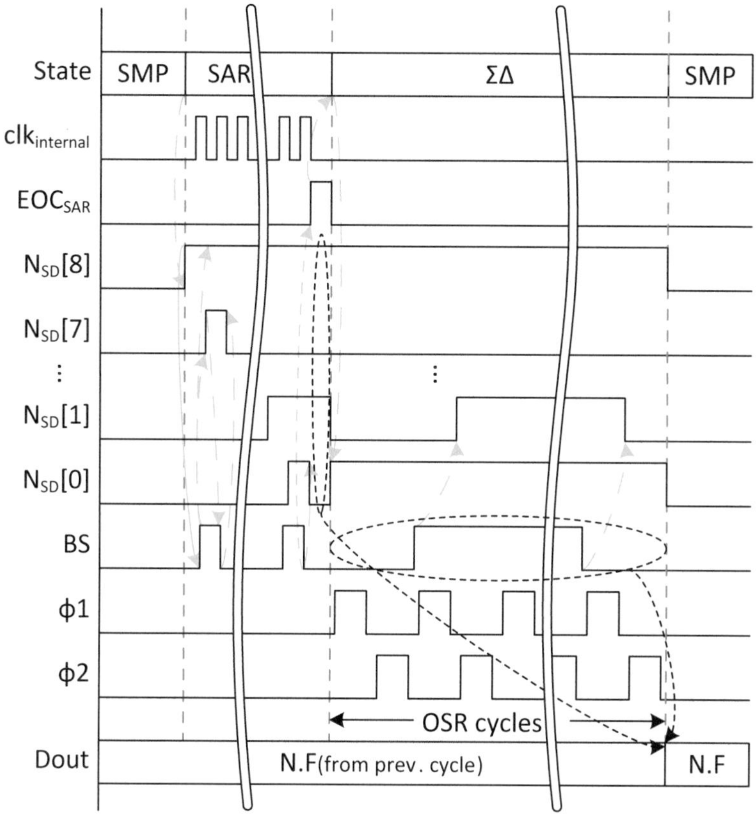

Fig. 7 CDC waveforms

After the SAR phase, the incremental converter provides added resolution based on the SAR result. The architecture is a second order feed-forward structure, similar to [17]. Dynamic element matching (DEM) is used to suppress CDAC mismatch. Either regular sequence or common centroid (CC) configuration can be assigned to the CDAC element indexes. $\varphi 1$ and $\varphi 2$ are 150 kHz non-overlapping clocks. The OTAs are cascoded inverter amplifiers as in [15]. OTA_1 and OTA_2 consume 12 μW and 1 μW, respectively.

3.2 *Measurement Results*

The proposed CDC is implemented in 180 nm CMOS and an active area is 0.46 mm^2. Figure 8 shows how output codes (N.F) are generated. The SAR output has only an 1 code error, and this error is tolerable in the $\Delta\Sigma$ converter. The CDC linearity test is performed by changing the input voltage applied to a fixed 24 pF

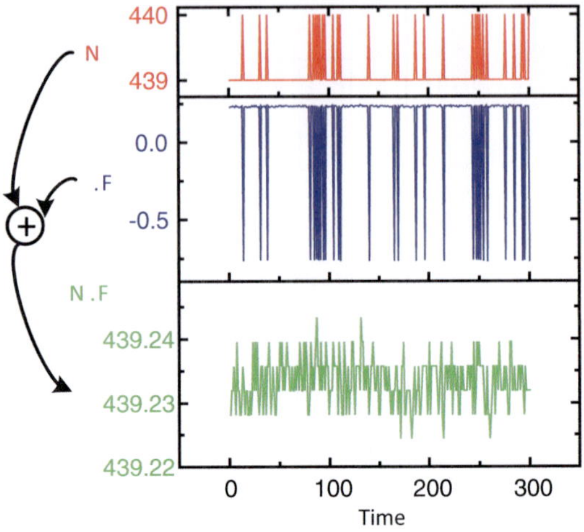

Fig. 8 Measured output code (N.F) generation

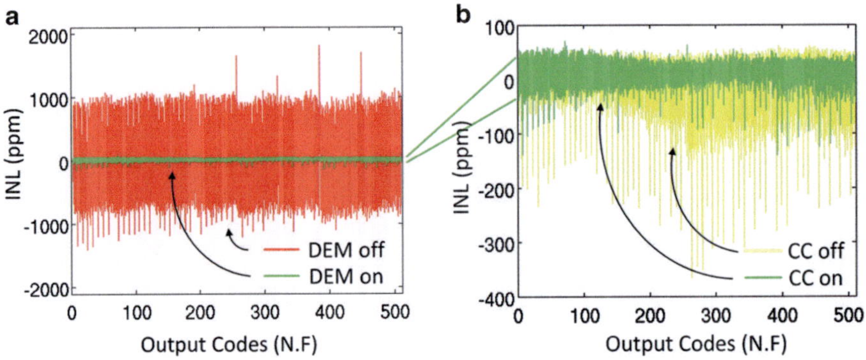

Fig. 9 INL across modes. (**a**) Dynamic element matching on/off with common centroid indexing on. (**b**) Common centroid indexing on/off with dynamic element matching on

sensor capacitor. In Fig. 9, almost all errors are within ±50 ppm (=14.3b) when DEM and CC configuration are used. When CC configuration is not used, the CDC deviates from the $\Delta\Sigma$ working range more often, resulting in more non-linearity. When DEM is OFF, SAR and $\Delta\Sigma$ use different CDAC elements, and it loses all bits from the $\Delta\Sigma$ operation because of capacitor mismatch. This work achieved 94.7 dB SNR, 0.16 fF resolution, and 175 fJ/c-s FoM at 32 OSR and 4.29 kS/s.

4 Dual Slope CDC with Energy-Efficient Charge Subtraction

The third design example presents an energy-efficient dual slope CDC (DS-CDC) for millimeter sensor nodes [2]. Iterative charge subtraction/accumulation using a configurable capacitor bank cancels base capacitance and zooms in to a variable

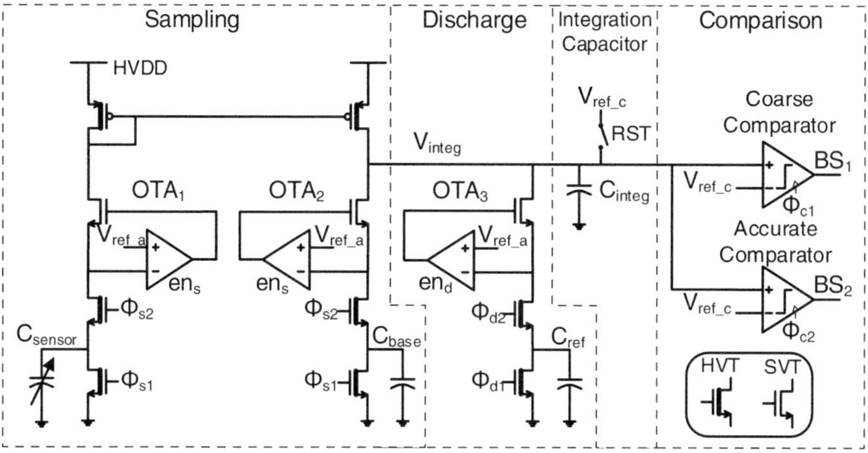

Fig. 10 Circuit diagram of the proposed CDC

input region, thereby reducing conversion time and energy for the DS-CDC. Dual-precision comparators are used in conjunction to achieve the high resolution of the accurate comparator and low energy of the coarse comparator. The CDC has a low power consumption of 110 nW with 8.7 fF resolution and 44.2 dB SNR at 6.4 ms conversion time.

4.1 CDC Operation

The DS-CDC consists of a current mirror, charge subtraction/accumulation devices, and two comparators as shown in Fig. 10, followed by a ripple carry counter and digital control logic. Figure 11 shows the waveforms of the DS-CDC. The sampled charge difference between C_{sensor} and C_{base} is transferred to C_{integ}, and the transferred charge is removed by iterative subtraction using C_{ref}. In the reset state, all the OTAs are disabled and C_{integ} voltage is set to V_{ref_c}. During the sampling state, OTA_1 and OTA_2 are enabled. While $\phi_{s1} = 1$, the charge is removed from C_{sensor} to C_{base} by shorting both nodes of the capacitors to ground. With $\phi_{s2} = 1$, the top plate of these capacitors is set to V_{ref_a} due to the feedback of the OTA and the device gated by ϕ_{s2}. Since $\phi_{s1} = 0$ in this phase, all current conducted by the source followers is accumulated on C_{sensor} or C_{base}. The full sampling operation consists of $4\phi_s$ cycles, thus providing $4\times$ charge amplification. In the following discharge state, OTA_3 and one of the two comparators are turned ON, and the charge in C_{integ} is subtracted by C_{ref} for each ϕ_c cycle. The discharge state ends when V_{integ} becomes smaller than V_{ref_c}; the total number of required cycles is recorded by a ripple carry counter as a digital code: $Code \sim 4\left(C_{sensor} - C_{base}\right)/C_{ref}$.

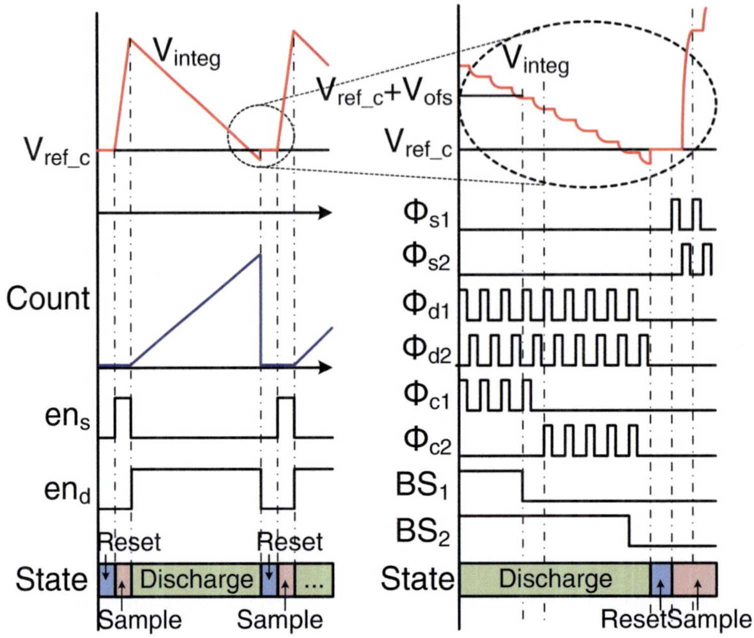

Fig. 11 Waveforms of the proposed CDC

4.2 Energy-Efficient Charge Subtraction

Settling time constant for an OTA feedback loop is $\tau = C_L/(\beta g_{m,OTA})$. In a conventional charge balancing approach (Fig. 12a) this time constant becomes $\left(C_{sensor} + C_{Ref\,fullrange}\right)/g_{m,OTA}$, whereas in the proposed method (Fig. 12b) it is reduced to $C_{ref}/g_{m,OTA}$ by Cref isolation. Since Cref is much smaller than Csensor, it allows for significantly lower OTA tail current for a fixed sampling rate. OTAs in the proposed design use a single stage design with 32 nW tail-currents. Although the OTAs use a 1.2 V supply, the current mirror uses 3.6 V, which is available in a battery powered complete microsystem [2]; this increases V_{integ} range while keeping power low.

4.3 Dual Comparators

The proposed CDC uses two clocked comparators; a coarse comparator with minimum size transistors (720 μV rms input referred noise, simulated) and a fine comparator with 10× larger transistors (100 μV rms noise, simulated). The coarse comparator is used for the most cycles during the discharge state. After the coarse comparator is flipped, the fine comparator is used to find the final

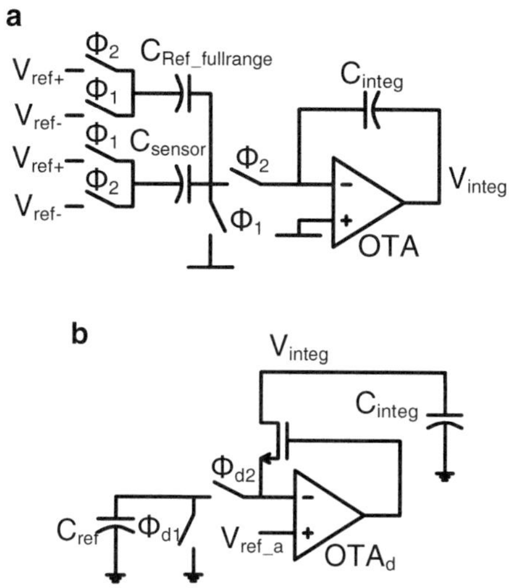

Fig. 12 (**a**) Conventional charge balancing CDC and (**b**) the proposed charge subtraction method

point that $V_{integ} < V_{ref\_c}$ (Fig. 11). To accomplish this, a reference voltage offset is required between the two comparators. In the proposed scheme, the lower power coarse comparator is used for the vast majority (>99 %) of the discharge cycles while the fine comparator increases the CDC resolution. Clocks ϕ_{s1}/ϕ_{s2} and ϕ_{d1}/ϕ_{d2} are non-overlapping 125 kHz clock pairs.

4.4 Measurement Results

The CDC is implemented in 180 nm CMOS and has an active area of 0.11 mm^2. By changing C_{base}, the CDC can measure capacitances ranging from 5 to 31 pF. Each C_{base} has 4 pF linear range and the capacitance ranges overlap to avoid missing codes. Maximum linearity error is 16.5 fF from nine different ranges calibrated by two points in each range. Power and resolution are measured at the worst-case, maximum input capacitance condition. The CDC uses three power domains: 0.6 V for digital control logic and non-overlapping clock generator, 1.2 V for most analog blocks, and 3.6 V for the current mirror. Total CDC power is 110 nW, consuming 90 nW from 1.2 V and the other 20 nW from 0.6 V. The measured resolution is 8.7 fF, resulting in 5.3 pJ/c-s FoM.

5 Iterative Delay-Chain Discharge CDC

The fourth design example illustrates an energy-efficient variant of a single-slope CDC [10]. This design concentrates on reducing conversion energy as low as possible while maintaining decent performance comparable to prior art.

With minimalized design excluding any complex analog circuit, this design results in a fully-digital CDC with >60 dB SNR which achieves the lowest energy consumption per conversion per capacitance, and also <0.06 % linearity error across a very wide capacitance range of 0.7 pF to over 10 nF.

5.1 Basic Operation Scheme

This design is based on the observation that when a ring oscillator (RO) is powered from a charged capacitance, the number of RO cycles needed to discharge the capacitance to a fixed voltage is naturally linear with the capacitance value. Figure 13 explains the concept of the conversion process. The top node of sensed capacitor *CT* is directly connected to the supply node of a ring oscillator. This node is initially charged to V_{HIGH}, and is then discharged gradually as the inverter RO oscillates. As signals in the RO transition, the RO draws some charge from C_{SENSE}, gradually lowering V_{CT}. As a result, the RO propagation delay increases, which is compared to a constant delay reference. The RO transition count until the period delay becomes longer than the reference delay is recorded by a counter, which becomes the output code D_{OUT}.

Since RO delay only depends on V_{CT} (neglecting noise initially), D_{OUT} is equal to the number of RO transitions while V_{CT} is discharged from V_{HIGH} to some constant voltage, V_{LOW}. As shown in Fig. 14, during conversion, at any particular V_{CT} value the amount of charge withdrawn per RO transition only depends on V_{CT} at that time. Therefore, the number of transitions required to reduce V_{CT} by a certain small voltage is proportional to input capacitance C_{SENSE}. As this is true at any V_{CT} level, the output code D_{OUT}, the sum of transition counts across all continuous small intervals from V_{HIGH} to V_{LOW}, is also proportional to C_{SENSE}. As the RO draws charge directly from C_{SENSE} without initial capacitance to voltage conversion, the CDC input capacitance range is essentially unlimited, constrained only by the

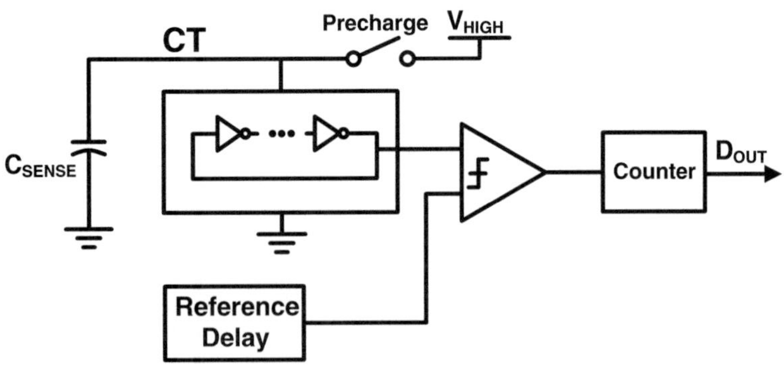

Fig. 13 Basic structure of the proposed CDC

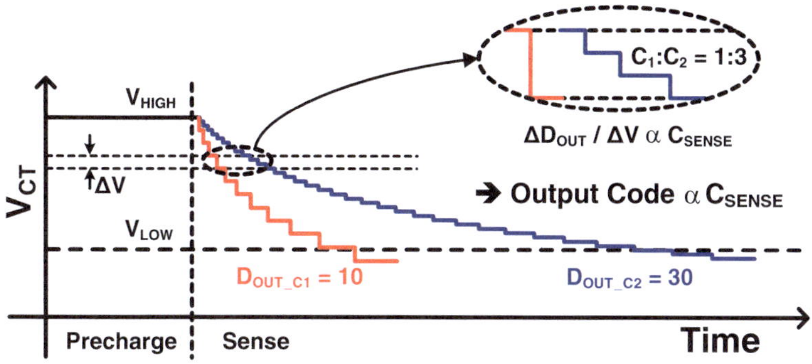

Fig. 14 Basic operation scheme of the proposed CDC

counter size. This is desirable when the C_{SENSE} range is uncertain at design time. Furthermore, energy used to charge C_{SENSE} is reused to oscillate the RO, reducing overall power consumption.

5.2 Detailed Implementation

Figure 15 shows the detailed implementation of the CDC circuit and its operation. Here an inverter chain is used in place of an RO to discharge C_{SENSE}—it is a 16-stage chain that is identical to the reference delay generator. Because of the identical structures, conversion stops when V_{CT} drops below V_{LOW}. The number of stages in the inverter chain is chosen for optimal SNR per conversion energy, where the energy to charge C_{SENSE} is balanced with the energy consumed by other blocks. The two propagation delays are compared by three delay comparators, which have a similar structure to an RS latch. The bottom comparator compares the propagation delay of falling edges, and the middle one compares the rising edges. Whenever the reference delay is shorter than the C_{SENSE} discharge delay chain, the comparators output pulses once, increasing counts stored in the *sub1* and *sub2* counters. A third counter tracks the main oscillation triggering signal. After each comparison, the next edge generator block triggers the next discharge and delay comparison, maintaining oscillation. All blocks except the C_{SENSE} delay chain operate at V_{LOW}, and a level converter drives the two delay chain inputs with V_{HIGH}.

As shown in the timing diagram of Fig. 16, conversion starts by precharging C_{SENSE} to V_{HIGH}. This is followed by *Sense* rising, triggering the first edge to propagate through the two delay chains. The top comparator takes in a slightly delayed version of the reference delay and determines when to finish the overall conversion, which occurs when V_{CT} becomes lower than V_{LOW} by some margin. As V_{CT} approaches V_{LOW}, the bottom two delay comparators pulse CK_1 and CK_2. They initially pulse sporadically due to noise, and then more frequently as V_{CT} crosses V_{LOW}. Just before conversion finishes, these two comparators pulse every cycle.

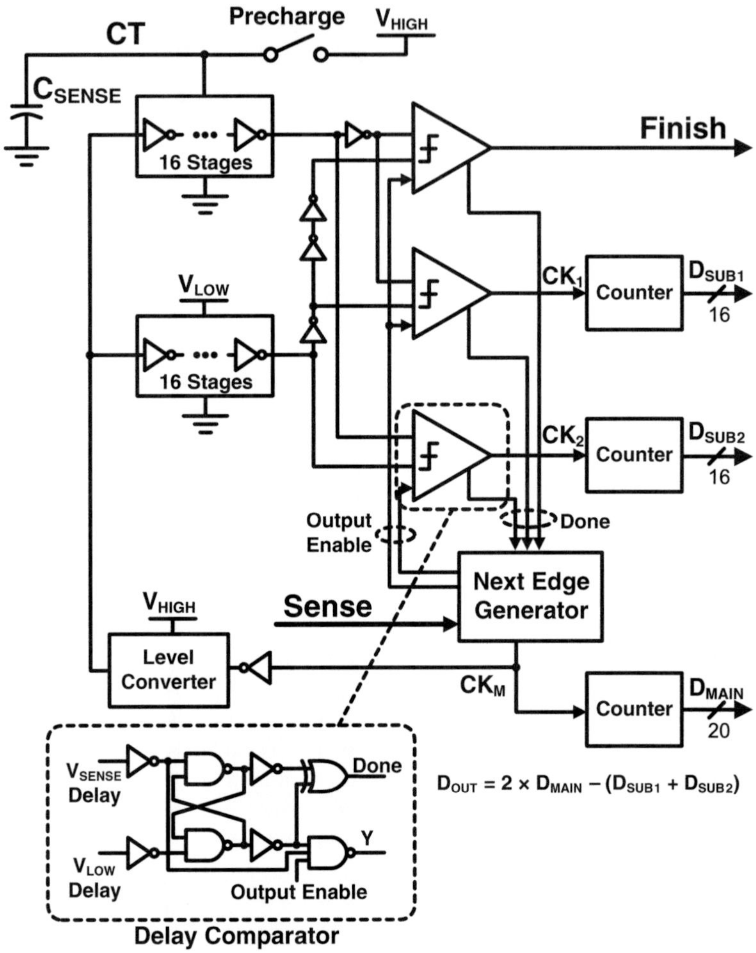

Fig. 15 Detailed implementation of the CDC

When the top comparator pulses *Finish*, *Sense* is turned off, and oscillation stops. Final D_{OUT} is the total count of comparator outputs for which $V_{CT} > V_{LOW}$, and is calculated as $2 \times D_{MAIN} - (D_{SUB1} + D_{SUB2})$.

The use of three comparators is designed to increase SNR by averaging noise over many comparisons when V_{CT} is near V_{LOW}. Comparing both rising and falling edges doubles the number of comparisons. By extending the conversion to where V_{CT} falls some margin below V_{LOW}, comparisons are performed through the whole noisy region around V_{LOW}, whereby false "$V_{CT} < V_{LOW}$" decisions above V_{LOW} are stochastically compensated by false "$V_{CT} > V_{LOW}$" decisions below V_{LOW}. The simulation shows that energy increases by 3 % compared to the standard approach of stopping conversion immediately after the first comparison triggers, while overall

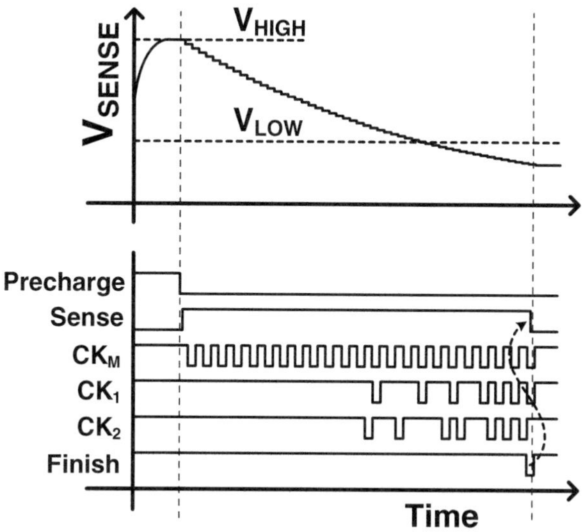

Fig. 16 Detailed timing diagram of the CDC

conversion noise is square rooted. In addition, the distribution of D_{OUT} using this scheme is centered at the number of exact counts from V_{HIGH} to V_{LOW}, thereby improving output code linearity.

5.3 Parasitic Capacitance Cancelation

The CDC measures the capacitance between one input node and ground, but some applications require the capacitance value between two input nodes, excluding parasitic capacitance to ground. We accomplish this through three conversions, as shown in Fig. 17. First, node B is connected to ground and the capacitance between node A and ground is measured, which includes parasitic capacitance C_{PA}. Second, nodes A and B are flipped and $C_{SENSE} + C_{PB}$ is measured. Finally, both A and B nodes are connected to V_{CT} to measure $C_{PA} + C_{PB}$. By adding the first two codes and subtracting the third, the parasitic capacitance is canceled out. While this requires three conversions, parasitic capacitance typically remains unchanged or changes slowly, and the parasitic cancelation can be performed infrequently, thus amortizing its overhead.

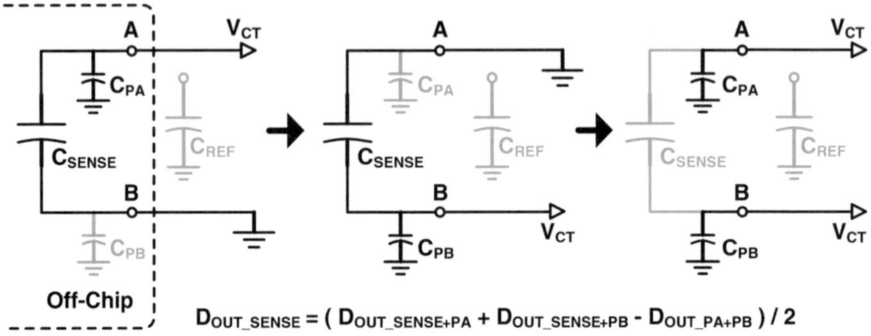

Fig. 17 Technique for parasitic capacitance cancelation

5.4 *Output Code Calibration*

The output code varies as temperature or supply voltage changes. One-point calibration removes this code deviation. In a calibration phase, V_{CT} is connected to an internal reference capacitor with known capacitance C_{REF} and the ratio of C_{REF} to corresponding D_{OUT} is stored. In subsequent normal conversion, digital output codes are converted to actual capacitance value by multiplying the code and the stored ratio. If the supply voltage changes sufficiently slowly, this calibration can occasionally be re-done.

5.5 *Measurement Results*

The CDC is fabricated in 40 nm CMOS and tested with $V_{HIGH} = 1.0$ V and $V_{LOW} = 0.45$ V. Core circuit area without testing circuits and internal capacitors is 0.0017 mm^2. This small area comes from simplicity of the CDC core circuit which consists of only a few hundreds of logic gates. Figure 18 shows the test chip has a very wide input capacitance range from 0.7 pF to 10 nF with a small linearity error of <0.06 %. The measured output noise percentage reduces as C_{SENSE} increases due to noise averaging. At 11.3 pF, the CDC has 0.109 % resolution, 35.1 pJ total conversion energy, and 141 fJ/c-s FoM.

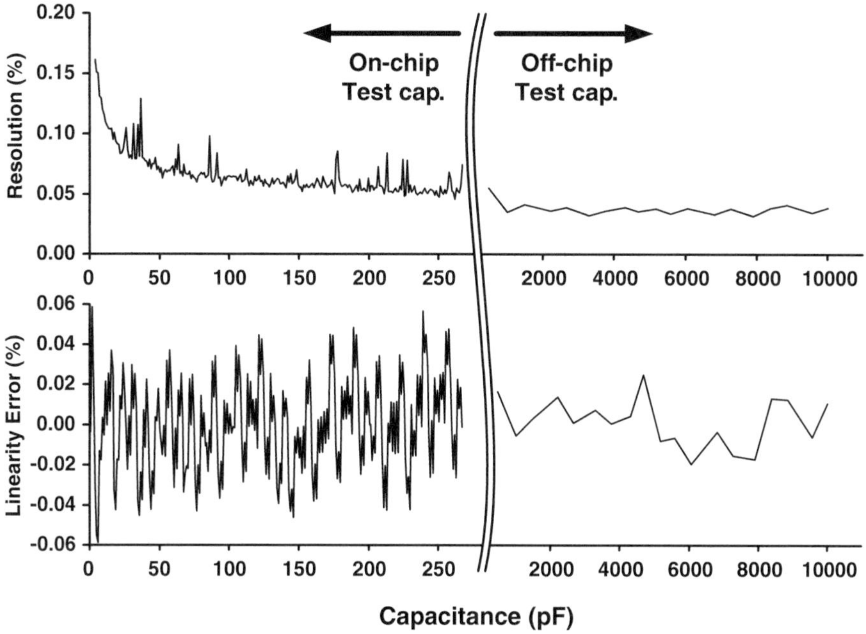

Fig. 18 Measured CDC resolution and linearity error

6 Conclusions

Energy-efficient CDC techniques have been described by means of several design examples. Because of their similarity to ADCs, a CDC FoM can be defined as $\frac{Power \times Meas.\ Time}{2^{(SNR-1.76)/6.02}}$, where $SNR = 20\ \log\left(\frac{Capacitance\ Range/2\sqrt{2}}{Capacitance\ Resolution}\right)$. Here $2\sqrt{2}$ is the crest factor [15] that is used to fairly compare DC-input CDCs to sinusoidal-input ADCs. This SNR definition imagines a sinusoidal continuous capacitance is given as input with the amplitude of *Capacitance Range*/2, and signal rms is regarded as $Capacitance\ Range/2\sqrt{2}$.

Recent CDCs performances are summarized and compared in Fig. 19. The SAR CDC introduced in Sect. 2 achieves a decent FoM, while the use of CDS suppresses offset and enables an increased input range. The highest SNR with an excellent FoM is achieved by the incremental CDC with zoom-in SAR converter. The dual slope CDC achieves a decent FoM with its energy-efficient charge subtraction. The iterative delay-chain discharge CDC obtains the lowest FoM, as well as a very wide input range. These energy-efficient CDCs are suitable for millimeter sensor nodes.

	Technology (nm)	Method	Input Range (pF)	Meas. Time (ms)	Power	SNR (dB)	FoM (pJ/c·s)
Sec.2 [7]	180	SAR	2.5- 75.3	4	120-160 nW	55.4-60.6	0.54-1.3[1]
Sec.3 [9]	180	IΔΣ	0 - 24	0.23	33.7 μW	94.7	0.18
Sec.4 [2]	180	DualSlope	5.3- 30.7	6.4	110 nW	44.2	5.3[1]
Sec.5 [10]	40	Iterative Delay-Chain Discharge	0.7 - 10000	0.019	1.84 μW	49.7	0.14[2]
[8]	180	SAR	N/A	0.004	240 μW	43.2	7.9
[3]	350	ΔΣ	8.4 - 11.6	0.02	14.9 mW	84.8	21
[4]	160	ΔΣ	0.54 - 1.06	0.8	10.3 μW	68.4	3.8
[11]	350	PM	1 - 6.8	7.6	210 μW	83	140
[12]	350	PM	0.8 - 1.2	0.05	15.8 mW	45.7	5000
[13]	320	PM	0.5 - 0.76	0.033	84 μW	40.9	98
[18]	130	Freq	6.0 - 6.3	1	270 nW	29.4	11

Fig. 19 Performance summary and comparison with recently published state-of-the-art CDCs. (*1*) One subrange FoM. (*2*) FoM at 11 pF sensor cap

References

1. Lee Y, Bang S, Lee I, Kim Y, Kim G, Ghaed H, Pannuto P, Dutta P, Sylvester D, Blaauw D (2013) A modular 1mm^3 die-stacked sensing platform with low power I^{2}C inter-die communication and multi-modal energy harvesting. IEEE J Solid-State Circuits 48(1):229–243
2. Oh S, Lee Y, Wang J, Foo Z, Kim Y, Blaauw D, Sylvester D (2014) Dual-slope capacitance to digital converter integrated in an implantable pressure sensing system. In: Proceedings of the European solid-state circuits conference, pp 295–298
3. Xia S, Makinwa K, Nihtianov S (2012) A capacitance-to-digital converter for displacement sensing with 17b resolution and 20μs conversion time. In: IEEE international solid-state circuits conference digest of technical papers, pp 198–200
4. Tan Z, Daamen R, Humbert A, Ponomarev Y, Chae Y, Pertijs M (2013) A 1.2-V 8.3-nJ CMOS humidity sensor for RFID applications. IEEE J Solid-State Circuits 48(10):2469–2477
5. Paavola M, Kamarainen M, Laulainen E, Saukoski M, Koskinen L, Kosunen M, Halonen K (2009) A micropower ΔΣ-based interface ASIC for a capacitive 3-axis micro-accelerometer. IEEE J Solid-State Circuits 44(11):3193–3210
6. Ghaed H, Chen G, Haque R, Wieckowski M, Kim Y, Kim G, Lee Y, Lee I, Fick D, Kim D, Seok M, Wise K, Blaauw D, Sylvester D (2013) Circuits for a cubic-millimeter energy-autonomous wireless intraocular pressure monitor. IEEE Trans Circuits Syst I 60(12):3152–3162
7. Ha H, Sylvester D, Blaauw D, Sim J (2014) 12.6 A 160nW 63.9fJ/conversion-step capacitance-to-digital converter for ultra-low-power wireless sensor nodes. In: IEEE international solid-state circuits conference digest of technical papers, pp 220–221
8. Tanaka K, Kuramochi Y, Kurashina T, Okada K, Matsuzawa A (2007) A 0.026mm^2 capacitance-to-digital converter for biotelemetry applications using a charge redistribution technique. In: Proceedings of the IEEE Asian solid-state circuits conference, pp 244–247
9. Oh S, Jung W, Yang K, Blaauw D, Sylvester D (2014) 15.4b incremental sigma-delta capacitance-to-digital converter with zoom-in 9b asynchronous SAR. In: Symposium on VLSI circuits digest of technical papers, pp 222–223

10. Jung W, Jeong S, Oh S, Sylvester D, Blaauw D (2015) A 0.7pF-to-10nF fully digital capacitance-to-digital converter using iterative delay-chain discharge. In: IEEE international solid-state circuits conference digest of technical papers
11. Tan Z, Shalmany S, Meijer G, Pertijs M (2012) An energy-efficient 15-bit capacitive-sensor interface based on period modulation. IEEE J Solid-State Circuits 47(7):1703–1711
12. Bruschi P, Nizza N, Piotto M (2007) A current-mode, dual slope, integrated capacitance-to-pulse duration converter. IEEE J Solid-State Circuits 42(9):1884–1891
13. Nizza N, Dei M, Butti F, Bruschi P (2013) A low-power interface for capacitive sensors with PWM output and intrinsic low pass characteristic. IEEE Trans Circuits Syst I 60(6):1419–1431
14. Pertijs M, Tan Z (2013) Nyquist AD converters, sensor interfaces, and robustness: Chapter 8. Energy-efficient capacitive sensor interface. Springer, New York, pp 129–147
15. Chae Y, Souri K, Makinwa K (2013) A 6.3 μW 20 bit incremental zoom-ADC with 6 ppm INL and 1 μV offset. IEEE J Solid-State Circuits 48(12):3019–3027
16. Harpe P, Zhou C, Bi Y, Meijs N, Wang X, Philips K, Dolmans G, Groot H (2011) A 26 μW 8 bit 10 MS/s asynchronous SAR ADC for low energy radios. IEEE J Solid-State Circuits 46(7):1585–1595
17. Markus J, Silva J, Temes G (2004) Theory and applications of incremental $\Delta\Sigma$ converters. IEEE Trans Circuits Syst I 51(4):678–690
18. Danneels H, Coddens K, Gielen G (2011) A fully-digital, 0.3V, 270 nW capacitive sensor interface without external references. In: Proceedings of the European solid-state circuits conference, pp 287–290

A Micro-Power Temperature-to-Digital Converter for Use in a MEMS-Based 32 kHz Oscillator

Samira Zaliasl, Jim Salvia, Terri Fiez, Kofi Makinwa, Aaron Partridge, and Vinod Menon

Abstract This paper describes the design of a low-power energy-efficient temperature-to-digital converter (TDC) intended for the temperature compensation of a 32 kHz MEMS-based oscillator (TCXO). The compensation scheme enables a frequency stability of ± 3 ppm over temperatures ranging from -40 to 85 °C. The TDC consists of an NPN-based temperature sensing element and a 15-bit second order $\Delta\Sigma$ modulator. A novel dynamic element matching (DEM) scheme ensures that DEM tones do not inter-modulate with the modulator's bit-stream, thus improving the TDC's accuracy without impacting its resolution. The TDC occupies 0.085 mm^2 in a 180 nm CMOS process, draws less than 4.5 μA from a 1.5 to 3.3 V supply, and achieves a resolution of 25 mK in a conversion time of 6 ms. This corresponds to a figure of merit of 24 pJ°C^2.

1 Introduction

Low power systems such as mobile and wearable devices often use micro power 32.768 kHz oscillators to track time of day and duty-cycle higher power circuitry. Key specifications for such oscillators are their PCB foot-print, frequency accuracy and power consumption.

These functions have previously been based on 32 kHz quartz-crystal tuning fork resonators. Further reducing the size of the crystals is challenging [1], while improving the frequency stability has significantly increased the required PCB area and power consumption [2, 3]. A promising alternative is the use of 32 kHz

S. Zaliasl (✉) • J. Salvia • A. Partridge • V. Menon
SiTime Corporation, Sunnyvale, CA, USA
e-mail: samira.zaliasl@gmail.com

T. Fiez
Oregon State University, Corvallis, OR, USA

K. Makinwa
Delft University of Technology, Delft, The Netherlands

K.A.A. Makinwa et al. (eds.), *Efficient Sensor Interfaces, Advanced Amplifiers and Low Power RF Systems*, DOI 10.1007/978-3-319-21185-5_4

oscillators (XOs) based on MEMS resonators [4], which offer a small foot-print (1.55 mm × 0.85 mm) together with moderate frequency stability (100 ppm from −40 to 85 °C).

To address smart metering and other precision applications, greater stability can be achieved with the help of temperature compensation engine. In a previous implementations of such a temperature-compensated oscillator (TCXO), however, the temperature compensation scheme was the most power hungry part of the system [3, 5]. As a result, they had to be heavily duty-cycled (1sample/min [3], 1sample/10 s [5]) to reduce their average current consumption, at the expense of limiting their ability to track rapid changes in ambient temperature.

In this paper, we describe the design of a temperature compensation engine used in a 32 kHz MEMS-based oscillator. It is optimized for small size, high stability and low power. By low power consumption, it can be operated at 3samples/s, thus dramatically improving the temperature tracking capability compared to previous works.

The temperature compensation engine consists of a temperature-to-digital converter (TDC), whose output is used to digitally scale the oscillator's output frequency via a third order polynomial function. The TDC employs a BJT-based sensing element and a second order delta-sigma ($\Delta\Sigma$) modulator which together produce a digitized representation of temperature. To improve its accuracy and stability, the TDC employs several dynamic correction techniques. These include correlated double sampling (CDS) in the modulator's first integrator, chopping at the system-level and dynamic element matching (DEM) of the BJT-biasing current sources.

Section 2 describes the top-level architecture of the 32 kHz MEMS-based oscillator. Section 3 discusses the implementation of the temperature sensor, while Sect. 4 describes the proposed DEM algorithm. Measurement results are shown in Sect. 5, followed by conclusions.

2 System Level Overview of 32 kHz MEMS-Based Oscillator

A simplified block diagram of the sub-μA MEMS-based oscillator is shown in Fig. 1. A 524 kHz MEMS resonator and sustaining amplifier provide a frequency reference to a programmable fractional-N synthesizer, which in turn generates an accurate 32 kHz output. This architecture addresses two challenges: (a) initial frequency offset due to process variations in the MEMS resonator, and (b) frequency variation of the resonator over temperature. In XO mode, a fractional-N synthesizer compensates frequency offset using a digital $\Delta\Sigma$ modulator (DSM), while in TCXO mode, the compensation engine corrects for the MEMS oscillator's temperature drift.

Figure 2 shows a detailed block diagram of the fractional-N PLL and the compensation path. In XO mode, the target frequency of 32.768 kHz is achieved by appropriately setting the fractional value of the second order digital $\Delta\Sigma$ modulator

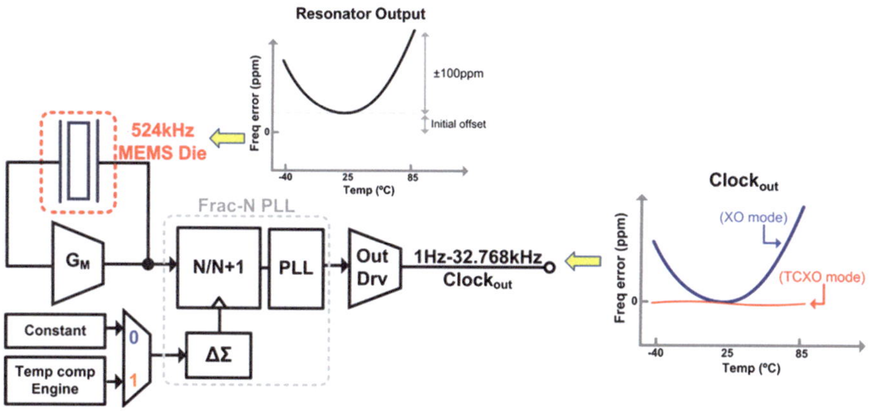

Fig. 1 Block diagram of the MEMS-based 32 kHz clock generator in XO or TCXO mode

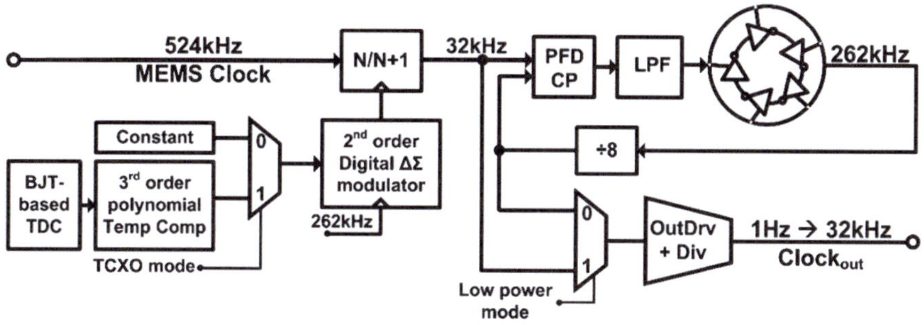

Fig. 2 Block diagram of PLL and its different configurations in XO, TCXO and low power mode

(DSM) during factory calibration. The following integer-N PLL then low-pass filters the quantization noise produced by the modulator. In TCXO mode, TDC and third order polynomial additionally compensate for frequency variation over temperature. The polynomial coefficients are determined during factory calibration and are stored in an on-chip non-volatile memory (NVM).

This fractional-N PLL differs from the classical $\Delta\Sigma$ fractional-N PLL in two specific points. First, the multi-modulus frequency division, under the control of a digital $\Delta\Sigma$ modulator, is performed in a pre-driver rather than in the PLL feedback path. Second, the output is tapped from the integer divider in the feedback path rather than from the VCO itself. Performing the fractional division in the reference path means that the pre-divider output is 32.768 kHz which allows the PLL to be disabled and bypassed in XO mode to reduce current consumption in low power mode. The output of the pre-divider is at the target frequency but it carries the $\Delta\Sigma$ modulator's quantization noise. This noise is not detrimental in applications that count pulses, e.g. 32,768 pulses to define 1 s. In applications where the jitter needs to be low and where a moderate power increase is acceptable, then integer-N PLL can be turned on to filter out this noise. As shown in Fig. 2, the integer-N PLL is a charge-pump based type II PLL [6] optimized for low power consumption.

In the described TCXO system, TDC noise is the dominant source of low frequency jitter. For some RF applications, the TDC noise must be below 50 mK/Conversion to keep long term jitter of 32 kHz output clock below a specified 2 μs in 2.5 s time period.

3 Temperature Sensor and Read-Out Implementation

To achieve TCXO frequency stability, a low-power energy-efficient TDC is required. For this, a NPN BJT-based front end is combined with a switched capacitor (SC) ΔΣ modulator as shown in Fig. 3. Such TDCs are known to achieve the best combination of accuracy and energy efficiency [7]. Inaccuracies of less than ±0.25 °C have been achieved from −55 to 125 °C, while dissipating less than 10 μW [8, 9].

As illustrated in Fig. 3, the heart of the TDC consists of two identical NPNs biased at a collector current ratio of p. The resulting base-emitter voltage, V_{BE}, has a negative temperature coefficient of approximately −2 mV/°C and can be expressed as follows:

$$V_{BE} = \eta \cdot \left({}^{kT}/_{q}\right) \cdot \ln\left({}^{I_C}/_{I_S}\right)$$

where η is a process dependent non-ideality factor (≈1), k is the Boltzmann constant, q is the electron charge, T is the temperature in Kelvin, I_C and I_S are the collector current and the NPN's saturation current, respectively. The difference between the two BJT's base-emitter voltages is proportional-to-absolute temperature (PTAT):

$$\Delta V_{BE} = V_{BE2} - V_{BE1} = \eta \cdot \left({}^{kT}/_{q}\right) \cdot \ln(p)$$

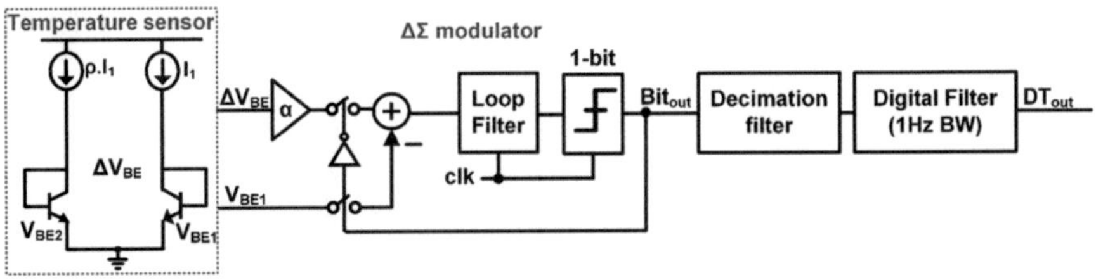

Fig. 3 Block diagram of the temperature sensor

The linear combination of V_{BE} (CTAT) and ΔV_{BE} (PTAT) generates a band-gap voltage of

$$V_{BG} = V_{BE1} + \alpha \cdot \Delta V_{BE}$$

where α is a fixed gain factor.

As in [10], both V_{BE} and ΔV_{BE} can then be applied to a charge-balancing second order $\Delta\Sigma$ modulator, which generates a bit-stream whose average value is proportional to $(\alpha\cdot\Delta V_{BE})/V_{BG}$. To save power, the modulator is operated in incremental mode, i.e. the integrators are initially reset and then the modulator is clocked for a fixed number of cycles. In this design, the following decimation filter produces a temperature reading with a resolution of 25 mK from 192 samples of the modulator's bit-stream. A 1 Hz digital filter is implemented at final stage to filter TDC noise further.

To achieve the best possible energy efficiency, the contributions of BJT-core, kT/C and quantization noise sources to the TDC's noise were carefully balanced with the help of the CppSim system-level simulator [11].

3.1 Circuit Implementation

Figure 4 shows the sensor's front-end. It consists of two identical NPNs biased at a 1:6 current ratio. NPNs were used instead of PNPs for two reasons. First, because their forward current gain is higher in the chosen process, which reduces errors due to voltage drops across their base resistances. Second, because their collector currents can be accurately defined, which translates into lower spread [8]. The bias circuit uses two NPN transistors, again biased at the same current ratio, to force the resulting ΔV_{BE} across a resistor (R_E), and thus generate an accurate PTAT current I_{Bias} with value of 10 nA at room temperature. Since the bias currents directly set the NPNs' collector currents, the temperature dependence of I_{Bias} doesn't impact the accuracy of ΔV_{BE}. However, it does impact the curvature of V_{BE}, and hence the sensor's non-linearity. Compared to the use of a temperature-independent bias current, the use of a PTAT current results in lower non-linearity [10].

The TDC is based on a single-bit feed-forward second order switched-capacitor (SC) $\Delta\Sigma$ modulator. As shown in Fig. 5, its main components are two SC integrators and a 1-bit quantizer. Both integrators are based on folded-cascode OTAs, with the first OTA gain-boosted to improve its DC gain. Depending on the quantizer's output (Bit_{out}), the modulator's next input is either $-V_{BE1}$ ($Bit_{out} = 1$) or $\alpha\cdot\Delta V_{BE}$ ($Bit_{out} = 0$) [10]. This charge-balancing scheme ensures that the average value of Bit_{out} is equal to the desired ratio $(\alpha\cdot\Delta V_{BE})/(V_{BE} + \alpha\cdot\Delta V_{BE})$. As shown, $-V_{BE}$ is integrated during a single clock cycle, while ΔV_{BE} is integrated over α clock cycles [12]. In other words, each $\Delta\Sigma$ cycle takes either 1 or α clock cycles, when Bit_{out} is 1 or 0, respectively. This scheme only requires one unit sampling capacitor (Cs), which avoids the extra area and mismatch errors associated with implementing

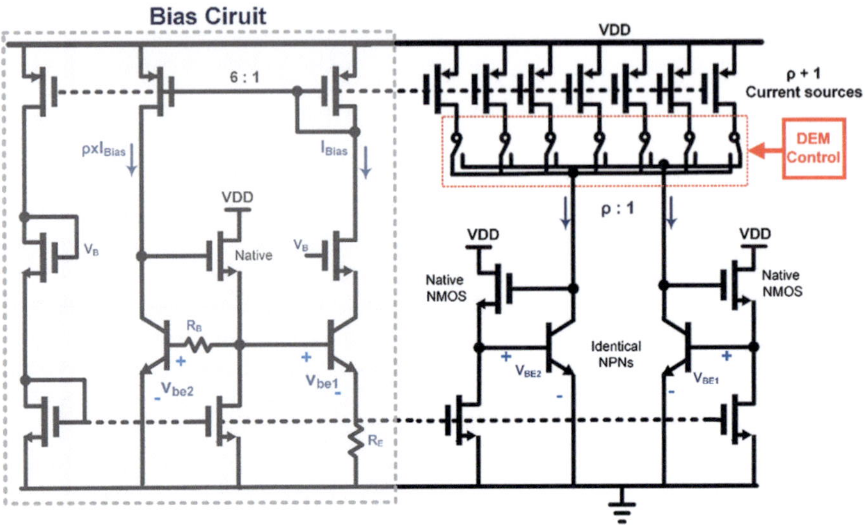

Fig. 4 Circuit diagram of the BJT core and its Bias circuit

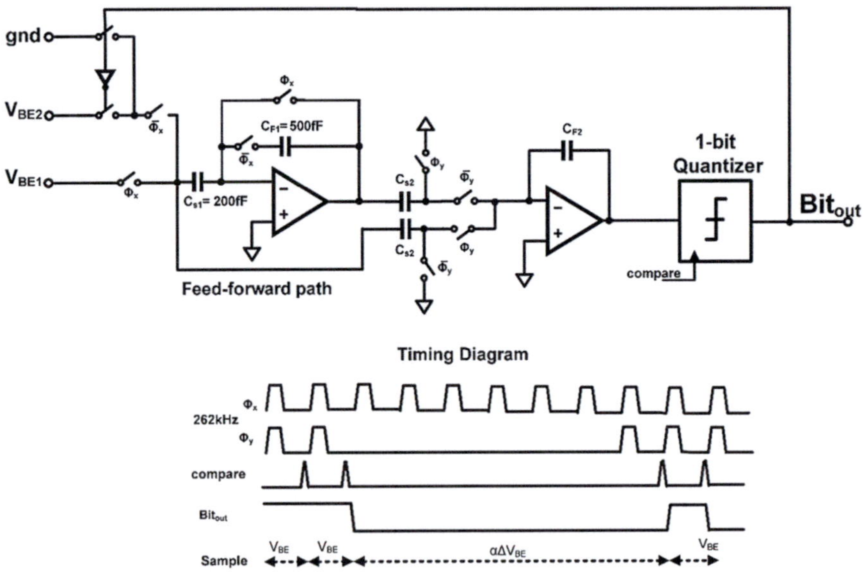

Fig. 5 Circuit diagram of the read-out circuit and timing diagram

α with multiple sampling capacitors [10]. It also ensures that the first integrator's closed-loop gain is fixed, irrespective of the modulator's state.

Several dynamic techniques are used to improve the TDC's accuracy. Mismatch in the current sources impacts the accuracy of ΔV_{BE}. Even with careful layout, a relative current ratio mismatch in the order of 0.1 % can be expected, which leads to an error of more than 100 mK at room temperature [13]. Therefore, dynamic element matching of the seven current sources in the bipolar core (see Fig. 4) is essential to mitigate their mismatch for an accurate ΔV_{BE}.

The first integrator's offset and flicker noise are mitigated by employing correlated double sampling (CDS) [14]. To mitigate charge injection errors, the first integrator employs a fully differential structure using minimum size switches. Any residual offset is then removed by system-level chopping, which is implemented by separating a full TDC conversion into two half conversions with requisite polarity inversions in the BJT core and at the quantizer output [9]. To provide a differential input to the first integrator, the positions of the current sources are swapped during the two phases of each $\Delta\Sigma$ clock cycle. This simultaneously averages out the mismatch between the two BJTs [10].

The TDC's overall timing diagram is shown in Fig. 6. The integrators of the modulator are reset at the start of each incremental $\Delta\Sigma$ conversion. To implement system level chopping, the TDC performs two consecutive conversions and averages the results. For each conversion, the modulator runs for 192 cycles, producing Bit_{out} which is fed into the decimation filter.

The total current consumption of the implemented TDC is 4.5 μA operating at a 262 kHz sampling clock. Of this, 1.5 μA is consumed by the front-end, 1.7 μA is dissipated by the loop filter and 0.8 μA is used by the decimation filter, the digital filter, and the clock generator. One full TDC conversion requires 6 ms at room temperature. The TDC achieves a resolution of 25 mK (rms) leading to a figure of merit (regarding Energy/Conversion and Resolution) of 24 pJ°C^2. This compares favorably with the state of the art [5].

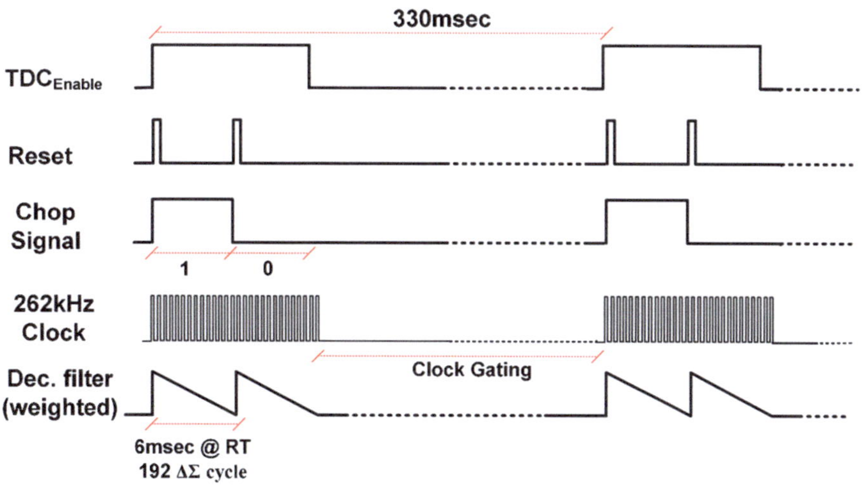

Fig. 6 TDC timing diagram in duty-cycled mode

To meet the 1 μA total-current-consumption target for TCXO part, the TDC is duty-cycled to reduce its average current consumption. However, the use of duty-cycling leads to a trade-off between temperature tracking accuracy and power consumption. Applications require a frequency error of <1 ppm in the presence of temperature ramps of ±1 °C/s. To meet this, the TDC's update rate is set to 3samples/s as shown in Fig. 6. This results in an average current of 150 nA. As shown in Fig. 6, clock gating is used to prevent power dissipation in digital circuits when the TDC is not being used.

4 Proposed Dynamic Element Matching Algorithm

Applying DEM to the front-end's current sources significantly improves their effective matching. However, the use of DEM translates any mismatch into periodic tones which, when applied to a ΔΣ modulator, can inter-modulate with its bitstream causing the well-known problem of quantization-noise folding [15–17]. This, in turn, can significantly increase the modulator's in-band noise floor. Previous designs mitigated this effect by randomization [16] or by the use of bitstream-controlled DEM [17]. However, these techniques do not guarantee that each conversion corresponds to a full DEM cycle and so suffer from residual mismatch error [13]. In this work, these shortcomings are addressed by a novel DEM implementation. The associated timing is illustrated in Fig. 7, which compares the timing of the proposed and traditional DEM schemes for the case when the current ratio $p = 6$. Since a complete DEM cycle requires $(p + 1)$ clock cycles, it can be combined with the α clock cycles required to integrate ΔV_{BE}. Choosing $\alpha = n \cdot (p + 1)$, where $n = 1, 2, 3, \ldots$, insures n full DEM cycles can be completed during exactly one ΔΣ cycle, thus avoiding any quantization-noise fold-back. Additionally, in one TDC conversion, there is no residual mismatch error since in such errors are averaged out during each ΔV_{BE} cycle.

To demonstrate the effectiveness of the proposed DEM algorithm, a 1 % random mismatch was intentionally added to the current sources in simulation. A transient noise simulation showed that with a conventional DEM scheme the TDC's in-band noise increased to 150 mK (rms), whereas with the proposed scheme the in-band noise was restored to 25 mK(rms). This is further highlighted in Fig. 8, in which the standard deviation of the random mismatch was set to 10 % for three cases: (1) DEM disabled, (2) according to the proposed algorithm and (3) a conventional DEM scheme. It can be seen that compared to the situation without DEM and the situation with the proposed scheme, which were at essentially the same level, the conventional DEM scheme increased the noise floor by 20 dB.

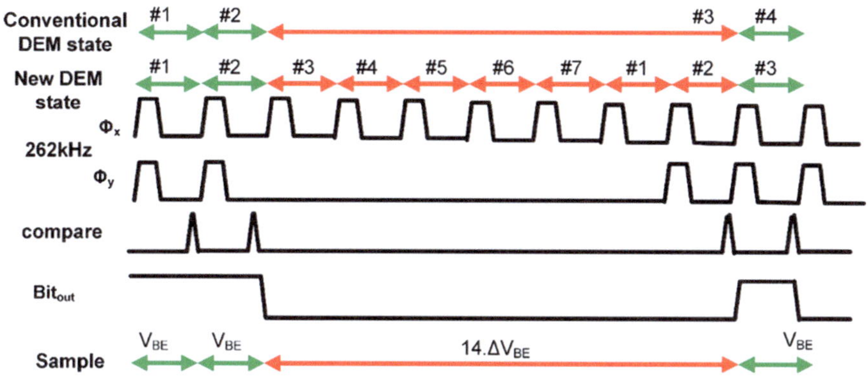

Fig. 7 Timing diagram of conventional and proposed DEM schemes

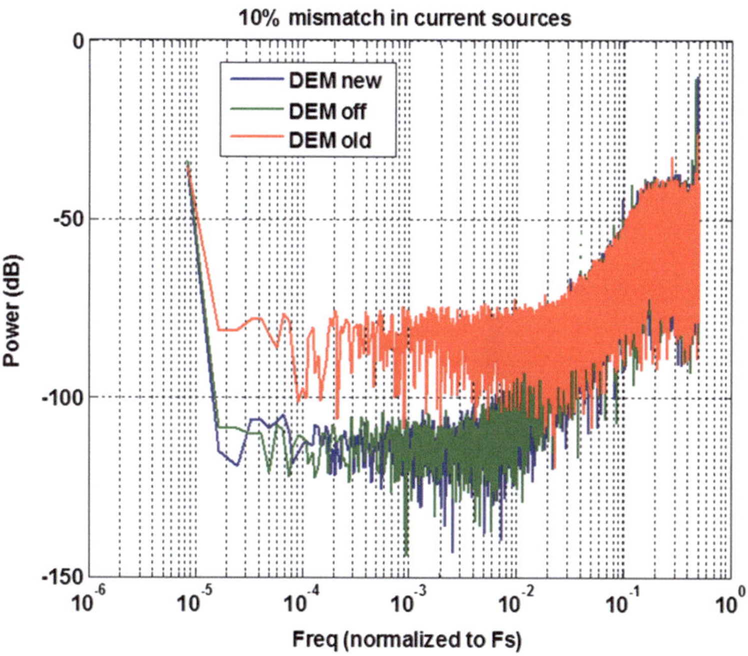

Fig. 8 Power spectrum density for three cases: (1) DEM off (2) proposed DEM (3) conventional DEM. As shown proposed DEM has similar noise performance as when DEM is disabled

5 Realization and Measurement

Figure 9a shows the 180-nm CMOS die with an area of 1.5 mm × 0.8 mm, and the MEMS die with an area of 0.42 mm × 0.42 mm. On the CMOS chip, TDC consumes 0.085 mm^2 and the PLL and OSC have area of 0.21 mm^2, while the

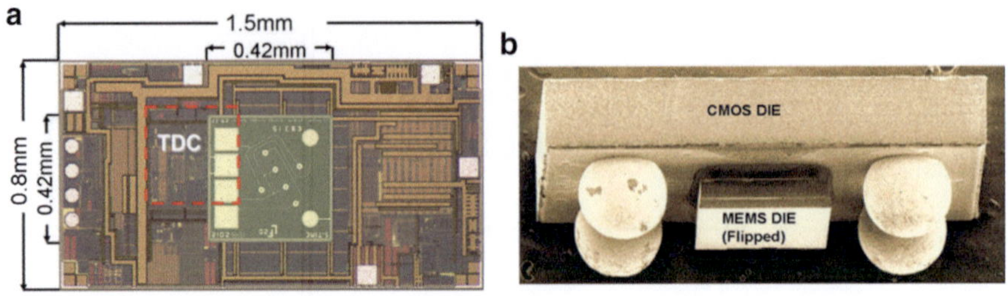

Fig. 9 (**a**) Die photograph of 524 kHz MEMS die and a 180-nm CMOS chip in (**b**) chip scale package

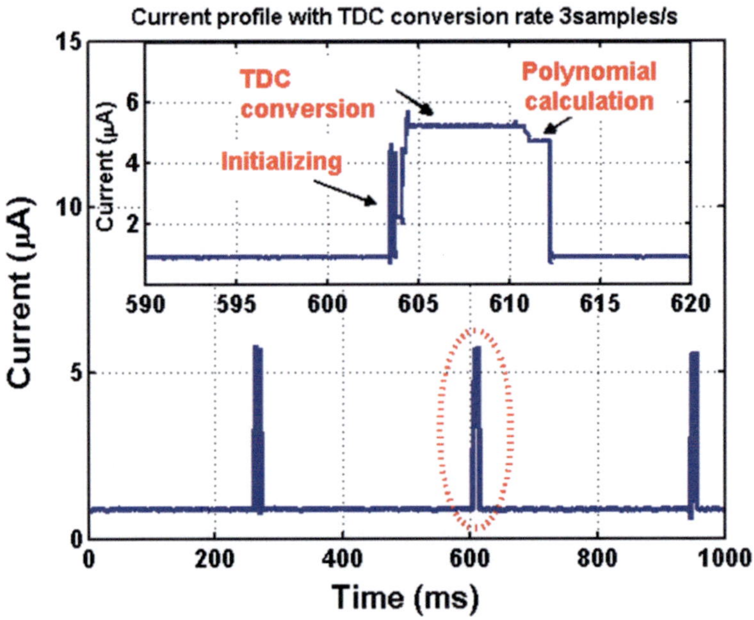

Fig. 10 Transient current profile of entire chip when TDC conversion rate is 3samples/s

digital including the third order polynomial correction logic occupies 0.5 mm^2. To minimize the package size, the MEMS resonator is flip-chip bonded to the CMOS die in a 1.55 mm × 0.85 mm chip-scale package (CSP), as shown in Fig. 9b. In this packaging, epoxy under-fill is applied between two dies to fully insulate the MEMS-CMOS interconnections which makes it fully compatible with standard lead-free PCB assembly processes.

The measured average current of the chip is 1.0 μA from 3.3 V supply with a 32 kHz output under no external load, of which the temperature compensation consumes less than 150 nA on average. Figure 10 displays the chip's dynamic current consumption, which can be seen to have a peak value of 5.5 μA. The implemented TDC consumes 4.5 μA operating at a 262 kHz sampling clock.

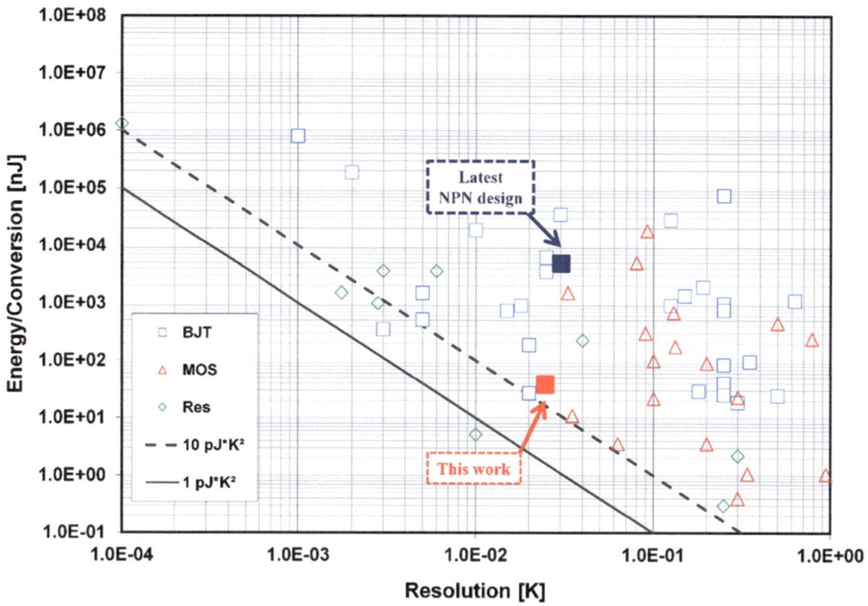

Fig. 11 Energy per conversion versus resolution for different smart temperature sensors using different sensing principles [5]

During each conversion, it takes 1 ms for the BJT core and modulator to initialize, 6 ms for two half-conversions at 25 °C, and 2 ms to evaluate the polynomial. After each conversion, the TDC and compensation engine sleep and are turned on every 330 ms (3samples/s) by a low power wake-up circuit that runs continuously. The TDC achieves a resolution of 25 mK (rms) leading to a figure of merit (Energy/Conversion $\times$ Resolution^2) of 24 pJ°C^2. Figure 11 compares the sensor's performance in terms of energy/conversion and resolution with several other smart temperature sensors [7]. The FoM is improved a lot compared to last NPN design [8].

As shown in Fig. 12a, the measured frequency stability over temperatures ranging from −40 to 85 °C is better than ±100 ppm for 28,000 XO devices. The initial accuracy of the XO devices is within ±3 ppm with one point calibration at 25 °C. From Fig. 12a, it can also be seen that the resonator frequency spread can be more than 50 ppm at the temperature extremes. This means that to realize a TCXO with a stability of a few ppm, each device should be individually trimmed at factory level. As shown in Fig. 12b, the measured frequency accuracy improves to ±3 ppm for 28,000 TCXO devices after individual trimming at five temperature points at wafer level. It should be noted that trimming will not compensate for the effects of ageing and the mechanical stress due to packaging. These are mitigated by the use of dynamic correction techniques such as DEM, CDS and chopping in the TDC. With these techniques in place, the oscillator's solder-down shift is less than ±1.5 ppm over the temperature range of −40 to 85 °C, despite the use of chip-scale packaging.

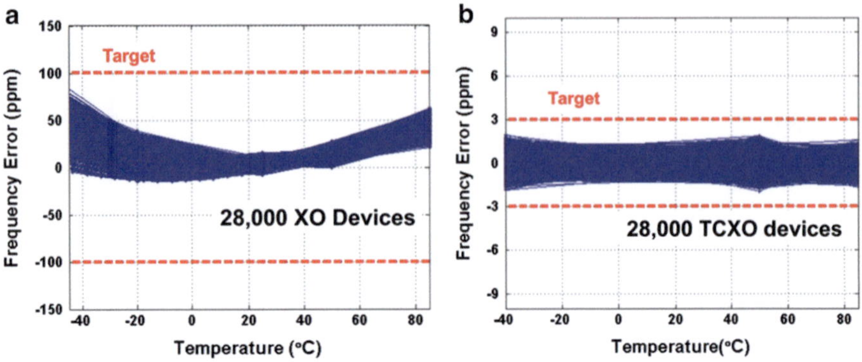

Fig. 12 Frequency stability in (**a**) XO and (**b**) TCXO configurations

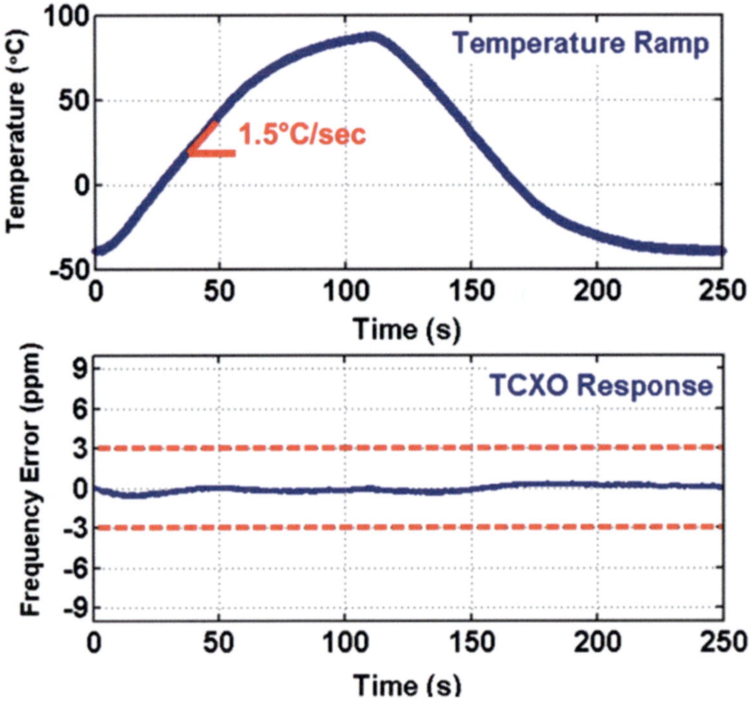

Fig. 13 TCXO response to temperature ramp

Figure 13 shows the temperature tracking performance in TCXO mode where a temperature transient with a slope as high as 1.5 °C/s is applied and the measured normalized frequency error is plotted. An evident from the figure, a TDC conversion rate of 3samples/s is sufficient to maintain the resulting frequency changes within ±3 ppm over temperature.

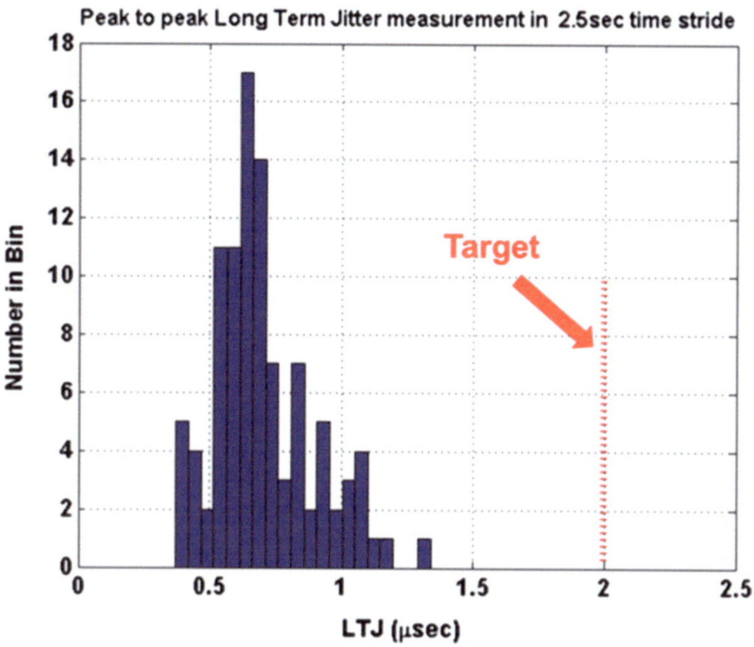

Fig. 14 Measured long term jitter in 2.5 s time stride for 100 TCXO devices

Noise performance of a 32 kHz clock is critical when it is used in wake-up circuitry for phones. Noise performance applicable for this application which demonstrates the wake-up time accuracy is shown in Fig. 14. It is the measured peak-to-peak long term jitter (LTJ) at stride of 2.5 s for 100 devices with a mean value of 0.7 μs. As shown, it comfortably meet the target noise performance suited for this application.

Table 1 summarizes the measured performance and compares it with existing quartz-based 32 kHz TCXO devices [2, 3, 18]. As shown, the described 32 kHz MEMS-based oscillator outperforms quartz crystal-based solutions in several aspects but most notably in current consumption (at least 2× less) and package area (at least 6× smaller package) while demonstrating ±3 ppm frequency stability over the industrial temperature range.

6 Conclusion

This paper presents a 32 kHz temperature-compensated MEMS-based programmable oscillator with footprint of 1.55 mm × 0.85 mm which is at least 6× smaller than previous works. In this system, the MEMS resonator and the CMOS circuitry are realized on separate dies and stacked together in a chip-scale package

Table 1 Performance summary of 32 kHz TCXO device and comparison table

Parameter	This work	Maxim DS32kHz	Epson TG-3530	Kyocera KT3225T
Supply voltage (V)	1.5–4.5	2.7–3.5	2.2–5.5	2–5.5
Temperature range (°C)	−40 to 85	−40 to 85	−20 to 70	−40 to 85
Frequency stability vs. temp (ppm)	±3	±7.5	±5	±5
Supply sensitivity (ppm/V)	±2.5	2.5	±1	±1
Start up (s)	0.2	1	3	3
Current (μA)	1 typ	1.85 typ	1.7 typ	1.5 typ
Clock enabled, no load	1.5 max	4 max	4 max	4 max
Package size (mm × mm)	1.55 × 0.85	18.5 × 6.35	5 × 10.1	3.2 × 2.5

(CSP). It achieves frequency stability of 3 ppm from −40 to 85 °C and long term jitter of less than 1.5 μs in 2.5 s time stride. Its overall current consumption is 1 μA. It is able to track temperature transient with a slope as high as 1.5 °C/s. Such performance is achieved by combining a temperature insensitive MEMS resonator with a stable energy-efficient temperature-to-digital converter and a low power fractional-N PLL, and by the extensive use of duty-cycling and clock gating techniques.

The temperature todigital converter uses NPN temperature-sensing elements followed by a second order SC delta-sigma modulator. It draws 4.5 μA current from 1.5 V supply voltage. It achieves a resolution of 25 mK in a conversion time of 6 ms which corresponds to a figure of merit of 24 pJ°C^2. Optimized design and the extensive use of dynamic techniques such as chopping, correlated-double-sampling (CDS) and dynamic element matching (DEM) enables this level of stability and energy efficiency compared to latest NPN designs.

A novel DEM algorithm is introduced which has no tonal behavior. It effectively reduces current-source mismatch without increasing the modulator's noise floor due to quantization noise folding. It insures full DEM cycles to be completed during one $\Delta\Sigma$ cycle, thus avoiding no residual error and noise folding.

References

1. Dalla Piazza S. Quartz tuning forks: a high-volume, low-cost, high-tech MEMS product. [online]. http://www.go4time.eu/publications/37-general.html
2. Datasheet. Epson TG-3530SA [online]. http://www.eea.epson.com/portal/pls/portal/docs/1/1547462.PDF
3. Datasheet. Maxim integrated DS32KHz [online]. http://datasheets.maximintegrated.com/en/ds/DS32kHz.pdf

4. Zaliasl S, Salvia JC et al (2015) A 3 ppm 1.5 × 0.8 mm^2 1.0μA 32.768 kHz MEMS-based oscillator. IEEE J Solid State Circuits 50:291–302
5. Ruffieux D, Krummenacher F, Pezous A, Spinola-Durante G (2010) Silicon resonator based 3.2μW real time clock with ±10ppm frequency accuracy. IEEE J Solid-State Circuits 45:224–234
6. Gardner F (1979) Phaselock techniques, 2nd edn. Wiley, New York
7. Makinwa KAA. Smart temperature sensor survey [online]. http://ei.ewi.tudelft.nl/docs/TSensor_survey.xls
8. Sebastiano F, Breems LJ, Makinwa KAA, Drago S, Leenaerts D, Nauta B (2010) A 1.2-V 10μW NPN-based temperature sensor in 65-nm CMOS with an inaccuracy of 0.2°C (3σ) from −70°C to 125°C. IEEE J Solid-State Circuits 45(12):2591–2601
9. Souri K, Makinwa KAA (2011) A 0.12mm^2 7.4μW micropower temperature sensor with an inaccuracy of 0.2°C (3-sigma) from −30°C to 125°C. IEEE J Solid-State Circuits 46(7):1693–1700
10. Pertijs MAP, Makinwa K, Huijsing J (2005) A CMOS smart temperature sensor with a 3σ inaccuracy of 0.1°C from 55°C to 125°C. IEEE J Solid-State Circuits 40(12):2805–2815
11. Perrott MH. CppSim system simulator package [online]. http://www.cppsim.com
12. Kashmiri SM, Pertijs MAP, Makinwa KAA (2010) A thermal-diffusivity-based frequency reference in standard CMOS with an absolute inaccuracy of ±0.1% from 55°C to 125°C. IEEE J Solid-State Circuits 45(12):2510–2520
13. Pertijs MAP, Huijsing JH (2006) Precision temperature sensors in CMOS technology. Springer, Dordrecht
14. Enz CC, Temes GC (1996) Circuit techniques for reducing the effects of op-amp imperfections: auto-zeroing, correlated double sampling, and chopper stabilization. Proc IEEE 84(11):1584–1614
15. Vadipour M (2000) Techniques for preventing tonal behavior of data weighted averaging algorithm in ΣΔ modulators. IEEE Trans Circuits Syst II Analog Digit Signal Process 47(11):1137–1144
16. Wang CB (2001) A 20-bit 25-kHz delta-sigma A/D converter utilizing a frequency-shaped chopper stabilization scheme. IEEE J Solid-State Circuits 36(3):566–569
17. Pertijs MAP, Huijsing JH (2004) A sigma-delta modulator with bitstream-controlled dynamic element matching. In: Solid-state circuit conference, ESSCIRC 2004, pp 184–190
18. Datasheet. Kyocera KT3225T [online]. http://global.kyocera.com/prdct/electro/pdf/khz/kt3225t_e.pdf

Low-Power Biomedical Interfaces

Refet Firat Yazicioglu, Jiawei Xu, Rachit Mohan, Bogdan Raducanu, Nick Van Helleputte, Carolina More Lopez, Srinjoy Mitra, Julia Pettine, Roland Van Wegberg, and Mario Konijnenburg

Abstract The design of energy efficient instrumentation has long been fueled by the mobile applications where low-power sensors and sensor interfaces have been used for continuous measurement of inertial measurements and environmental parameters. On the other hand, during the last decade, together with the increasing interest on continuous measurements of physiological and neural signals, new generations of energy efficient instrumentation amplifiers have emerged. This paper presents the state of the art of instrumentation architectures in the field of biomedical instrumentation and discusses their use in wearable and implantable biomedical signal acquisition systems.

1 Introduction to Biomedical Interfaces

The measurement of physiological signals and vital signs is a common procedure in modern clinical practice. The growing interest towards continuous measurement of physiological signals has fueled the miniaturization of medical instrumentation and lead to the development of wearable and implantable biomedical systems. These miniature biomedical systems, due to the limited available volume, have to consume minimal energy. These energy critical systems then have been the main driver for the development of novel components and systems with extremely low energy consumption.

This paper describes energy efficient instrumentation techniques. First, we will present different instrumentation amplifiers that have been custom developed for

R.F. Yazicioglu (✉)
imec, Leuven, Belgium

Holst Center, Eindhoven, The Netherlands
e-mail: firatyazicioglu@gmail.com

J. Xu • J. Pettine • R. Van Wegberg • M. Konijnenburg
Holst Center, Eindhoven, The Netherlands

R. Mohan • B. Raducanu • N. Van Helleputte • C.M. Lopez • S. Mitra
imec, Leuven, Belgium

K.A.A. Makinwa et al. (eds.), *Efficient Sensor Interfaces, Advanced Amplifiers and Low Power RF Systems*, DOI 10.1007/978-3-319-21185-5_5

acquiring high precision biopotential signals. Later we will discuss how these instrumentation amplifiers can be combined with different sensor interfaces to result in complete biomedical signal acquisition systems.

2 Fundamentals of Biopotential Signal Acquisition

Acquiring biopotential signals require two main components; an electrode and an instrumentation amplifier (Fig. 1). An electrode is a transducer that converts ionic current in the biology into electronic current. As a common practice, electrodes are modeled as complex impedance in series with a voltage source, which represents the polarization effect between skin and electrode material [1]. This polarization voltage appears due to the ion concentration mismatch between the tissue and the electrode, and is dependent on the electrode material as well as skin conditions. Instrumentation amplifiers connected to the electrodes, amplify the microvolt level biopotential voltage appearing differentially between the electrodes. This measurement scheme also helps rejecting common mode signals such as mains interference.

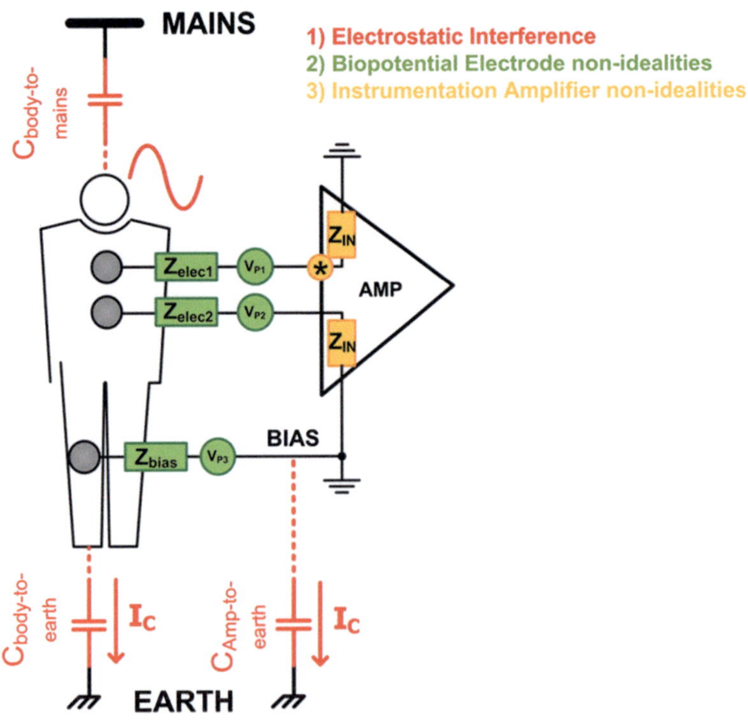

Fig. 1 Acquiring biopotential signals using electrodes and an instrumentation amplifier

There are several aggressors in biopotential measurements, which define the performance requirements of instrumentation amplifiers; *(i)* According to the IEC standards [2], biopotential acquisition systems need to accurately amplify biopotential signals even in the presence of ± 300 mV polarization voltage. This either necessitates the use of high-pass filters prior to amplification or the presence of large dynamic range signal acquisition, *(ii)* The mains interference, appearing as a common mode signal between the acquisition electrodes, has to be rejected by the amplifier. This requires high performance common-mode rejection-ratio (CMRR). While for implantable applications, 70–80 dB is sufficient, for wearable applications performance higher than 100 dB is required, *(iii)* the input impedance of the amplifier can affect the quality of signal acquisition. The mismatch between impedance of the two recording electrodes converts the common mode interference into differential at the input of the amplifier due to the finite input impedance of instrumentation amplifier.

In this paper, we will particularly discuss the advances in the design of instrumentation amplifiers for biomedical signal acquisition. We will look into different applications such as wearable and implantable biomedical applications, and present emerging instrumentation techniques that implement energy efficient signal acquisition even under the presence of aggressors.

3 Instrumentation Techniques for Wearable Applications

3.1 Fully-Differential Instrumentation Amplifiers

In most biopotential measurements, the signals are measured differentially between two electrodes (as shown Fig. 1). The main advantage of this measurement scheme is the capability to reject common mode interference. There are different instrumentation amplifier topologies that can be used for differential amplification of biopotential signals.

The most commonly utilized instrumentation amplifier is the classical 3-opamp IA [3] (Fig. 2a). This architecture consists of two-stage amplification. However, matching of the resistors, R2 and R3, defines the CMRR and the presence of resistor feedback does not yield good power-noise trade-off. The primary noise contributors are the operation amplifiers (OP1 and OP2) and the feedback resistors R and RG. Hence the operational amplifiers, especially OP1 and OP2, need large-power input stages as well as low-impedance output stages. Furthermore since it is DC-coupled in nature, this instrumentation amplifier cannot implement high gain without avoiding signal saturation due to the presence of the electrode polarization voltage.

Current-Feedback Instrumentation Amplifiers (CBFIA in Fig. 2b) have emerged as an interesting alternative to obtain a high performance with low-power dissipation [4, 5]. The input stage and resistor R1 determine the noise of this architecture. Contrary to the 3-opamp IA, the input stage actually drives this resistor. Hence, CBFIAs don't require high-power output stages to drive small value resistors saving

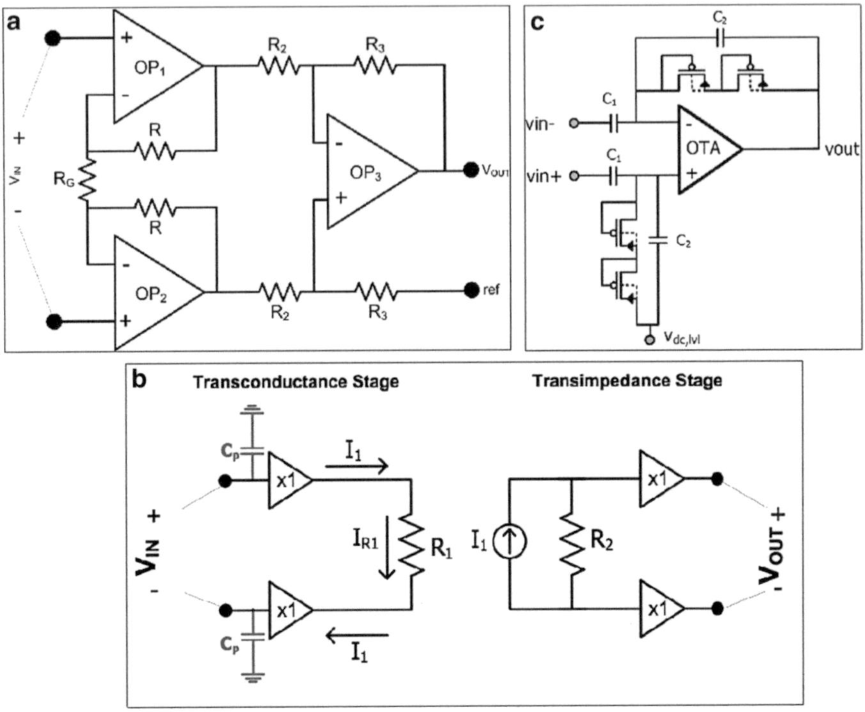

Fig. 2 Commonly used IA architectures like the 3-opamp (**a**), current-feedback (**b**), or capacitive IA (**c**)

power dissipation. To deal with the polarization voltage, either passive high-pass filtering or active DC-servo loops (DSL) can be used (by injecting an offset current into R1). However, both of these solutions have drawbacks; *(i)* DSLs usually have a limited range of a few 10s of mV [5, 6]. Increasing this range typically results in increased noise or power. In addition, a large DC-offset will result in a significant operating point mismatch in the input differential pair which would negatively impact the CMRR. While filtering the DC-offset before the actual IA, like in capacitive feedback IAs (Fig. 2c) [7, 8] or AC-coupled IAs [4], usually allows to deal with large polarization voltages, it also tends to result in poor CMRR [8] due to mismatch in the capacitors unless large (external) capacitors [10] or trimming are employed [4].

Since the bandwidth of interest in typical biopotential signals can be quite low (sub-Hz region), *1/f* noise is usually a major aggressor that may dominate the total integrated noise of the amplifier. Chopper modulation is a common-technique to mitigate the effects of *1/f* noise. In our research, we have proposed a large dynamic range chopper-modulated IA with DSL to improve upon the DC Electrode Offset (DEO) rejection range while maintaining the benefits of the other architectures [11] (Fig. 3). This architecture consists of two internal CFBIAs similar to [5] for the forward signal path and a feedback loop, which behaves differently for

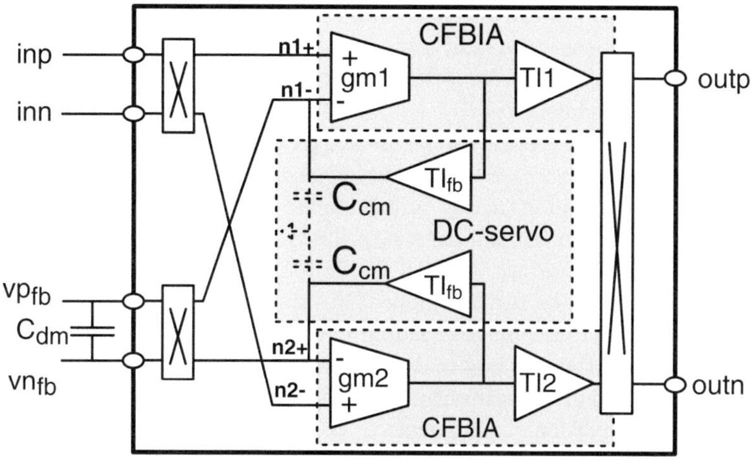

Fig. 3 Architecture of current balancing instrumentation amplifier

common-mode signals than for differential mode signals and is responsible for rejecting the polarization voltage between the electrodes. The input signals of the IA contain three main components: *(i)* desired differential signal of interest (ECG); *(ii)* undesired polarization voltage; *(iii)* undesired common-mode interference. The proposed IA rejects the two undesired components by replicating both of them at the output of the feedback loop and feeding those signals to the negative inputs of the CFBIAs where they will be rejected by the inherent high CMRR of the CFBIA. It should be noticed that it is not sufficient to provide only the polarization voltage at the output of the feedback loop as traditional DC-servo loops do. To achieve a good CMRR, it is necessary to ensure that the differential input pairs of each of the individual forward amplifiers will see exactly the same common-mode signals.

3.2 Active Electrode Instrumentation Amplifiers

Many scalp biopotential acquisition systems use gel electrodes that help reducing the electrode impedance. However, the use of gel prevents long-term recordings and increases set-up time, especially in applications where a large number of electrodes are required, such as scalp EEG measurements. Furthermore, the gel dries out over time and degrades the signal quality, and thus requires frequent intervention from medical professionals.

Dry electrodes solve usability problems of gel electrodes with reduced set-up time and increased user comfort, at the expense of significantly higher contact impedance. Higher impedance increases noise pick-up from environment and mains interference, resulting in poor signal quality. An electrode with a co-integrated amplifier, namely Active Electrode (AE), can reduce noise pick-up by minimizing the length of cabling between the electrode and the amplifier's input. However, there are significant challenges to yield a medical-grade system with low-noise, high-input impedance, large electrode offset tolerance, and high CMRR while utilizing Active Electrodes architecture.

In prior-art, AEs are typically implemented by using analog buffers, i.e. voltage followers. Analog buffers have low output impedance and low gain variation across process corners. In addition, analog buffers require only three connections (V_{dd}, V_{ss} and V_{out}) between each AE and the rest of the system, minimizing the number of cables. However, a major drawback of the buffer AEs is their noise-power trade-off. Analog buffers only perform impedance conversion without providing any voltage gain. As a result, the succeeding back-end circuit also needs low-noise performance, increasing the power dissipation of the overall system. An AE can be also implemented using instrumentation amplifiers. By amplifying signals using AEs, noise and precision requirements of the following stages are significantly relaxed, leading to a power-efficient system [10].

An AC-coupled inverting amplifier with capacitive feedback (Fig. 2c) [12] has been a widely used architecture for wearable and implantable applications. This is because of the rail-to-rail offset rejection capability and area efficiency. A single-ended version of this amplifier has been used as an AE in reference [10]. However, the major drawback of this architecture is its sensitivity to $1/f$ noise. Different architectures, mostly employing chopper modulations techniques, have been proposed to minimize the $1/f$ noise. Chopper modulation can be implemented before the input coupling capacitors (Fig. 4), as the cost of reduce input impedance and limited offset tolerance [9]. Chopper modulation approach at the virtual ground has been proposed too (Fig. 4). However, it has been shown that this architecture suffers from the low frequency $1/f^2$ noise due to excess current noise [13].

A non-inverting amplifier with single-end input (Fig. 5) can also be used as an Active Electrode amplifier. This architecture provides higher input-impedance, simply because the input impedance is only defined by the parasitic gate capacitance.

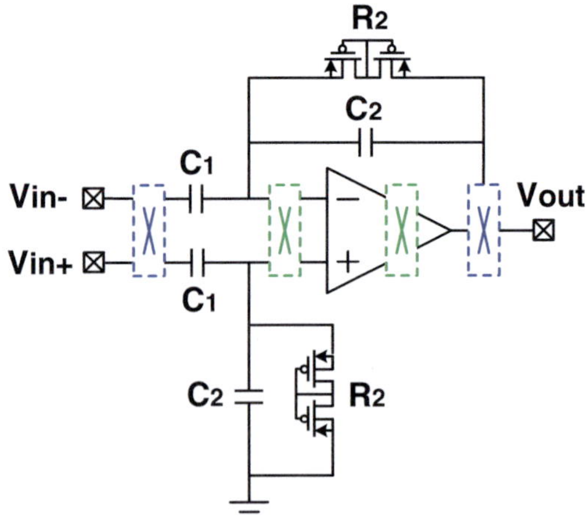

Fig. 4 An AC-coupled inverting amplifier with alternative chopping schemes

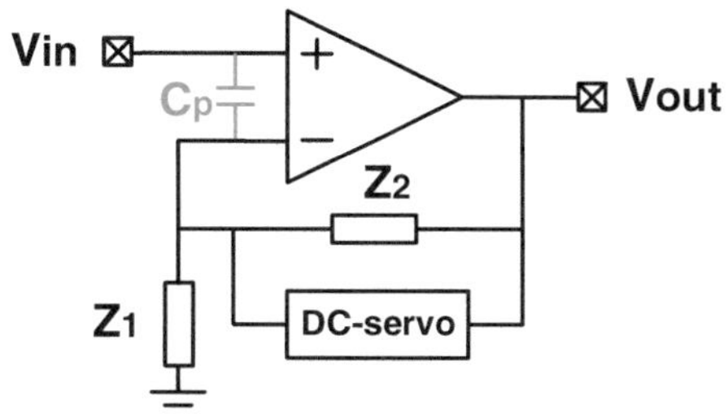

Fig. 5 A non-inverting amplifier with electrode offset compensation

Figure 5 shows the architecture of the capacitive feedback AE [14, 15]. To further reduce $1/f$ noise, chopping technique is utilized at the cost of $1/f^2$ noise at very low frequency [15]. In addition, a DC-servo loop (DSL) has been used to filter the electrode offset [14, 15].

In some applications of EEG measurements, such as extremely low-frequency surface potentials, the measurement of very low-frequency (<0.5 Hz) is necessary. This requires a DC-coupled amplification. However, as discussed earlier, the presence of polarization voltage requires large dynamic range increasing power dissipation. A functionally DC-coupled amplifier [16], used as an AE, combines the advantages of the AC-coupled and the DC-coupled IAs, i.e. compensating large electrode offset with low power while still preserving DC information. The functionally DC-coupled amplifier utilizes a voltage-to-voltage DC servo loop for electrode offset compensation (Fig. 6). The DC servo loop can be implemented with a gm-C integrator [16] that tracks the output offset and then cancels it by driving the IA's inverting input. The biggest advantage of voltage-to-voltage feedback based on gm-C architecture is that it can compensate up to hundreds of mV electrode offset with low power. The functionally DC-coupled amplifier has the same transfer function as a true DC-coupled amplifier. Even though other AC-coupled amplifiers with voltage-to-current feedback [17, 18] can also realize such characteristic, they usually suffer from limited offset tolerance (<100 mV) due to a limited amount of feedback current.

4 Instrumentation Techniques for Implantable Applications

Among implantable applications, *in vivo* neural probes have drawn increasing attention during the last 10 years. The increasing research focus on de-cyphering the functioning of the brain has made neural probes the most important tool for enabling neuroscientists to monitor, acutely or chronically, the extracellular activity of individual neurons or groups of neurons. Current neural interfaces, capable of monitoring the activity of large numbers of neurons, are composed of a neural probe connected to an integrated circuit for recording neural signals from multiple electrodes and transmitting the recorded data to external tools. Emerging technologies are trying to address small form-factor requirements and

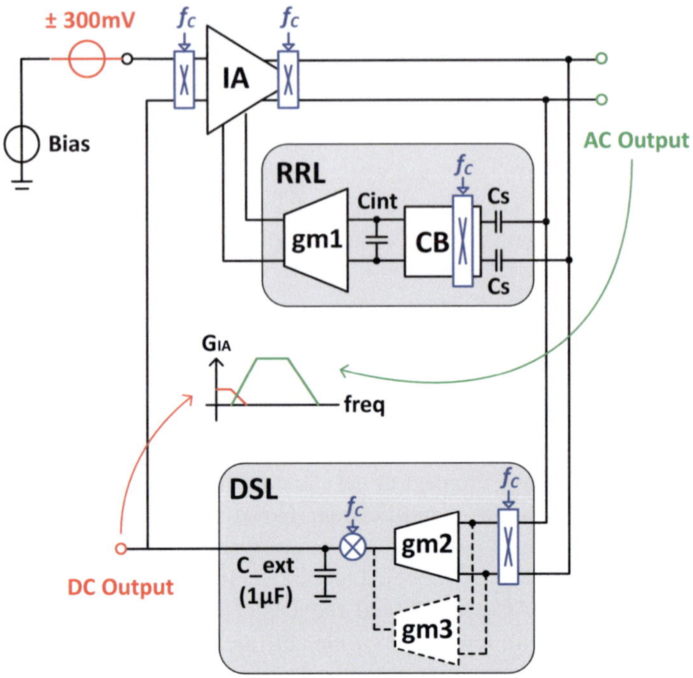

Fig. 6 A functionally DC-coupled amplifier with DC servo loop (DSL)

low power-consumption constraints, while providing high spatial and temporal resolution. Furthermore, implantable neural interfaces are pursuing two main goals: (*i*) replacing hardwired connections with flexible cables or wireless links in order to reduce cable tethering and infection risks, and (*ii*) enabling local processing of neural signals in order to reduce noise coupling and improve signal integrity [19].

The interconnection of neural probes with application-specific integrated circuits (ASICs) to form fully implantable devices poses important power and area limitations to the circuit design and creates several tradeoffs among different circuit blocks and specifications. For instance, implantable systems may dissipate only very low power in order to avoid heating of the surrounding tissue [20], but low-power telemetry usually achieves only limited bandwidths, making the transmission of many recording channels difficult. In the last years, researchers have proposed different kinds of neural interface architectures to deal with such tradeoffs [21–27]. The main targets of these designs are the development of low-power circuit techniques, efficient data management and smart power scheduling [19]. Wireless communication has received great attention due to its benefits in fully-implantable applications [18, 22, 25]. However, this technology is still relying on data reduction techniques in order to be able to transmit the information of large number of channels, which is not useful for applications that require the full raw data.

In a typical neural signal acquisition system, the signal chain is comprised of an instrumentation amplifier and subsequent stages, an analog to digital converter, signal processing and telemetry [28]. Due to the chemical processes happening at the electrode-tissue interface, the recorded signal may exhibit a large DC offset [29], larger than the polarization voltage of electrodes used in wearable applications, in comparison to the useful neural information [28]. This has led to a broad use of AC-coupled instrumentation amplifiers such as the one shown in Fig. 2c. However, the need for amplifying low-frequency signals requires the use of large coupling capacitors, causing the neural amplifier to occupy a large chip area. For instance, commonly-used neural amplifiers [28, 30] employ a capacitive-coupled instrumentation amplifier implemented with an OTA, requiring chip areas in the order of 0.1 mm^2 per channel. While area is not a limiting factor for other neural- or cell-interfacing applications [31–33], it is an important design constraint in implantable devices due to its direct influence on tissue damage [34, 35]. Due to these restrictions, the number of channels in neural recording systems has increased in a slow rate [36], while the systems have been traditionally implemented as a hybrid assemblage of passive neural probes and ASICs (see Fig. 7) [21, 37, 38].

Some attempts to implement active neural probes have been reported in the last years. Some of them are limited to a switch matrix providing the selection of a few recording electrodes from a large electrode array [39], or to a small number of analog recording channels providing signal processing [40, 41]. The most recent approach [26, 27] includes active electrodes, which enables the integration of a

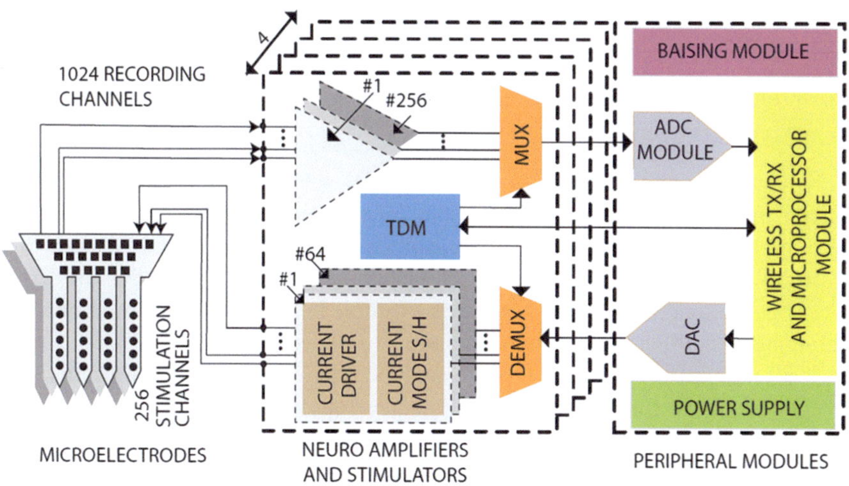

Fig. 7 Typical neural recording system shows separation between probe and electronics [34]

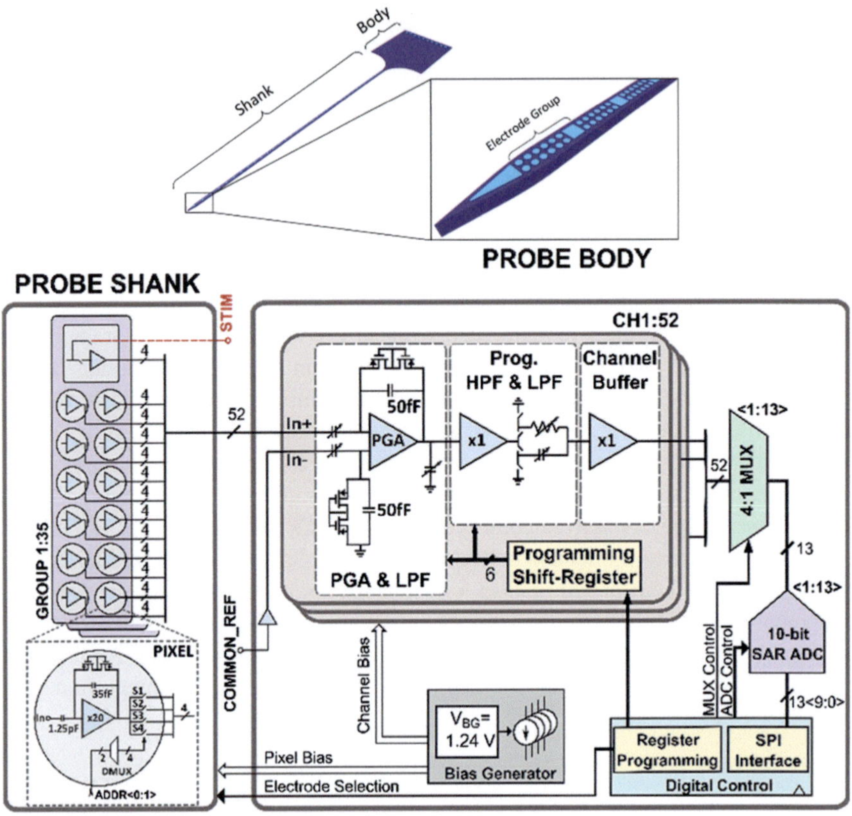

Fig. 8 An implantable active-electrode CMOS neural probe [26, 27]

very large array of electrodes in a single shank (see Fig. 8). The active electrode is implemented using capacitive feedback instrumentation amplifier similar to Fig. 5. The single ended outputs of the active electrodes are further amplified using another capacitive feedback architecture (Fig. 2c) positioned at the probe body. The active electrodes provide a low-impedance connection between the electrode and the subsequent amplifier, thus minimizing the crosstalk between channels and the noise pick-up and artifacts. In this design, amplification, filtering and digitization are integrated on the same probe substrate, while following roughly the same block architecture as previous systems. Future implementations may explore innovative circuit techniques to enable the integration of even larger arrays of active electrodes. In order to reduce the device area, more compact and lower-power analog-to-digital conversion methods may be used to achieve a completely digital electrode, thus eliminating the need for large analog processing circuits.

5 Future Directions in Instrumentation Amplifiers Architectures

The primary focus of instrumentation amplifier design during the past two decades has been noise power efficiency trade-off [11, 42, 51]. However, with the advent of new applications like large array sensor readout and multi-sensor readout systems, there has also been a growing interest in developing area efficient instrumentation amplifier architectures.

Thanks to the advancements in low-power instrumentation, the power consumption of large multi-sensor systems, such as wearable electronics and wireless sensor nodes, is dominated by the on-chip DSP or the RF, and not by the instrumentation amplifier anymore [18, 43]. To reduce the power consumption, digital and RF systems have started employing smaller technologies [44]. However, the instrumentation amplifier consumes significant amount of chip area, and due to the transistor mismatch and noise sensitivity of sensor interfaces, the area of instrumentation amplifier does not scale with technology. Similar to the multi-sensor systems, neural probe applications benefited from advancements in energy-efficient instrumentation amplifier technologies. However, the increasing channel count in neural interfacing technologies and stringent constraints on power dissipation, have also lead to the exploration of miniature area instrumentation amplifiers with ever reducing supply voltage.

One of the main challenges in low-voltage and low-area area instrumentation amplifier design is the dynamic range. Low voltage supply limits the maximum input signal, while low-area design generally increases noise due to the increasing kT/C and flicker noise. In addition, in scaling CMOS technologies, there are additional challenges such as reducing intrinsic gain and increasing the gate-oxide leakage [45], both of which negatively affect the accuracy of the instrumentation. Hence, the instrumentation architectures have to adopt additional compensation technique to overcome these non-idealities, which will in turn require additional power and area.

To tackle the challenge of a large input dynamic range at low supply voltages (<0.5 V) requires architecture innovations. Reference [46] uses a dynamic range folding scheme "fold" the large voltage dynamic range into a smaller range. On a separate track, in our work [47] (Fig. 9), we use a time-domain instrumentation amplifier architecture to handle a large dynamic range as well as obtaining a large intrinsic gain at 0.35 V.

To obtain noise-efficient designs, a capacitive-based IA design is preferred as capacitors are noise-less elements. However, to obtain a low input-referred noise, often, large capacitors are required which in turn dominate the total chip area consumed [12]. Reference [48] uses a T-capacitive network to reduce the area of the feedback capacitor. Such an approach can be beneficial for designs whose area

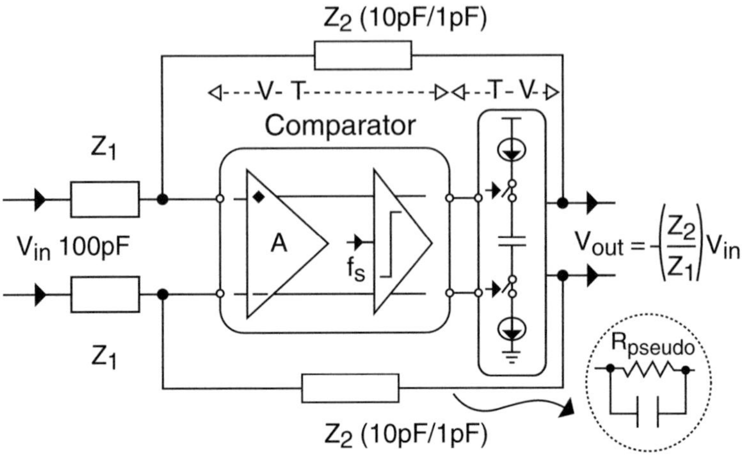

Fig. 9 Time-domain instrumentation amplifier [47]

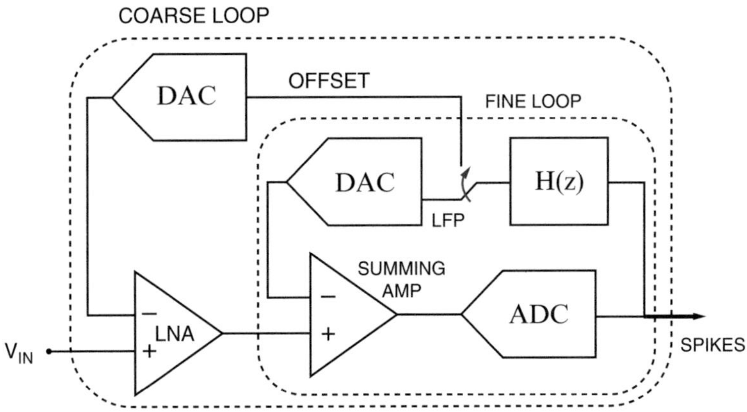

Fig. 10 DC-coupled mixed-signal feedback [18]

is limited due to the gain capacitors. Reference [18] presents DC-coupled readout architecture (Fig. 10). It eliminates usage of capacitors and utilizes a mixed-signal feedback to implement the required bandpass transfer function.

Although, as discussed in the brief literature survey above, low voltage and low area designs have been presented in literature in the past few years, design of instrumentation amplifiers in 40 nm and lower technologies is almost non-existent [49] and yet to be explored.

6 Using Instrumentation Amplifiers for Biomedical Signal Acquisition

6.1 Multi-Sensor Interfaces for Wearable Signal Acquisition

Figure 11 shows the architecture of our multi-sensor interface system-on-chip and the die micrograph. The SoC was fabricated in 0.18 μm CMOS technology and measuring 7 mm × 7 mm. The core supply voltage is 1.2 V and IO supply voltage can be varied from 1.8 to 3.3 V. The analog-front-end of this mixed-signal IC consists of three-channel differential instrumentation amplifiers as presented in Sect. 3.1. These three channels are used for ECG signal acquisition.

In addition to ECG, there is a single channel bio-impedance measurement circuit. The bio-impedance measurement circuit also utilizes the same instrumentation amplifier architecture. However, the instrumentation amplifier is adjusted such that the bio-impedance signals can be demodulated at the output of the instrumentation amplifier to yield complex impedance measurements.

Bio-impedance measurements consist of injecting a differential current into the body by use of electrodes and measuring the resulting voltage, which is proportional to the tissue impedance. This electrical measurement method has been explored actively and numerous physiological parameters can be monitored including respiration rate, heart rate, cardiac output, body fat and fluids, which are relevant for wearable device applications [50]. The choice of the injected current amplitude and frequency is governed by the target applications and the safety standards to guarantee that, for instance in the case of cardiac devices, the current levels are a few orders of magnitude lower than defibrillation current during a system failure [2].

The current injection circuit includes a waveform synthesizer and a 5 bit pseudo-sine DAC current generator (CG) with on-chip clock module and quadrature synchronous demodulation readout [11] as shown in Fig. 12. To support a large range of electrode types (gel and dry) while driving current levels in the range of 25–100 μApp, the CG is designed with cascode current sources and sinks to combine large dynamic range and large output impedance. The bio-impedance readout performs synchronous demodulation of the I and Q components at $f_{bioz}(0°)$ and $f_{bioz}(90°)$. This demodulation is simply achieved by removing the first chopped modulator of the instrumentation amplifier in Fig. 3 and let the output demodulator perform the synchronous demodulator function. Once demodulated, the output bio-impedance signal is low-pass filtered to separate spurious components from the base-band bio-impedance signal.

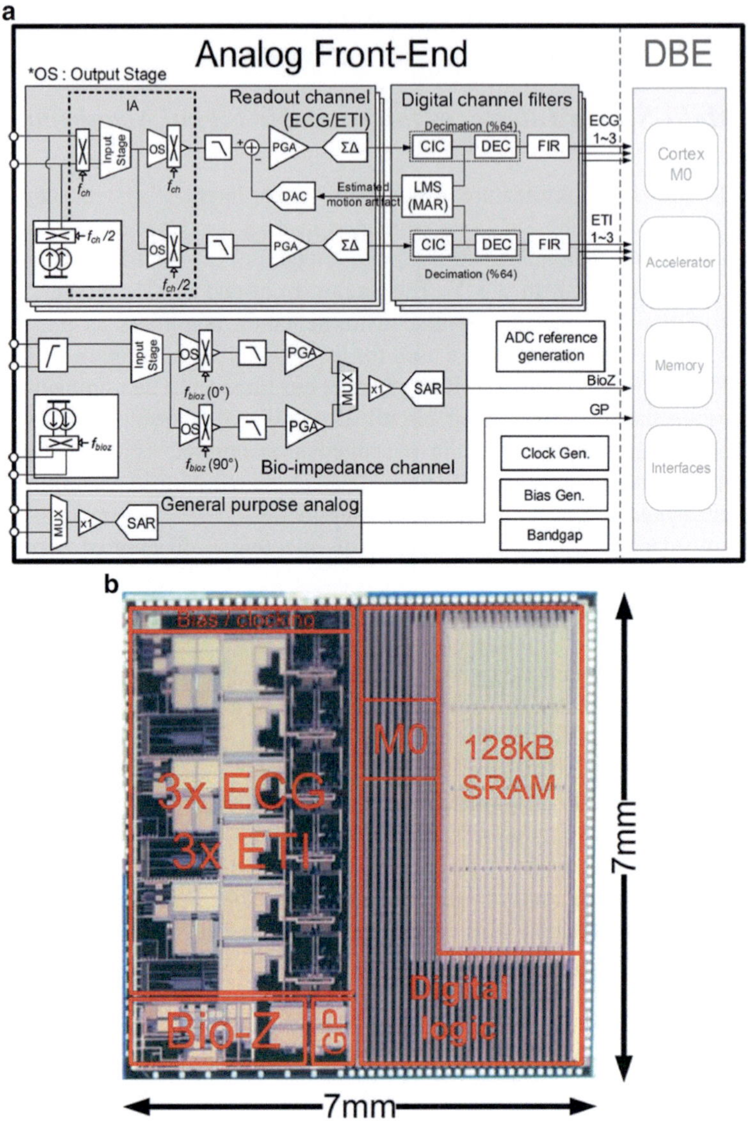

Fig. 11 Die photo of the multi-sensor interface circuit (7 mm × 7 mm)

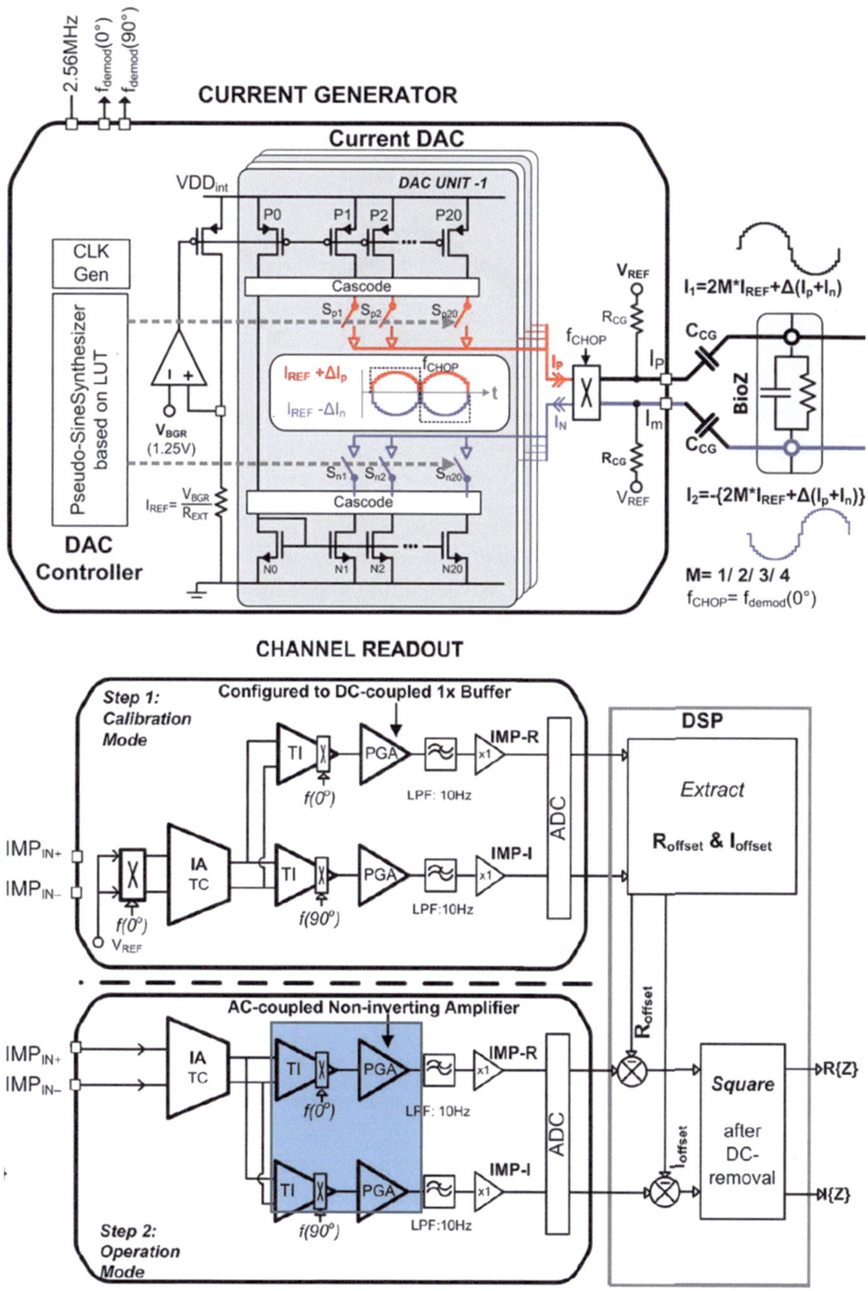

Fig. 12 Bio-impedance measurement circuit, making use of the instrumentation amplifier architecture presented in Sect. 3.1

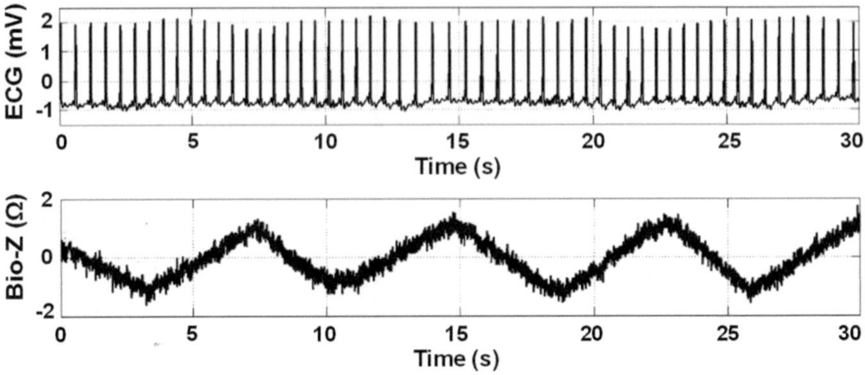

Fig. 13 Concurrent ECG and bio-impedance recording for respiration monitoring using the instrumentation of the multi-sensor interface IC

Figure 13 shows the measurement results from the multi-sensor ASIC. The instrumentation techniques presented in the earlier sections measure ECG and bio-impedance information simultaneously, enabling applications where heart rate and respiration rate needs to be continuously monitored, for instance fitness applications.

6.2 High-Density Neural Probes

A CMOS probe including active electrodes was recently reported by Lopez et al. [26, 27], which integrates the largest array of electrodes in a single shank (i.e. 455). In this design, 52 electrodes can be recorded simultaneously and all the readout circuits are integrated on the same probe substrate (see Fig. 14). The complete functionality of this active probe was validated by successful *in vivo* experiments in anesthetized rats. For this, a battery-powered portable headstage for chronic use was designed, which provides real-time monitoring and acquisition of 52 selected electrode signals (Fig. 15, left). The headstage provides USB connection to the computer and its dimensions are $17 \times 27 \times 8$ mm^3, with a weight of 3.8 g. Figure 16 shows the recorded spontaneous activity from a set of 26 electrodes positioned in the cortex and the thalamus. Chronic experiments, carried out by using the setup shown in Fig. 15 (right), have also been made to further validate this technology in awake animals performing behavioral tasks.

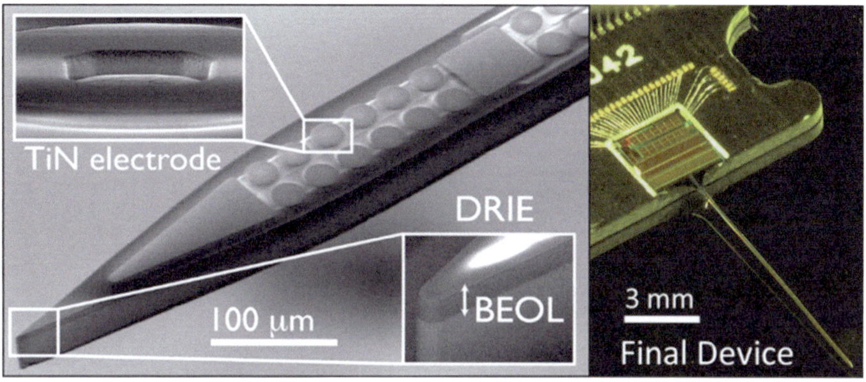

Fig. 14 Active-electrode probe containing the largest electrode array in a single shank [26, 27]

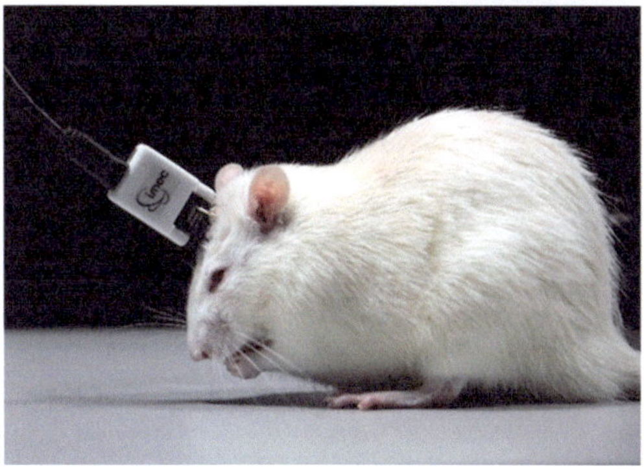

Fig. 15 Picture of the packaged active neural probe and the custom-designed headstage (*left*). Picture of the setup used for chronic in vivo experiments (*right*) [26, 27]

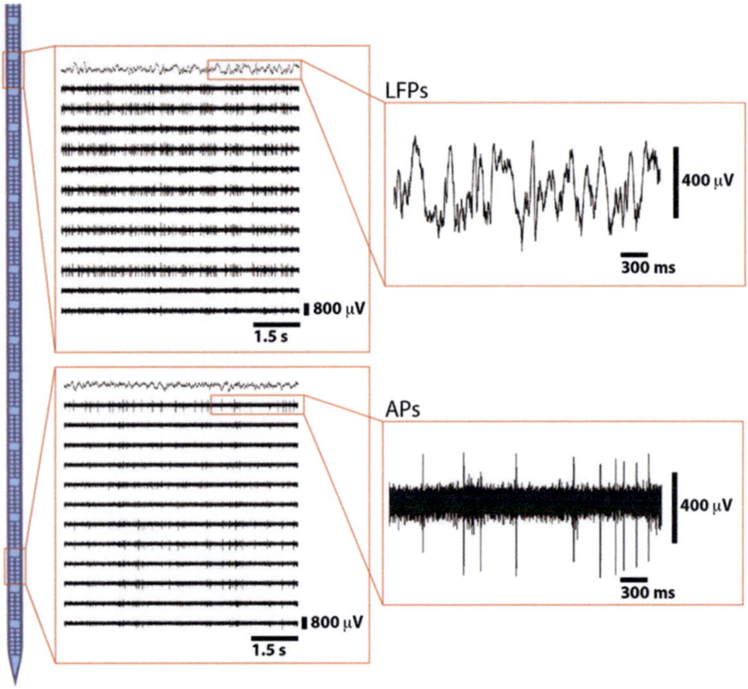

Fig. 16 In vivo recording from 26 parallel channels in an implanted neural probe located in the thalamus and the cortex [26, 27]

References

1. Webster J (1992) Medical instrumentation: application and design, 2nd edn. Houghton Mifflin, Boston
2. IEC60601-1 (2005) Medical electrical equipment—part 1: general requirements for basic safety and essential performance. International Standard
3. Texas Instruments ADS1292R (2012) Low-power, 8-channel, 24-bit AFE for biopotential measurements. SBAS502B
4. Van Helleputte N, Kim S, Kim H, Kim JP, Van Hoof C, Yazicioglu RF (2012) A 160μA biopotential acquisition IC with fully integrated IA and motion artifact suppression. IEEE Trans Biomed Circuits Syst 6(6):552–561
5. Yazicioglu RF, Merken P, Puers R, Van Hoof C (2008) A 200μW eight-channel acquisition ASIC for ambulatory EEG systems. In: Digest ISSCC, pp 164–603
6. Fan Q, Sebastiano F, Huijsing JH, Makinwa KAA (2011) A 1.8 μW 60 nV/$\sqrt{}$Hz capacitively-coupled chopper instrumentation amplifier in 65 nm CMOS for wireless sensor nodes. IEEE J Solid-State Circuits 46(7):1534–1543
7. Zou X, Liew W-S, Yao L, Lian Y (2010) A 1V 22μW 32-channel implantable EEG recording IC. In: IEEE international solid-state circuits conference, pp 126–127
8. Qian C, Parramon J, Sanchez-Sinencio E (2011) A micropower low-noise neural recording front-end circuit for epileptic seizure detection. IEEE J Solid-State Circuits 46(6):1392–1405

9. Denison T, Consoer K, Kelly A, Hachenburg A, Santa W (2007) A 2.2μW 94nV/√Hz, chopper-stabilized instrumentation amplifier for EEG detection in chronic implants. In: ISSCC, pp 162–594
10. Xu J, Yazicioglu RF et al (2011) A 160μW 8-channel active electrode system for EEG monitoring. IEEE Trans Biomed Circuits Syst 5(6):555–567
11. Van Helleputte N, Konijnenburg M, Pettine J, Jee D-W, Kim H, Morgado A, Van Wegberg R, Torfs T, Mohan R, Breeschoten A, de Groot AH, Van Hoof C, Yazicioglu RF (2015) A 345 μW multi-sensor biomedical SoC with bio-impedance, 3-channel ECG, motion artifact reduction, and integrated DSP. IEEE J Solid-State Circuits 50(1):230–244
12. Harrison RR, Charles C (2003) A low-power low-noise CMOS amplifier for neural recording applications. IEEE J Solid-State Circuits 38(6):958–965
13. Xu J et al (2013) Measurement and analysis of current noise in chopper amplifiers. IEEE J Solid-State Circuits 48(7):1575–1584
14. Guermandi M, Cardu R, et al (2011) Active electrode IC combining EEG, electrical impedance tomography, continuous contact impedance measurement and power supply on a single wire. In: Proceedings of ESSCIRC, pp 335–338
15. Xu J, Mitra S, Matsumoto A, Patki S, Makinwa KAA, Van Hoof C, Yazicioglu RF (2014) A wearable 8-channel active-electrode EEG/ETI acquisition system for body area networks. IEEE J Solid-State Circuits 49(9):2005–2016
16. Xu J, Büsze B, Kim H, Makinwa KAA, Van Hoof C, Yazicioglu RF (2014) A 60nv/rt(Hz) 15-channel digital active electrode system for portable biopotential monitoring. In: Digest ISSCC, pp 424–425
17. Yazicioglu RF, Merken P et al (2007) A 60μW 60 nV/√Hz readout front-end for portable biopotential acquisition systems. IEEE J Solid-State Circuits 42(5):1100–1110
18. Muller R et al (2012) A 0.013 mm^2 2.5 μW, DC-coupled neural signal acquisition IC with 0.5 V supply. IEEE J Solid-State Circuits 47(1):232–243
19. Gosselin B (2011) Recent advances in neural recording microsystems. Sensors 11(5):4572–4597
20. Harrison RR (2008) The design of integrated circuits to observe brain activity. Proc IEEE 96(7):1203–1216
21. Gosselin B, Ayoub AE, Roy J-F, Sawan M, Lepore F, Chaudhuri A, Guitton D (2009) A mixed-signal multichip neural recording interface with bandwidth reduction. IEEE Trans Biomed Circuits Syst 3(3):129–141
22. Chae M-S, Yang Z, Yuce MR, Hoang L, Liu W (2009) A 128-channel 6 mW wireless neural recording IC with spike feature extraction and UWB transmitter. IEEE Trans Neural Syst Rehabil Eng 17(4):312–321
23. Shahrokhi F, Abdelhalim K, Serletis D, Carlen PL, Genov R (2010) The 128-channel fully differential digital integrated neural recording and stimulation interface. IEEE Trans Biomed Circuits Syst 4(3):149–161
24. Gao H, Walker RM, Nuyujukian P, Makinwa KAA, Shenoy KV, Murmann B, Meng TH (2012) HermesE: a 96-channel full data rate direct neural interface in 0.13 μm CMOS. IEEE J Solid-State Circuits 47(4):1043–1055
25. Biederman W, Yeager DJ, Narevsky N, Koralek AC, Carmena JM, Alon E, Rabaey JM (2013) A fully-integrated, miniaturized (0.125 mm^2) 10.5 μW wireless neural sensor. IEEE J Solid-State Circuits 48(4):960–970
26. Lopez CM, Andrei A, Mitra S, Welkenhuysen M, Eberle W, Bartic C, Puers R, Yazicioglu RF, Gielen G (2014) An implantable 455-active-electrode 52-channel CMOS neural probe. IEEE J Solid-State Circuits 49(1):248–261
27. Tung-Chien Chen, Kuanfu Chen, Zhi Yang, Cockerham, K., Wentai Liu (2009), A biomedical multiprocessor SoC for closed-loop neuroprosthetic applications, IEEE International Solid-State Circuits Conference (ISSCC), pp. 434–435, 435
28. Jochum T, Denison T, et al (2009) Integrated circuit amplifiers for multi-electrode intracortical recording. J Neural Eng 6

29. Franks W, Schenker I, Schmutz P, Hierlemann A (2005) Impedance characterization and modeling of electrodes for biomedical applications. IEEE Trans Biomed Eng 52(7):1295–1302
30. Majidzadeh V, Leblebici Y (2009) A micropower neural recording amplifier with improved noise efficiency factor. In: European conference circuit theory and design (ECCTD), pp 319–322
31. Muller J, Livi P, Chen Y (2013) Conferring flexibility and reconfigurability to a 26,400 microelectrode CMOS array for high throughput neural recordings. In: Solid-state sensors, actuators and microsystems transducers, pp 744–747
32. Eversmann B, Jenkner M, Hofmann F, Paulus C (2003) A 128 × 128 CMOS biosensor array for extracellular recording of neural activity. IEEE J Solid-State Circuits 38(12):2306–2317
33. Lei N, MacLean J, Yuste R (2008) A 256Ã—256 CMOS microelectrode array for extracellular neural stimulation of acute brain slices. In: ISSCC digest of technical papers, pp 148–603
34. Bagheri AGS, Salam M, Velazquez J (2012) 1024-channel-scalable wireless neuromonitoring and neurostimulation rodent headset with nanotextured flexible microelectrodes. In: IEEE biomedical circuits and systems conference (BioCAS), pp 184–187
35. Liu L, Yao L, Zou X, Goh WL (2013) Neural recording front-end IC using action potential detection and analog buffer with digital delay for data compression. Conf Proc IEEE Eng Med Biol Soc 2013:747–750
36. Stevenson IH, Kording KP (2011) How advances in neural recording affect data analysis. Nat Neurosci 14:139–142
37. Harrison R, Watkins P, Kier R, Lovejoy R (2007) A low-power integrated circuit for a wireless 100-electrode neural recording system. IEEE J Solid-State Circuits 42(1):123–133
38. Perlin GSA, Wise K (2006) Neural recording front-end designs for fully implantable neuroscience applications and neural prosthetic microsystems. Conf Proc IEEE Eng Med Biol Soc 1:2982–2985
39. Torfs T et al (2011) Two-dimensional multi-channel neural probes with electronic depth control. IEEE Trans Biomed Circuits Syst 5(5):403–412
40. Olsson R, Wise K (2005) A three-dimensional neural recording microsystem with implantable data compression circuitry. In: Solid-state circuits conference, 2005. Digest of technical papers (ISSCC), pp 558–559
41. Olsson RH III, Buhl DL, Sirota AM, Buzsaki G, Wise KD (2005) Band-tunable and multiplexed integrated circuits for simultaneous recording and stimulation with microelectrode arrays. IEEE Trans Biomed Eng 52(7):1303–1311
42. Steyaert MSJ, Sansen WMC (1987) A micropower low-noise monolithic instrumentation amplifier for medical purposes. IEEE J Solid-State Circuits 22(6):1163–1168
43. Verma N, Shoeb A, Bohorquez J, Dawson J, Guttag J, Chandrakasan AP (2010) A micropower EEG acquisition SoC with integrated feature extraction processor for a chronic seizure detection system. IEEE J Solid-State Circuits 45(4):804–816
44. Ashouei M, Hulzink J, Konijnenburg M, Zhou J, Duarte F, Breeschoten A, Huisken J, Stuyt J, de Groot H, Barat F, David J, Van Ginderdeuren J (2011) A voltage-scalable biomedical signal processor running ECG using 13pJ/cycle at 1MHz and 0.4V. In: Solid-state circuits conference digest of technical papers (ISSCC), 2011 IEEE international, pp 332–334
45. Annema A-J, Nauta B, van Langevelde R, Tuinhout H (2005) Analog circuits in ultra-deep-submicron CMOS. IEEE J Solid-State Circuits 40(1):132–143
46. Han D, Zheng Y, Rajkumar R, Dawe G, Je M (2013) A 0.45V 100-channel neural-recording IC with sub-μW/channel consumption in 0.18μm CMOS. In: Solid-state circuits conference digest of technical papers (ISSCC), 2013 IEEE international, pp 290–291
47. Mohan R, Yan L, Gielen G, Van Hoof C, Yazicioglu RF (2014) 0.35 V time-domain-based instrumentation amplifier. Electron Lett 50(21):1513–1514
48. Ng KA, Xu YP (2012) A compact, low input capacitance neural recording amplifier with C in/gain of 20fF.V/V. In: Biomedical circuits and systems conference (BioCAS), 2012 IEEE, pp 328–331

49. De Smedt V, Gielen G, Dehaene W (2013) A 40nm-CMOS, 18 μW, temperature and supply voltage independent sensor interface for RFID tags. In: Solid-state circuits conference (A-SSCC), 2013 IEEE Asian, pp 113–116
50. Grimnes S, Martinsen OG (2015) Bioimpedance and bioelectricity basics, 3rd edn. Elsevier, Amsterdam
51. Wattanapanitch W, Fee M, Sarpeshkar R (2007) An energy-efficient micropower neural recording amplifier. IEEE Trans Biomed Circuits Syst 1(2):136–147

A Power-Efficient Compressive Sensing Platform for Cortical Implants

Mahsa Shoaran and Alexandre Schmid

Abstract Smart and miniaturized implantable microsystems with diagnostic and therapeutic capabilities are becoming increasingly important for patients suffering from neurological disorders such as epilepsy. Recent developments in microfabrication technology have provided new insights into seizure generation at an unprecedented spatial scale. Based on these findings, designing powerful acquisition systems capable of probing the wide-range spatiotemporal activities within the brain holds a great promise to improve the quality of life of epileptic patients. As a major technological barrier, the high overall data rate of digitized neural signals recorded by dense electrode arrays can drastically increase the power consumption of the wireless transmission module. Consequently, extensive system-level design improvement is needed to meet the requirements of the implantable device, while preserving the high-resolution monitoring capability. In this context, low-power circuit and system design techniques for data compression and seizure detection in multichannel cortical implants are presented. The first fully-integrated circuit that addresses the multichannel compressed-domain feature extraction is proposed, consuming sub-μW of power within an effective area of 250 μm $\times$ 250 μm per channel.

1 Introduction

Emerging applications in brain-machine interface systems require high-resolution, chronic multisite cortical recordings, which cannot be obtained with conventional technologies due to high power consumption and high invasiveness [1]. Despite substantial improvements in today's implantable medical device technology, several requirements remain to be addressed such as intimate integration of the implant with human body, high-resolution mapping of biological signals, and reliable detection of

M. Shoaran (✉)
Mixed-mode Integrated Circuits and Systems Laboratory, California Institute of Technology, Pasadena, CA, USA

A. Schmid
Microelectronic Systems Laboratory, Swiss Federal Institute of Technology in Lausanne (EPFL), Lausanne, Switzerland
e-mail: mshoaran@caltech.edu

K.A.A. Makinwa et al. (eds.), *Efficient Sensor Interfaces, Advanced Amplifiers and Low Power RF Systems*, DOI 10.1007/978-3-319-21185-5_6

symptoms by developing ultra-low-power and miniaturized devices with enhanced functionality and biocompatibility.

Epilepsy is a primary target of the neuroengineering field, along with movement disorders, stroke, affective disorders, head trauma, chronic pain and paralysis [2]. Frequent seizures are typical manifestations of epilepsy. Approximately one third of epileptic patients exhibit seizures that are not controlled by medication. Despite substantial innovations in anti-seizure drug therapy over the past 15 years, the proportion of patients with uncontrolled epilepsy has not changed, highlighting the need for new treatment strategies [2].

The development of new devices capable of performing a rapid and reliable seizure detection followed by brain stimulation is a promising solution. In some cases, an ablative surgery consisting in removing the areas of brain producing seizures may provide a favorable outcome. The most critical phase in both therapy methods pertains to the efficient and accurate detection of seizure onset or epilepsy focus, which is tackled in this chapter.

1.1 High-Resolution Electrophysiology

The high density of neurons in neurobiological tissue requires a large number of electrodes to obtain the most accurate representation of neural activity and provide better control over the location of the stimulation sites or resected tissue. In clinical practice, intracranial EEG (iEEG) used for epilepsy surgery has been traditionally provided by large and widely-spaced electrodes operated over a relatively narrow bandwidth, which offer only cumbersome wired operation. This practice, however, is largely based upon tradition and limits of sensor technology when iEEG was first being recorded, rather than our exact knowledge of human brain. These technological limitations often frustrate epileptologists looking for discrete, functional lesions to remove during epilepsy surgery [3].

In addition, the controversial performance of the seizure detection methods to date is partially due to the inefficiency of the current anti-seizure devices in capturing the high-resolution dynamics of the epileptic brain. The performance of several seizure detection methods at different sampling rates and for different electrode types shows that all the studied metrics perform significantly better at higher sampling rates, using higher resolution electrode arrays [4, 5]. Using the hybrid subdural strips containing both clinical macroelectrodes and microelectrodes as presented in Fig. 1, the seizure generation is shown to be preceded by a build-up of seizure-like activity on the microelectrodes [5]. Consequently, a more rapid seizure onset detection is expected at a higher spatial resolution.

The main goal of the presented research is to achieve a higher integration capability and lower power consumption compared to the existing systems, by introducing a new multichannel signal acquisition platform (Fig. 2a) for the recently developed dense arrays of electrodes [6]. Such highly flexible arrays of subdural electrodes (Fig. 2b) have unique advantages over penetrating microelectrode arrays

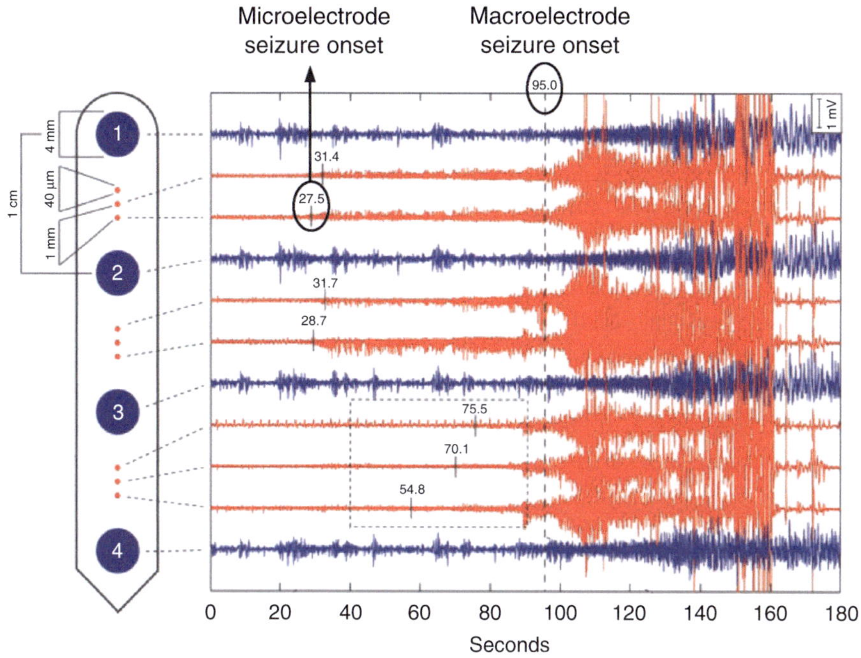

Fig. 1 Hybrid subdural strip containing clinical macroelectrodes (*blue*) and microelectrodes (*red*). Independent, asynchronous microseizure activity is recorded on microelectrodes (printed as a number superposed to the waveform at the moment of detection, e.g. 31.4, 28.7) significantly before the seizure becomes apparent on the clinical macroelectrodes [5]

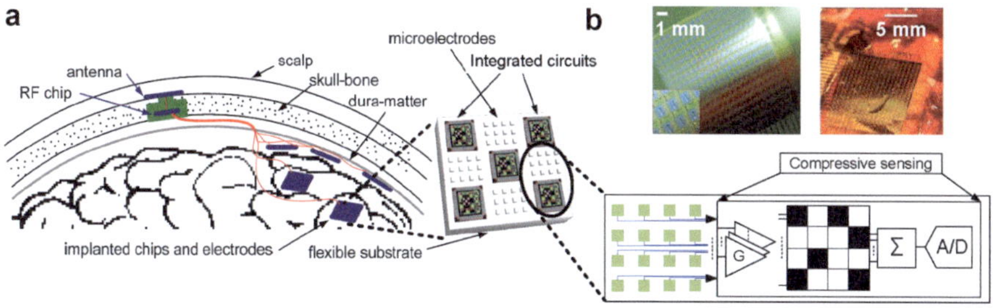

Fig. 2 (**a**) Schematic representation of the proposed distributed monitoring system. (**b**) Flexible, high-resolution electrode array, adapted from [6]. The electrode size and spacing are 300 µm × 300 µm and 500 µm, respectively. The array includes 360 electrodes. A compressive sensing solution shown in the *bottom inset* is proposed to limit the transmission data rate

(such as Utah array) in that they are able to maintain signal quality over extended periods of time with minimized irritation and injury to brain tissue [6].

The general solution proposed in this chapter consists of implanting highly flexible and thin substrates including the chip-electrode combinations (Fig. 2a). The proposed active electrode arrays combine a large number of microelectrodes with integrated circuits and are distributed over a relatively large area of the cortex. By placing these recording units on the potentially epileptic parts of the cortex previously indicated by standard non-invasive methods, sufficient information pertaining to the localization of the epileptic foci and/or detection of seizure onset with high spatial resolution quality can be recorded.

The long-term implantation is enabled by the presence of an RF chipset located in a burr hole in the skull for remote powering and wireless data transmission.

2 Compressive Sensing for Neural Signal Acquisition

Compressive sensing (CS) [7, 8] is an emerging compression method with interesting advantages over traditional methods thanks to its low encoder complexity and universality with respect to the signal model. A compressive sensing system samples a high-dimensional signal into a smaller number of linear measurements than dictated by the Nyquist sampling theorem. Many biological signals such as action potentials, EEG and ECG have an information rate much smaller than the rate dictated by the Nyquist sampling theorem. Thus, CS can be applied to these signals. Compared to spike detection and activity-dependent recording, CS has the advantage of preserving the temporal information and morphology of the neural signal for the entire recording period.

2.1 Conventional Architectures

The common microelectronic approach to CS is shown in Fig. 3a, b. The amplified and filtered signal of each channel passes through m multiplication and integration paths, performing the matrix multiplication $y = \mathbf{\Phi}x$ by which the d-dimensional signal x is mapped into an m-dimensional measurement vector y with a compression ratio of d/m, i.e., $m \ll d$.

The multiplication function can be implemented either in the analog domain prior to digitization (Analog CS or ACS shown in Fig. 3a) [9], or downstream the ADC (Digital CS or DCS shown in Fig. 3b) [10].

Employing CS leads to a significant data reduction of the system. However, a microelectronic architecture based on single-channel CS occupies a considerable die area when compared to the Nyquist-sampling alternatives. The power analysis in [10] shows the superior performance of a digital (Fig. 3b) over the analog implementation. However, including m multiplication and accumulation blocks in each channel results in a large area which summed up to the area of the low-noise amplifier, the Nyquist rate ADC and the random sequence generators, disqualifies the digital approach for a multichannel recording interface which should include the circuits supporting many channels in a limited die area.

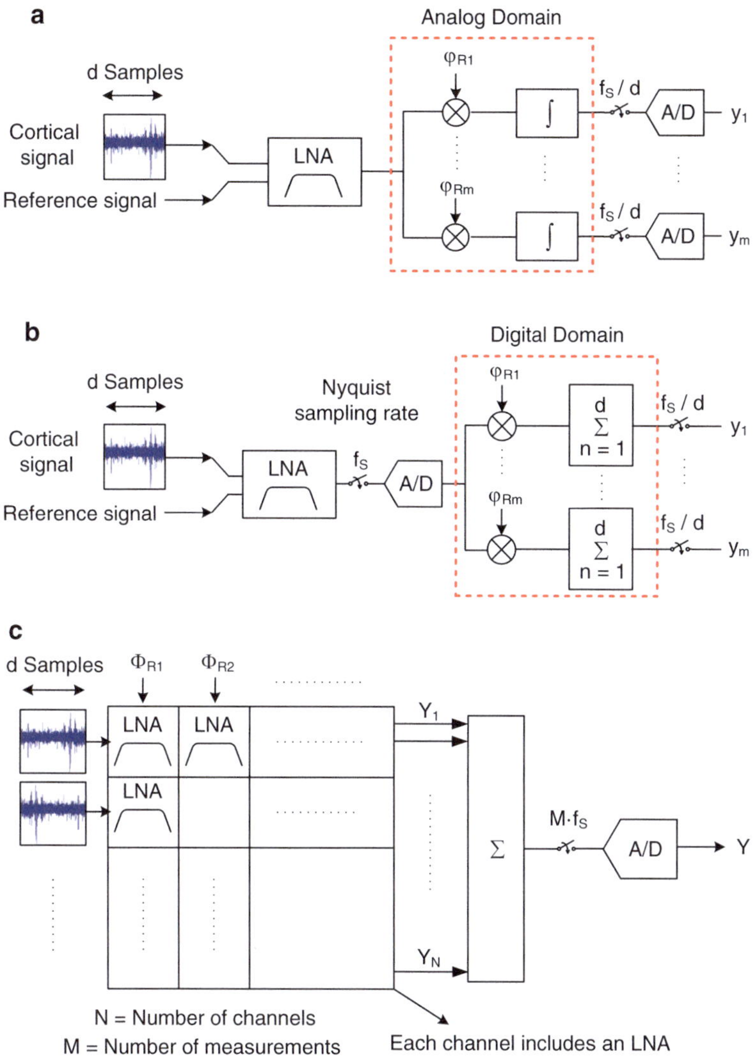

Fig. 3 Block diagram of (**a**) the analog single-channel CS, (**b**) the digital single-channel CS, and (**c**) the proposed multichannel CS (MCS) architectures

2.2 *Proposed Spatial-Domain Compressive Acquisition System*

To overcome these issues, a new multichannel measurement scheme (Fig. 3c) along with an appropriate recovery scheme are proposed [11], which encode the entire array into a single compressed data stream. In the proposed approach, the compression is carried out in the analog domain, and in a multichannel fashion. This technique circumvents the need to place one ADC per channel and results in a significant area saving. Based on this approach, a wireless monitoring system consisting of several recording/compressing units is proposed (Fig. 2). Taking benefit of the area-efficient implementation of CS, the number of recording units that are implantable on the cortex and satisfy the energy constraints of the system is scaled up by a factor equal to the compression ratio.

Let $\mathbf{X} \in \mathbb{R}^{d \times N}$ represent the multichannel iEEG signal where d is the dimension of the signal in each channel and in a defined time-window called *compression block* and N is the number of channels. We define a reshaping operator $\mathcal{P} : \mathbb{R}^{d \times N} \rightarrow \mathbb{R}^{d \cdot N}$ which transposes the input matrix and vectorizes the resulting matrix by concatenating its columns after each other [11].

The linear compressive measurements are obtained by acquiring $M = p/d$ measurements from columns of $\mathbf{X}$, i.e., M measurement at each time-sample from the array, where $p \ll d \times N$ is the total number of measurements. Hence, the multichannel linear map can be represented in matrix form $\mathbf{\Phi}_{MC} \in \mathbb{R}^{(Md) \times (dN)}$ as follows

$$\mathbf{\Phi}_{\mathrm{MC}} = \begin{bmatrix} \mathbf{\Phi}_1 & 0 & \cdots & 0 \\ 0 & \mathbf{\Phi}_2 & \cdots & 0 \\ \vdots & \vdots & \ddots & \vdots \\ 0 & 0 & \cdots & \mathbf{\Phi}_d \end{bmatrix}, \mathbf{\Phi}_i = \begin{bmatrix} \phi_i^{1,1} & \cdots & \phi_i^{1,N} \\ \phi_i^{2,1} & \cdots & \phi_i^{2,N} \\ \vdots & \ddots & \vdots \\ \phi_i^{M,1} & \cdots & \phi_i^{M,N} \end{bmatrix} \quad (1)$$

where $\mathbf{\Phi}_i \in \mathbb{R}^{M \times N}$ and $\phi_i^{k,l}$ is uniformly selected from $\{0, 1\}$ to approximate a measurement matrix similar to the Bernoulli matrix. The multichannel measurement vector $Y \in \mathbb{R}^p$ is defined as

$$Y = \mathbf{\Phi}_{\mathrm{MC}}\, \mathcal{P}(\boldsymbol{X}) \quad (2)$$

A comprehensive analysis of power and area overhead of several neural recording architectures is presented in [12]. The area efficiency of multichannel CS (MCS) architecture justifies its superior performance over other CS methods in a high-density recording system. As opposed to ACS and DCS, the die area of an MCS-based system is independent of the compression ratio (*CR*). The compression ratio is only tuned by adapting the sampling rate of the ADC and the frequency of the random sequence generators, without additional area overhead.

2.2.1 Microelectronic Architecture

The proposed MCS architecture based on an active switched capacitor integration method is shown in Fig. 4. Each of the N channels of the array consist of an area-efficient T-network based low-noise amplifier with a band-pass transfer function, an additional low-pass filter to limit the high cut-off frequency, a second capacitively coupled gain stage, and a buffered sample-and-hold circuit. The sampled signals of N channels of the array connect to the summing stage at the sampling phase (Φ_{S3}) which follows the two in-phase events Φ_{S1} and Φ_{S2}.

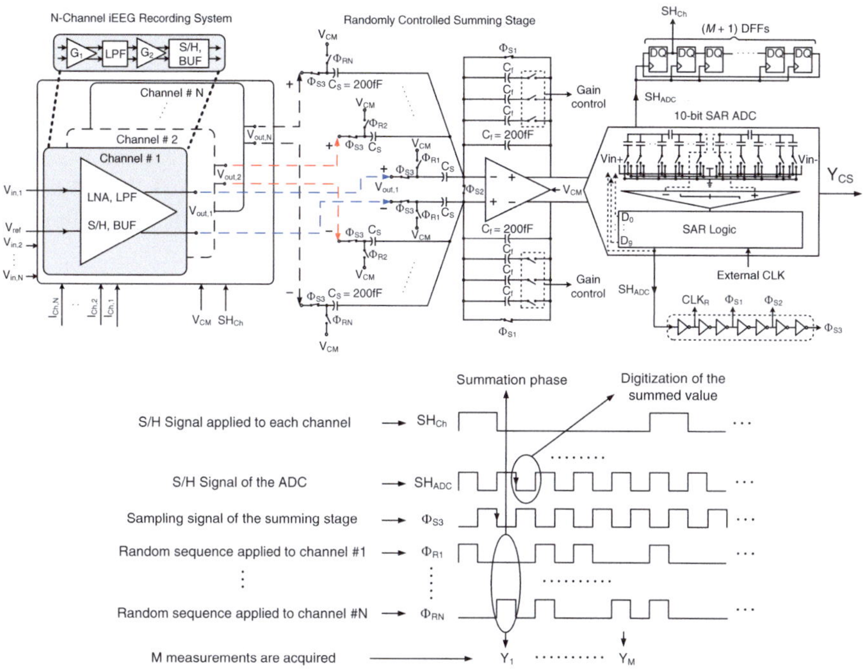

Fig. 4 Circuit implementation and timing diagram of the proposed multichannel CS (MCS) architecture including the individual channels, the randomly controlled summing stage and a binary-weighted SAR ADC with attenuation capacitor [11]

In sampling mode, Φ_{S1}, Φ_{S2} and Φ_{S3} are turned on, allowing the differential voltage across the two sampling capacitors (C_S) at the output of each channel to track the differential output voltage of the channel. In summation mode, the charge stored on the sampling capacitors of those channels configured with a random value equal to one are transferred to the capacitors in the feedback path (C_f).

In the proposed scheme, the data sampled from a channel is kept constant during M randomized summations and consecutive digitizations by the ADC (Fig. 4). Multi-output random sequence generation ($\Phi_{R1}, \ldots, \Phi_{RN}$) is achieved by XORing the multiple outputs of maximal-length pseudo-random binary sequence (PRBS) generators. All the required signals are generated using a single external clock.

A binary-weighted capacitive array with attenuation capacitor is used which enables the compact implementation of the ADC. The required resolution of the ADC in the proposed system is approximated [11] by $B_y \approx B_{sig} + \log_2 \sqrt{N} - 0.58$. Considering a background neuronal noise of 5–10 μV_{rms} and a typical amplitude of surface cortical signals of up to 1 mV and adding some extra margin, B_{sig} and B_y are set to 8 and 10 bits for 16 channels.

2.3 Experimental Results

The proposed multichannel compressive sensing system is designed and implemented in a 0.18 μm CMOS technology. The total current consumption of the chip including the buffers and bias generators is 140 μA drawn from a 1.2 V power supply, corresponding to an effective current of 8.75 μA per channel. The achieved power density of the system is 7.2 mW/cm^2, significantly below the safety limit of 80 mW/cm^2 for an implantable system.

2.3.1 Recovery Performance

In order to demonstrate the effectiveness of the proposed acquisition model, a long segment of multichannel iEEG signal recorded from subdural strip and grid electrodes implanted on the left temporal lobe of a patient with medically refractory epilepsy has been used as the input (Fig. 5a). The signals are recorded during an invasive pre-surgical evaluation phase to pinpoint the areas of the brain involved in seizure generation and to study the feasibility of a resection surgery. This data includes minutes of pre-ictal, ictal and post-ictal activities sampled at 32 kS/s, using Neuralynx.

The signals recorded by 16 adjacent channels of electrodes (arranged in grid formation) are applied to the proposed CS system. The iEEG signal is generated using an FPGA based platform. The FPGA transfers preloaded serial iEEG data to two off-chip eight-channel D/A converters and the outputs of the D/As are applied into the chip. The recovery SNR of the reconstructed signal ($\hat{x}$) with respect to the original signal (x) is calculated from

$$\mathrm{SNR} = -20\log_{10}\|x - \hat{x}\|_2 / \|x\|_2 \tag{3}$$

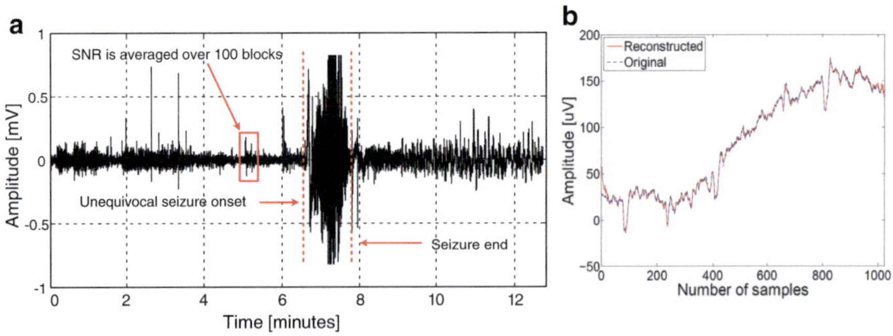

Fig. 5 (**a**) One channel of human intracranial EEG recorded by strip and grid electrodes implanted on the left temporal lobe. The recovery SNR is calculated by averaging over 100 blocks of signal in the low-voltage fast activity region. (**b**) Recovery performance for a block of length 1024 and compression ratio of 4; reconstruction $SNR_{CH1} = 28.04$ dB

for each recording channel (e.g. SNR_{CH1} represents the recovery SNR of first channel). The mean SNR of 16 channels are averaged over 100 blocks of the signal, as shown in Fig. 5a. The performance of the circuit is validated for low-voltage fast activities which are shown to be associated with seizure onset.

The achieved averaged SNR of the proposed system is 21.80 dB. Based on the statistical analysis reported in [13], a minimum SNR of 10.45 dB is acceptable to maintain the diagnostically important data in the recovered signal, e.g., for successful seizure detection. Reducing the number of measurements to $M = 1$, i.e., $CR = 16$, results in an average SNR of 13.72 dB. Thus, the system is able to successfully recover low-voltage intracranial EEG signals compressed by a ratio as high as 16.

The reconstructed signals versus the original signals corresponding to one block of an arbitrary channel using a mixed norm recovery [11] are shown in Fig. 5b. The length of each compression block (d) is equal to 1024 samples and is equivalent to 256 ms at a 4 kHz sampling frequency. The digitized data after ADC is used for recovery.

2.3.2 Discussion and Comparison

Table 1 summarizes the performance of the system and presents a comparison with published works. In this table, compression power and area refer to the extra power consumption and area usage of the signal digitization, compression and thresholding blocks, which are commonly added to the total power consumption and area of the analog front-end.

The authors in [10] apply compressive sensing on a single-channel pre-recorded EEG data by acquiring measurements in the digital domain. The power saving is significant while the area overhead is not addressed.

Due to the youthfulness of the field and the lack of similar electronic architectures that use CS in brain implants, the results are compared to a Discrete Wavelet Transform (DWT)-based design [17] for intra-cortical implants and several additional systems based on spike/AP detection [14–16, 18] for implantable neural recording applications.

The design in [17] addresses the area-efficiency of the implantable system and proposes an architecture that sequentially evaluates the DWT of the multichannel data in digital domain. Our results outperform this approach in terms of area and power efficiency. In addition, high compression ratios are achieved by means of the following thresholding and redundancy removal stages, while the DWT by itself does not result in any data compression. Thresholding, however, results in a significant loss of the signal in non-spiking regions while a more precise recovery is achieved at much lower compression ratios (e.g. at $CR = 2$ in [17]). The chip includes several memory registers containing threshold values of different channels and additional blocks such as controllers, address generator and buffer units which degrade the power and area efficiency of the system.

Table 1 Comparison of the proposed CS system performance with published literature

Parameter	[10]	[14]	[15]	[16]	[17]	[18]	This work
Technology (μm CMOS)	0.09	0.13	0.5	0.18	0.5	0.5	0.18
Power supply (V)	0.6	1.2	3.3	1.8	N.A.	3	1.2
Compression method	DCS	PWL Spike det.	Spike det.	AP det.	DWT Spike det.	Spike det.	MCS
Number of channels	1	1	100	16	32	32	16
Comp. area/chann. (mm^2)	0.103	0.080	<0.160	>0.0475	0.18	0.12	0.008
Comp. power/chann. (μW)	1.9	1.18	27	>96	95	75	0.95
Sample rate/chann. (kS/s)	≤20	90	15	30	25	20	4
Compression ratio	≤10	125	150	48	≤20	12.5	≤16

Some of the reported spike detector systems achieve significant data reduction [14, 15] with negligible overhead in terms of compression power and area [14]. However, the patient-specific threshold setting in such systems can result in design complexity in a real neural interface in addition to the loss of signal in non-spiking regions. Furthermore, the transmitted signal may not be acceptable to the clinicians who usually prefer to have access to the entire iEEG data, even though somewhat lossy, for a thorough neurological examination.

3 Feature Extraction for Epileptic Seizure Detection

The sudden occurrence of seizures without any forewarns is one of the most disabling aspects of epilepsy. Several researchers have therefore focused on developing methods that anticipate or detect impending seizures to enable a closed-loop therapeutic intervention such as stimulation or drug delivery. Recording from the surface of the cortex with high temporal and spatial resolution is a fundamental key to implement powerful seizure detector devices.

In this chapter, a new feature extraction method applied to the compressed signal of several channels is proposed [19], as an efficient method to limit the amount of data acquired by a seizure detector device (Fig. 6). The spatial evolution of the intracranial EEG (iEEG) signals recorded by the adjacent electrodes and the corresponding behaviour in the compressed domain are exploited for detecting seizure onset.

This approach enables the real-time, compact, low-power and low hardware complexity implementation of the seizure detection algorithm, as a part of an implantable neuroprosthetic device for the treatment of epilepsy.

3.1 Commercially Available Anti-Seizure Devices

The Vagus Nerve Stimulator (VNS) is the first FDA-approved device for treating epilepsy. The VNS reduces the number of seizures in an individual by an average rate of 30–40 %. The VNS operates by periodic electrical stimulation of the left vagus nerve by a contact wrapped around the nerve trunk in the neck. The VNS is an open-loop anti-seizure device lacking a direct feedback capability to modulate therapy. The design and implementation of responsive closed-loop devices hold the promise of better seizure control, lower side effects on nervous system and lower power consumption. Analogous to the feedback control in automatic implantable cardiac pacemakers, closed-loop devices actively record neural signals (intracranial EEG signals), process these signals in real time to detect evidence of imminent seizure onset, and then trigger an intervention. The Responsive Neurostimulator (RNS) system is a first generation closed-loop device currently being under pivotal clinical trial. The RNS utilizes the coastline or line-length as one of the parameters

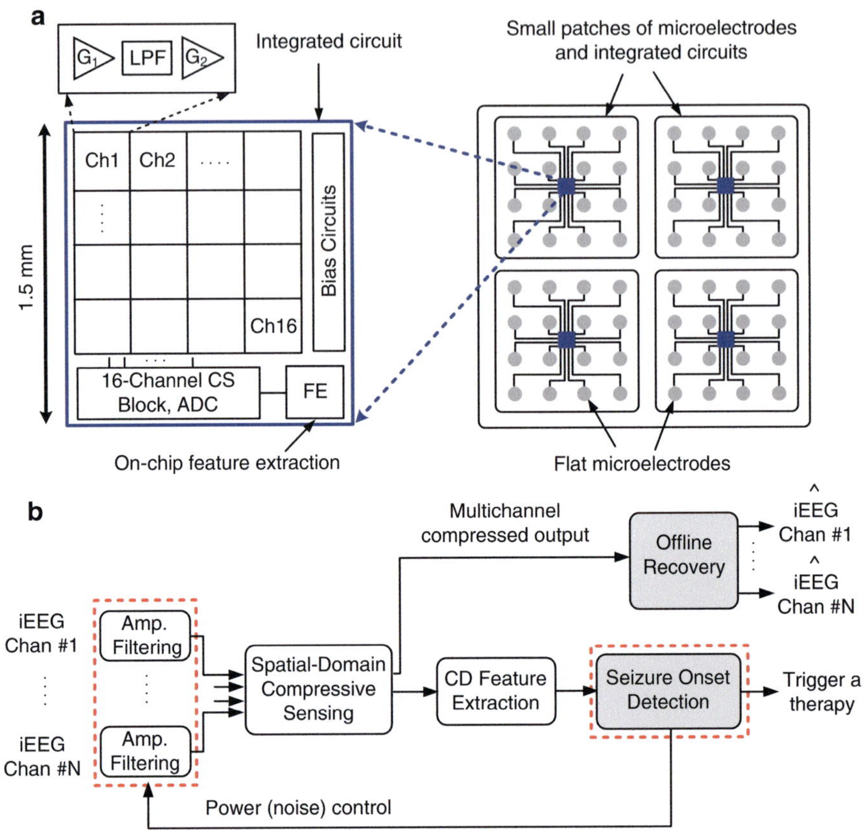

Fig. 6 System-level view of the proposed method [19]. (**a**) Every 16 electrodes are processed by a microelectronic chip, performing the amplification, filtering and compressed sensing on the outputs. Then features are extracted from the output data stream in the compressed domain. (**b**) Block-diagram of the proposed system [19]. Signals recorded by every 16 electrodes are processed by a spatial-domain compressive sensing and a feature extraction block. Offline recovery of the compressed data provides full access to the original signals

for seizure detection. A second detector implemented in RNS detects changes in signal power. The physician can program the RNS to use one or both detection tools.

The pulse generator of an RNS device is approximately 4 cm wide, 6 cm long, and 7 mm thick, housing the seizure detection electronics, a battery, and the connection ports to subdural and depth electrodes. Battery depletion necessitates recurring generator replacement and a large craniotomy is needed to place the device in the skull. As a potential risk, intracranial hemorrhage may occur when implanting the RNS System.

The RNS neurostimulator can sense the brain activity from a maximum of four amplification channels. A maximum of four leads may be implanted, two of which may be depth leads. Due to the small number of recording and stimulation channels, the efficiency of the RNS device in seizure suppression is highly limited.

3.2 *Existing On-Chip Solutions and Proposed Architecture*

In an emerging effort to miniaturize the size of the responsive neurostimulator device, improve the seizure detection performance and lower the power consumption of the system, novel integrated seizure detectors have been proposed. The majority of the existing systems extract specific biomarkers of the digitized neural signal (Fig. 7a) [20, 21]. These feature extraction methods are separately applied to all channels (Fig. 7a). In large recording arrays, the complexity of seizure detection based on the outcome of single-channel detections, the area and power overhead of implementing the in-channel feature extraction circuitry and the increased size of the extracted feature vectors necessitate an improved seizure detection strategy.

We present a computationally efficient compressed-domain feature extraction algorithm (Fig. 7b) suitable for seizure onset detection in dense iEEG recording arrays.

3.2.1 Reconstruction Overhead of Compressive Sensing

As opposed to the simple compression model in CS, the signal recovery is commonly achieved by nonlinear and relatively expensive optimization-based or iterative algorithms [22], generally shifting the following signal analysis required

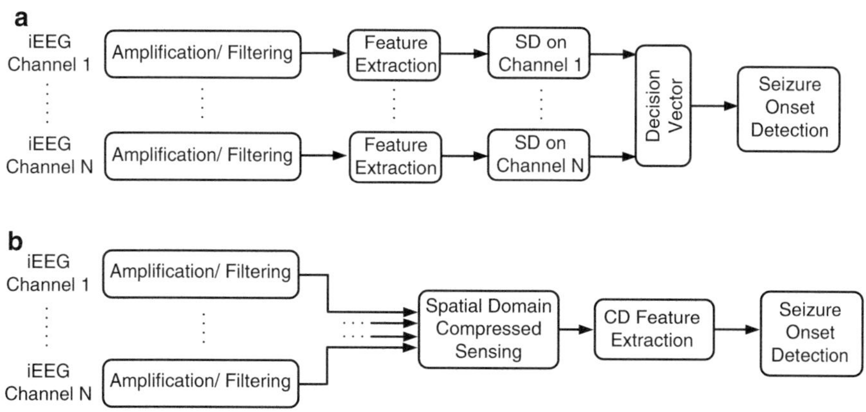

Fig. 7 Block diagram of seizure detection for (**a**) standard Nyquist-sampled system and (**b**) proposed multichannel compressive sensing system. In (**b**), features are extracted in the compressed domain and a multichannel decision for a group of adjacent channels is made

on the data into a base station. Although most of the CS literature has focused on improving the speed and accuracy of the recovery algorithms, the state-of-the-art is still lacking the capacity of real-time operation for many applications (e.g. in Fig. 5, 1.12 s is needed for the reconstruction of 256 ms of the original iEEG signal).

However, in many signal processing applications, only the specific features of the signal are of interest and the exact recovery is not necessary [22]. In order to leverage the benefits of data compression, developing algorithms which do not require the full data recovery and perform feature extraction and processing directly in the compressed domain is potentially interesting.

3.2.2 Compressed-Domain Feature Extraction

The possibility of applying feature extraction for seizure detection on the measurements acquired using the multichannel compressive sensing method (Figs. 6 and 7b) is studied in the following.

As an alternative to applying the feature extraction on every single channel, the algorithm receives the compressed output from 16-channel subblocks of an iEEG recording array and extracts the relevant features to distinguish between the seizure and non-seizure states. For dense recording arrays sampling iEEG signals at relatively high data rates, the size of feature vectors extracted from the single channels and the power and area cost of applying the feature extraction on each channel is significant, both in the compressed and non-compressed domains. The proposed approach to this problem consists of dividing the array into several subblocks of 4×4 channels, compressing the data recorded by each subblock and applying the feature extraction on the compressed output.

Multichannel iEEG signals exhibit significant correlation among spatially adjacent channels and are also correlated over time. An efficient compressive sensing system should exploit the spatial correlations among nearby channels over time. Similarly, an optimal seizure detection method should be sensitive to the spatial and temporal evolution of the iEEG signals in order to extract the most applicable features. We have implemented this concept in a low-dimensional domain, where the signals are compressed in the spatial domain and the features are extracted over time, exploiting both temporal and spatial characteristics of the iEEG signals. The proposed method is suitable for high-resolution recording arrays, where adjacent channels are expected to capture highly correlated neural activity. In such arrays, it is more beneficial to make decisions for small subsets of the electrodes representing the activity of a group of neurons, rather than for every single electrode. Thanks to the high sampling resolution in the spatial and temporal domains, the single-neuron activities contributing to the progressive evolution of neuronal networks are efficiently exploited by applying seizure detection in the compressed domain.

In general, many software-based seizure detection approaches employ sophisticated algorithms to increase the detection accuracy. However, the limited power and area budget of an implantable device impose strict constraints on the choice of appropriate seizure detection features. Extracting frequency-based features involves computationally intensive algorithms such as filtering and fast Fourier transform (FFT) evaluation, which is hardly suitable for an on-chip implementation. Considering the hardware cost and detection accuracy, the coastline feature has been selected and extracted from the multichannel iEEG signal prior to, and following the compression. Coastline achieves the best seizure detection performance among more than 65 different time and frequency domain features [23]. This feature is a measure of the line length between successive samples and provides an appropriate characteristic of epileptiform iEEG since it increases at low-amplitude fast activities as well as high-amplitude slow activities. In our circuit implementation, the coastline feature of block i is computed as:

$$C_i = \frac{1}{256} \sum_{256} |x[n] - x[n-1]| \tag{4}$$

where x[n] is the value of signal x at time $t_n = n \times T_{S/H}$ with $T_{S/H}$ being the sample and hold time of the analog-to-digital converter (ADC). Normalization is done by shifting the digital feature output by 8 bits towards the LSB. The choice of the division factor (i.e., the summation window length) depends on the signal amplitude, the detection accuracy and required latency of detection.

3.2.3 Microelectronic Architecture of Seizure Detector Chip

The proposed feature extraction method builds upon the MCS system initially presented in Sect. 2. Block diagram and circuit-level implementation of the proposed architecture are shown in Figs. 6 and 8.

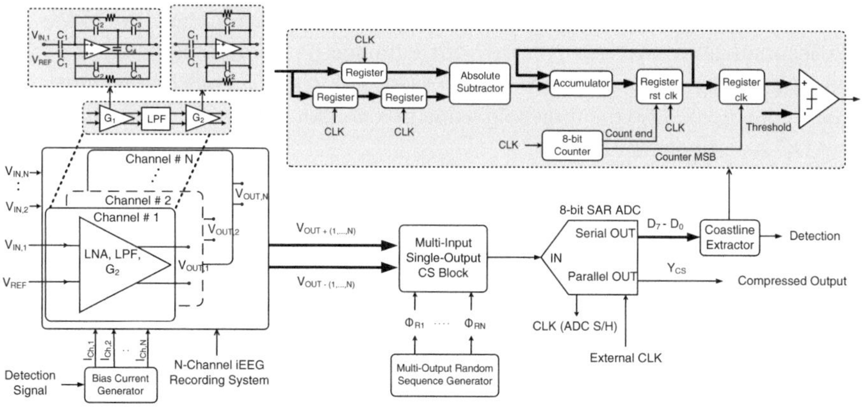

Fig. 8 Schematic of the proposed multichannel feature extractor system including the individual channels, a multichannel CS block with 16 differential inputs and a single output, an 8-bit successive approximation ADC and a digital coastline extractor block [19]

The neural signals recorded by every N electrodes of the array (N is equal to 16 in this design) are processed through amplification and filtering stages located inside the channels. The differential outputs of the channels connect to the multichannel compressive sensing block and randomly accumulate at the single-ended output of this stage. As shown in [11], the compressed sensing system is very robust against circuit nonidealities such as noise. Thus, the compressed data stream is digitized using a moderate-resolution successive approximation ADC. The serial output of the ADC can be used for a full recovery of the original iEEG signal at the receiver side.

The employed compressive sensing technique enables sixteen times data reduction and power saving of the transmitter, compared to the conventional non-compressed design. The 8-bit digital output is delivered to the feature extractor block in parallel form. To improve the power efficiency of the system, the bias current of the front-end amplifiers is doubled upon seizure detection at the output comparator of the feature extraction block. This technique avoids unnecessary power dissipation while normal (i.e., seizure-free) iEEG is being recorded. Upon seizure detection, the current of the low-noise amplifier (LNA) increases which simultaneously reduces the input-referred noise of the channel and enhances the achievable SNR at the output.

An 8-bit feature extractor circuit computes the coastline parameter at the output of the ADC. The parallel output of the ADC is passed through register stages which capture two successive compressed values. The absolute difference of the two inputs is generated at the output of the subtractor and accumulated over a window of length 256. Extra bits are employed in the arithmetic hardware to prevent overflow during the accumulation. An 8-bit synchronous binary counter controls the accumulation window and resets the register after 256 consecutive additions. Within the coastline generation chain, the absolute differential value of the successive samples is shifted towards the least significant bit by 8 bits, enabling the division operation in (4). At the rising edge of the counter MSB output (with a period equal to $256/f_{CLK}$), the accumulated value is compared with a trained threshold input using a digital comparator stage. The detection signal associated with the coastline parameter is raised upon exceeding the threshold value in a window of length 64 ms (Fig. 9), at an ADC sampling rate of 4 kS/s.

3.3 *Experimental Results and Performance Comparison*

In order to assess the overall performance of the compressed-domain seizure detector, 103 subblocks containing 23 seizures from 4 patients have been processed. The total length of the analyzed iEEG data is 420 h recorded from 16 channels and includes pre-ictal, ictal and post-ictal activities.

The average sensitivity in seizure detection, false alarm rate and latency are reported in Table 2. Sensitivity is computed as the ratio of correct seizure predictions to the total number of registered seizures. False alarm rate is measured by the number of false positive detections in 1 h of recording.

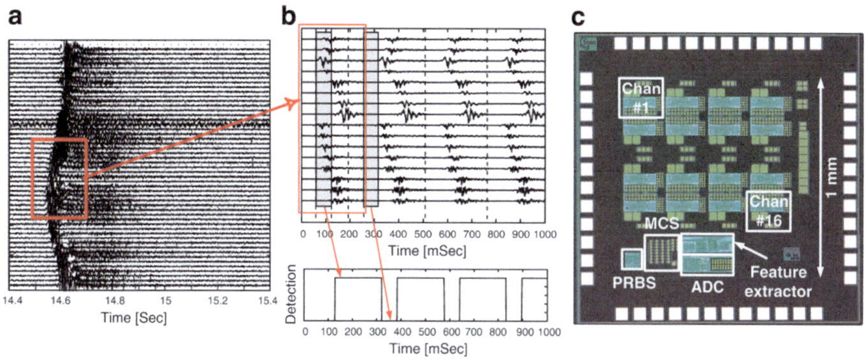

Fig. 9 (**a**) Epileptiform burst recorded of a rat's somatosensory cortex. (**b**) 16 Successive channels of the array are periodically applied to the chip. The seizure detection output is raised associated with the second subwindow of the signal. (**c**) Die micrograph of the fabricated seizure detector chip

Table 2 Performance comparison with published work

	Technology (μm CMOS)	Area/chann. (mm^2)	Power/chann. (μW)	Sens. (%)	FAR (#/h)	Latency (s)
This work	0.18	0.0625	0.85	100	0.34	−0.27
[24]	0.18	1.68[a]	162[b]	> 92	N.A.	0.8
[21]	0.18	3.125[c]	>66[d]	84.4	> 0.04	<2
[20]	0.18	6.25	6.7	N.A.	N.A.	N.A.

[a]Including telemetry, classification and stimulation/channel count
[b]Feature extraction and classification power/channel count
[c]Including classification engine and divided by channel count
[d]Analog front-end power

Latency is defined as the amount of time after (or before if negative) electrographic seizure onset taken by the proposed feature to cross the threshold and trigger a detection.

The seizure detection performance and die micrograph of the fabricated chip are shown in Fig. 9. A window of length 256 ms at the beginning of the epileptic burst (recorded from a rat somatosensory cortex) is periodically applied into the circuit. Every 64 ms of the iEEG signal is mapped into a seizure or non-seizure event, at the rising edge of the counter MSB signal. A seizure is detected after 128 ms, which correctly represents the seizure onset time falling within the second sub-window of the signal. The performance summary and a comparison with existing works are shown in Table 2. The total current consumption of the proposed system is 17 μA which is drawn from a 0.8-V power supply.

The proposed solution incurs lower computational complexity in the feature extraction stage, thanks to the reduction in the number of samples processed in the compressed domain. This computational advantage could be further enhanced by the fact that even fewer samples are required for detection purposes in the compressed domain, than for signal reconstruction [22]. In addition, the spatial compressive sensing enables proper offline data recovery of the original physiological signal at the receiver. This feature is highly desired by physicians who prefer to have access to the entire iEEG data for a thorough clinical evaluation.

4 Conclusions

The main objective of this chapter is to explore the design tradeoffs and minimize the implementation cost of implantable neural recording systems to achieve a powerful design in terms of power efficiency, compactness, functionality and scalability. The ultimate clinical goal is targeted at epilepsy, although similar challenges exist in smart IC design for monitoring several other biological signals. Several design solutions are proposed to enable high-density recording of intracranial EEG signals in an anti-seizure device.

In summary, applying the compressed sensing method to the multichannel iEEG data enables the wireless implantable system to accommodate a higher number of channels into the acquisition array, compared to traditional systems transmitting the raw data. The area and power costs of implementing the proposed multichannel CS hardware are negligible. Assuming a limit of 1 Mbit/s of the transmitter and wireless link, 8 bits of resolution of the digitized data and a sampling rate of 4 kS/s within the channels, the maximum allowable size of the array is equal to 31 channels in the traditional and 400 in a compressed sensing system with $CR = 16$ and 10 bits of resolution at the output. Transmitting the compressed-domain seizure detector features rather than the compressed data as discussed in Sect. 3, enables the same wireless link to transmit 10 different features from more than 21,000 channels at a rate of 4 Hz.

As a significant advantage, the proposed system can be easily scaled up to larger overall array sizes, while retaining the spatial resolution and power efficiency of the implantable device. The significant improvement in the allowable array size (e.g., 400 versus 31) enables the implementation of very high-density implantable arrays capturing the seizure-related information which originates from the fine-scale spatial and temporal activity during a seizure. Depending on the application, the system can be configured to transmit the compressed data for further processing and reconstruction upon detecting an imminent seizure.

References

1. Muller R, Le H-P, Li W, Ledochowitsch P, Gambini S, Bjorninen T, Koralek A, Carmena JM, Maharbiz MM, Alon E et al (2015) A minimally invasive 64-channel wireless μecog implant. IEEE J Solid-State Circuits (JSSC) 4:344–359
2. Stacey WC, Litt B (2008) Technology insight: neuroengineering and epilepsy-designing devices for seizure control. Nat Clin Pract Neurol 4(4):190–201

3. Pollo C, Shoaran M, Leblebici Y, Mercanzini A, Dehollain C, Schmid A (2012) The future of intracranial eeg recording in epilepsy: a technological issue. Epileptologie 29:114–119
4. Talathi SS, Hwang D-U, Spano ML, Simonotto J, Furman MD, Myers SM, Winters JT, Ditto WL, Carney PR (2008) Non-parametric early seizure detection in an animal model of temporal lobe epilepsy. J Neural Eng 5(1):85–98
5. Stead M, Bower M, Brinkmann BH, Lee K, Marsh WR, Meyer FB, Litt B, Van Gompel J, Worrell GA (2010) Microseizures and the spatiotemporal scales of human partial epilepsy. Brain 133:2789–2797
6. Viventi J, Kim D-H, Vigeland L, Frechette ES, Blanco JA, Kim Y-S, Avrin AE, Tiruvadi VR, Hwang S-W, Vanleer AC et al. (2011) Flexible, foldable, actively multiplexed, high-density electrode array for mapping brain activity in vivo. Nat Neurosci 14(12):1599–1605
7. Donoho DL (2006) Compressed sensing. IEEE Trans Inf Theory 52(4):1289–1306
8. Candès EJ (2006) Compressive sampling. In: Proceedings of the international congress of mathematicians, pp 1433–1452
9. Laska JN, Kirolos S, Duarte MF, Ragheb TS, Baraniuk RG, Massoud Y (2007) Theory and implementation of analog-to-information converter using random demodulation. In: Proceedings of the international symposium on signals, circuits and systems (ISCAS), 2007, pp 1959–1962
10. Chen F, Chandrakasan AP, Stojanovic VM (2012) Design and analysis of a hardware-efficient compressed sensing architecture for data compression in wireless sensors. IEEE J Solid-State Circuits 47(3):744–756
11. Shoaran M, Kamal M, Pollo C, Vandergheynst P, Schmid A (2014) Compact low-power cortical recording architecture for compressive multichannel data acquisition. IEEE Trans Biomed Circuits Syst 8(6):857–870
12. Shoaran M, Afshari H, Schmid A (2014) A novel compressive sensing architecture for high-density biological signal recording. In: IEEE biomedical circuits and systems conference (BioCAS), pp 13–16
13. Higgins G, Faul S, McEvoy RP, McGinley B, Glavin M, Marnane WP, Jones E (2010) Eeg compression using jpeg2000: how much loss is too much. In: International conference of the IEEE engineering in medicine and biology society (EMBC), pp 614–617
14. Rodriguez-Perez A, Ruiz-Amaya J, Delgado-Restituto M, Rodriguez-Vazquez A (2012) A low-power programmable neural spike detection channel with embedded calibration and data compression. IEEE Trans Biomed Circuits Syst 6(2):87–100
15. Harrison RR, Watkins PT, Kier RJ, Lovejoy RO, Black DJ, Greger B, Solzbacher F (2007) A low-power integrated circuit for a wireless 100-electrode neural recording system. IEEE J Solid-State Circuits 42(1):123–133
16. Gosselin B, Ayoub AE, Roy J-F, Sawan M, Lepore F, Chaudhuri A, Guitton D (2009) A mixed-signal multichip neural recording interface with bandwidth reduction. IEEE Trans Biomed Circuits Syst 3(3):129–141
17. Kamboh AM, Oweiss KG, Mason AJ (2009) Resource constrained vlsi architecture for implantable neural data compression systems. In: IEEE international symposium on circuits and systems (ISCAS), pp 1481–1484
18. Olsson RH, Wise KD (2005) A three-dimensional neural recording microsystem with implantable data compression circuitry. IEEE J Solid-State Circuits 40(12):2796–2804
19. Shoaran M, Pollo C, Schindler K, Schmid A (2015) A fully-integrated ic with 0.85-μW/channel consumption for epileptic ieeg detection. IEEE Trans Circuits Syst II Express Briefs 62(2):114–118
20. Verma N, Shoeb A, Bohorquez J, Dawson J, Guttag J, Chandrakasan AP (2010) A micro-power eeg acquisition soc with integrated feature extraction processor for a chronic seizure detection system. IEEE J Solid-State Circuits 45(4):804–816
21. Yoo J, Yan L, El-Damak D, Altaf MAB, Shoeb AH, Chandrakasan AP (2013) An 8-channel scalable eeg acquisition soc with patient-specific seizure classification and recording processor. IEEE J Solid-State Circuits 48(1):214–228

22. Davenport MA, Boufounos PT, Wakin MB, Baraniuk RG (2010) Signal processing with compressive measurements. IEEE J Sel Top Sign Proces 4(2):445–460
23. Logesparan L, Casson AJ, Rodriguez-Villegas E (2012) Optimal features for online seizure detection. Med Biol Eng Comput 50(7):659–669
24. Chen W-M, Chiueh H, Chen T-J, Ho C-L, Jeng C, Chang S-T, Ker M-D, Lin C-Y, Huang Y-C, Chou C-W et al. (2013) A fully integrated 8-channel closed-loop neural-prosthetic soc for real-time epileptic seizure control. In: 2013 IEEE international solid-state circuits conference digest of technical papers (ISSCC), pp 286–287

Part II
Advanced Amplifiers

The second part of the book deals with the design of one of the most important blocks in analog design: the operational amplifiers (opamps). This topic is addressed by six papers: some of them deals with more theoretical aspects, like with the optimum opamp design in different conditions, some others discuss more practical aspects presenting design issues for specific opamp embedded in target applications.

The first paper by Willy Sanses look at the future and discusses how will be the opamp design in the next technology nodes. Technology scaling will deeply change opamp design and this paper focuses on the different opamp design strategies. Opamps are always in competition with transconductance (gm) blocks for higher frequencies despite their higher linearity. The author oversees that both of them are now gradually being replaced by CMOS inverters. The first paper focuses on the merits and advantages of all three of them.

The second paper from Rinado Castello et al. deals with the design of amplifiers that need to drive heavy loads (low resistances and/or large capacitances) with good efficiency. In such cases, a push-pull output stage creates large open-loop distortion components that need to be compressed through feedback to insure high closed-loop linearity. Minimizing close loop residual distortion involves three steps. First, eliminate all open-loop source of distortion not intrinsic to the proper operation of the push pull structure. Second, choose the amplifier topology that gives the maximum close loop compression of the open-loop distortion components for a given bandwidth. Third, maximize the open-loop gain in the signal band and/or the unity gain bandwidth of the amplifier for a given topology while insuring stability in the presence of variable loads.

The third paper from Olivier Nys et al., discusses the design strategy of an acquisition chain for microphone input signal. For a set of target specifications, the overall strategy for the acquisition of the signal, pre-amplification and A/D conversion is then discussed, together with the main constraints, and a new architecture for the loop. The most critical analog blocks, such as first amplification stage, loop filter and biasing circuitry, are then investigated in more details, followed by measurement results.

The fourth paper from Giulio Ricotti et al., addresses the design of High Voltage Operational Amplifiers and in particular the aspects oriented to obtain high linearity. The first aspects is related to extend the bandwidth working on the transconductance as functions of the strongly nonlinear parasitic capacitance. The second aspects describes a technique to draw the integrated feedback network based on a resistive voltage divider with particular focus to low power consumption and high linearity.

The fifth paper from Qinwen Fan et al., focuses on the design of amplifiers that achieve micro-volt offset by employing dynamic offset reduction techniques, like chopping, auto-zeroing and chopper stabilization. The working principles and non-idealities of these techniques are described. The up-modulated offset associated with chopping causes ripple, which can be a significant source of error if not filtered effectively. Thus, various ripple reduction techniques are introduced to suppress this ripple to the micro-volt level. The paper presents also the ping-pong architecture, which enables the realization of auto-zeroed amplifiers with continuous-time behavior. Examples of chopped amplifiers, auto-zeroed amplifiers and chopper-stabilized amplifiers are presented, as well as designs in which multiple techniques are combined.

Finally the sixth paper from Jan Kaplon et al., deals with the design of front-end amplifiers used in the recent discovery of the Higgs boson by the ATLAS and CMS experiments at the Large Hadron Collider (LHC) at the European Laboratory for High Energy Physics (CERN) in Geneva. Particles are accelerated and brought into collision at well-defined interaction points. Detectors, giant cameras of about 40 m long by 20 m in diameter, constructed around these interaction points take pictures of the collision products as they fly away from the interaction point. They contain millions of channels often generating a small (1 fC) electric charge upon particle traversals. Integrated circuits provide the readout in a very aggressive radiation environment and accept collision rates of about 40 MHz with on-line selection of potentially interesting events before data storage. Power consumption directly impacts the measurement quality as it governs the amount of material present in the detector, and often the fraction of the power consumed by the front-end amplifiers is significant if not predominant. Basic architectures and a selection of front-end amplifiers as a representative overview for various types of particle detectors operated at LHC are proposed.

Opamps, Gm-Blocks or Inverters?

Willy Sansen

Abstract This Operational amplifiers have been the backbone of most amplifiers and filters in communication applications and ADCs. They are in competition with Gm blocks for higher frequencies despite their higher linearity. Both of them are now gradually being replaced by CMOS inverters. This text focuses on the merits and advantages of all three of them.

1 Introduction

Operational amplifiers use external feedback to increase linearity. They thus need high gain, which is easy to achieve at low frequencies. At high frequencies however, simpler configurations are required such as differential pairs, for voltage mode operation, or current amplifiers for current-mode operation. They achieve gain at much higher frequencies but need internal feedback to improve linearity. Moreover they can be tuned. All three are now compared.

The next paragraph looks at FOM's (Figure of Merit) to be able to measure and compare performance. The third paragraph gives an overview of the circuit tricks, which allow to extend the high-frequency region of operational amplifiers. The fourth paragraph then shows how to improve the linearity of Gm blocks, so that they can provide similar performance as operational amplifiers. The last paragraph looks at alternative configurations, such as VCO-based and Ring Amplifiers, which involve CMOS inverters only.

W. Sansen (✉)
KU Leuven, Leuven, Belgium
e-mail: willy.sansen@esat.kuleuven.be

K.A.A. Makinwa et al. (eds.), *Efficient Sensor Interfaces, Advanced Amplifiers and Low Power RF Systems*, DOI 10.1007/978-3-319-21185-5_7

2 Comparison with FOM's

The most used measure of the performance of operational amplifiers [1] includes their Gain-Bandwidth product GBW (in Hz), the load capacitance C_L (in pF), which represents the integrated noise at the same time, and the supply current (in mA). For example when a single MOST is biased at $V_{GS} - V_T = 0.2$ V, which corresponds to an Inversion Coefficient (IC) of about 10, then its FOM is about 1500 MHz pF/mA. This value increases if this MOST is biased deeper in weak inversion, as shown in Fig. 1. It can be increased by a factor of about three times.

How deep a MOST can be biased in weak inversion, solely depends on the speed required. A single MOST amplifier needs an f_T value of about ten times its GBW. The current and the corresponding f_T can thus be reduced accordingly. Low-frequency amplifiers and filters all end up in the weak-inversion region, especially for small channel lengths.

Indeed for smaller channel lengths, the corresponding values of the Inversion Coefficients decrease, as shown in Fig. 2 for two-stage Miller OTA.

A differential amplifier can only achieve half of this FOM, for the same biasing point of IC = 10. This also applies to a telescopic cascode. A folded cascode is half of this value again, which is now 375 MHz pF/mA. A conventional two-stage Miller OTA provides a value which is very close i.e. 350 MHz pF/mA [1]. A CMOS inverter doubles the value of a single MOST up to 3000 MHz pF/mA. This is a clear advantage of a CMOS inverter amplifier.

A full operational amplifier can do much better however, as explained next.

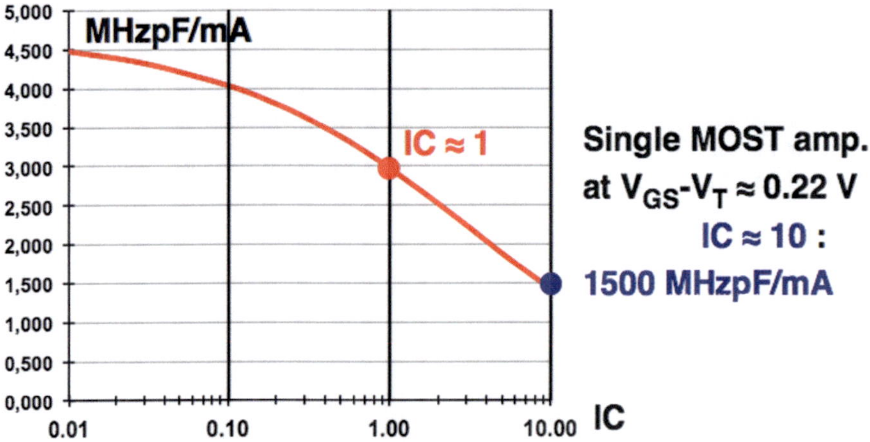

Fig. 1 FOM vs. Inversion Coefficient for a single-transistor amplifier

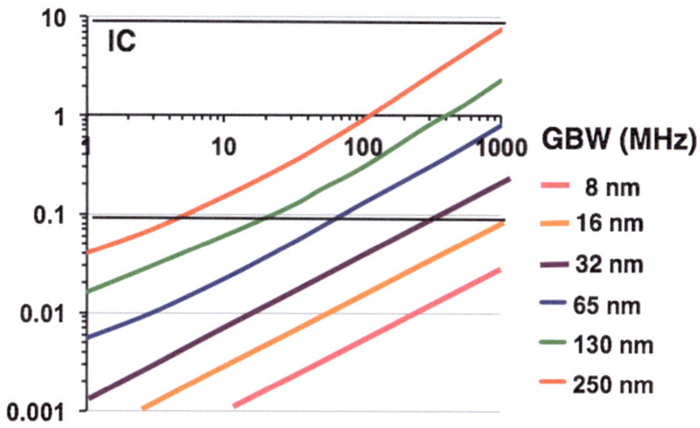

Fig. 2 IC for different channel lengths and GBW values, for two-stage Miller OTA's [2]

3 Opamp Gain at High Frequencies

Several circuit tricks are available to realize more GBW for the same current levels. The most used ones are the cancellation of the resistances and the capacitances and straight pole-zero cancellation. They are described next.

3.1 Cancellation of Resistance

The inclusion of negative resistance (or cross-coupling) has been around for a long time, especially to realize oscillators. They have been used in the Drains of a differential pair to increase the gain at low frequencies [3]. The GBW is not increased however (Fig. 3).

The GBW can be increased as well by including negative resistances in the sources, as shown for the three elementary configurations of Fig. 4. They will be used to realize high-performance filter blocks, as described later.

A very attractive OTA with negative resistances is the symmetrical OTA shown in Fig. 5 [3]. Without the negative resistances (or $A = 0$), its FOM is only 1260 MHz pF/mA, for $IC = 0.5$. It increases to 6090 MHz pF/mA for $A = 0.7$.

For larger values of A, mismatch may come in, which may cause unstability.

3.2 Cancellation of Capacitance

Cross-coupling capacitors provides negative capacitance which is able to compensate positive capacitance, and thus increases bandwidth, as shown in Fig. 6.

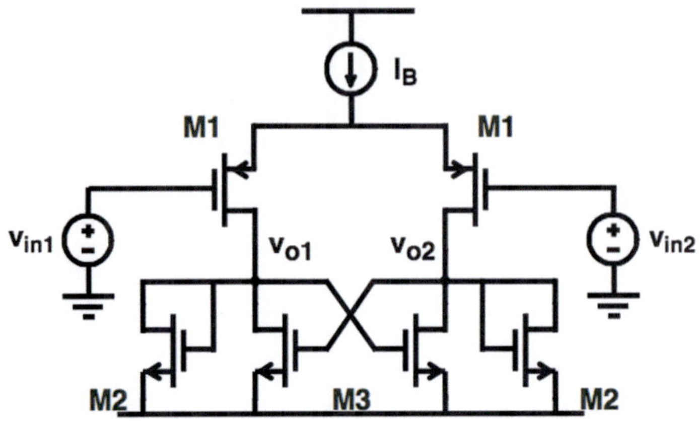

Fig. 3 Negative resistance in the drains

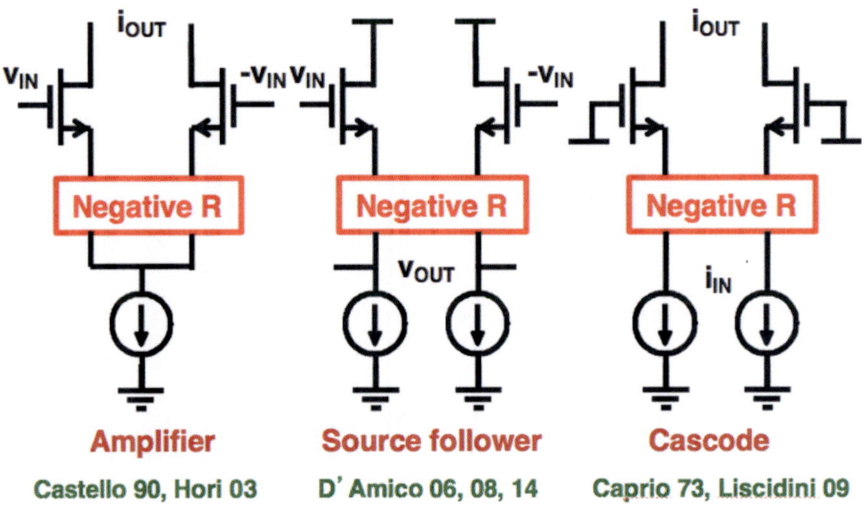

Fig. 4 Negative resistance in the sources

The exact compensation capacitance depends on six transistor and circuit parameters, which makes it quite difficult to realize. Nevertheless this compensation technique is applied in a multitude of wide-bandwidth amplifiers and transimpedance amplifiers.

This technique also uses pole-zero cancellation, as explained next.

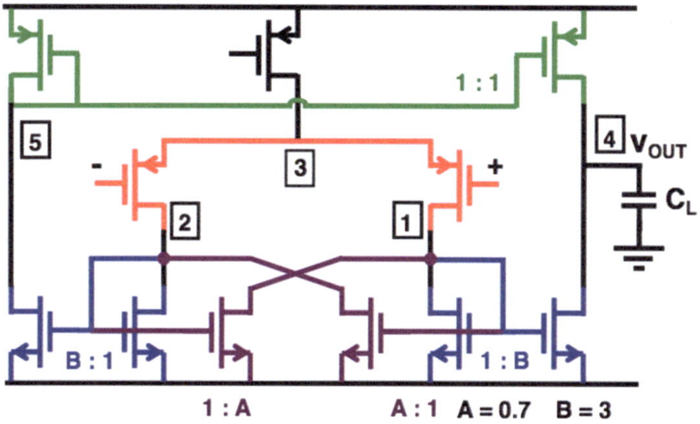

Fig. 5 Negative resistance in a symmetrical OTA

Fig. 6 Negative capacitance in a differential pair [4]

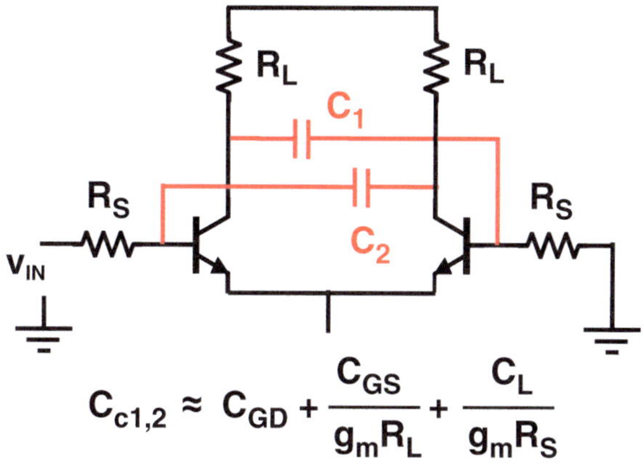

3.3 *Pole-Zero Cancellation*

Early operational amplifiers needed pole-zero cancellation avoid the slow lateral pnp transistors in the signal path [4, 5]. Recent examples in CMOS use pole-zero cancellation which reduces the power consumption drastically.

For example the three-stage amplifier in Fig. 7 realizes the cancellation of the first two non-dominant poles by means of zeros [6]. This is set by the ratio of the feedback gmt to gm2. Its FOM is 14,250 MHz pF/mA, which is about ten times better than what a single-transistor amplifier can provide.

Reverse Nested Miller compensation, shown in Fig. 8 [7], provides similar results with 15,700 MHz pF/mA.

Another good example is the four-stage amplifier shown in Fig. 9 [8]. A local feedback filter block is inserted in the middle to provide pole-zero compensation. Its FOM is 95,700 MHz pF/mA !!!

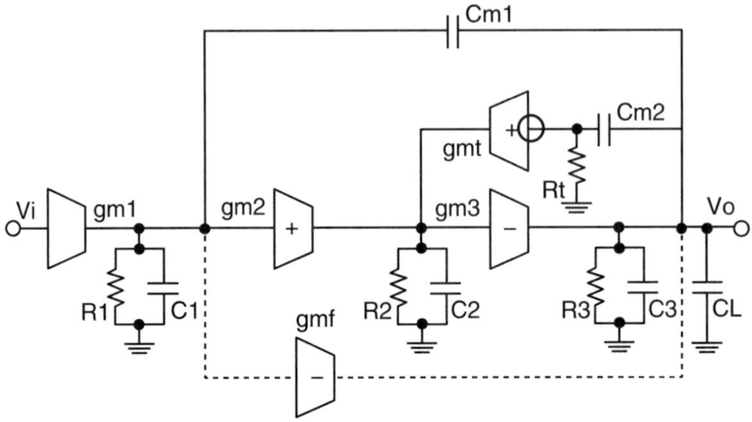

Fig. 7 Pole-zero compensation in a three-stage Miller OTA [6]

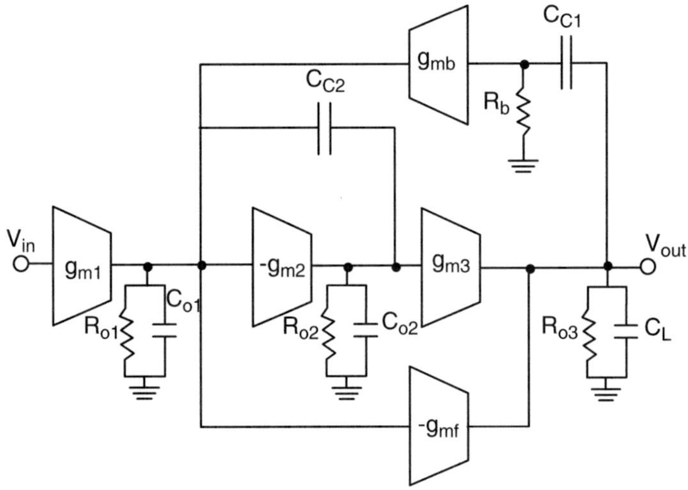

Fig. 8 Pole-zero compensation in Reverse Nested Miller OTA [7]

4 Linearization of Gm Blocks

At higher frequencies, simple structures need to be used such as differential amplifiers. However in order to maintain linearity, internal feedback is required. Many forms of feedback have been published but two basic configurations can be distinguished. The first one has additional gain **before** the input of the differential pair whereas the other one has the feedback amplifier **after** the input transistor, with feedback to the Source, the Bulk or the current source [9]. This is shown in Fig. 10 for half a differential pair.

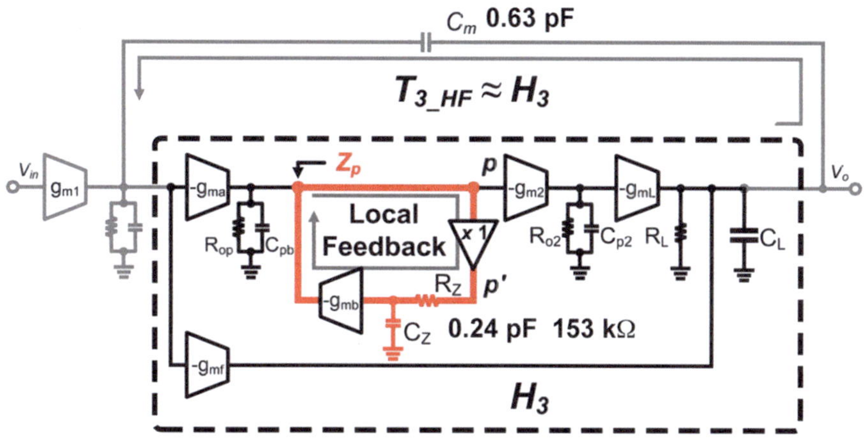

Fig. 9 Pole-zero compensation in a four-stage amplifier [8]

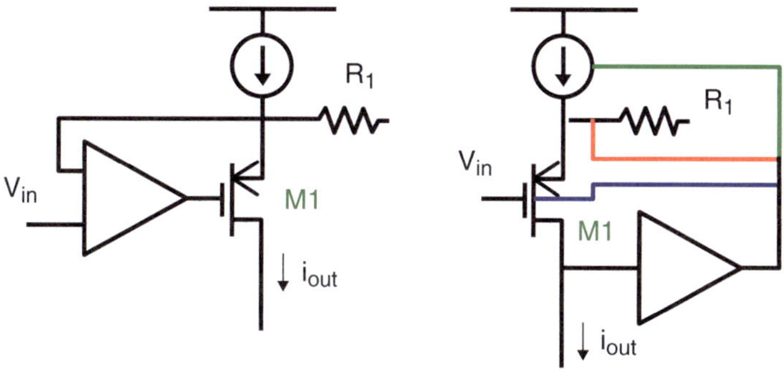

Fig. 10 Feedback in a differential pair [9]

An excellent example of the second configuration is shown in Fig. 11a. The additional loop gain reduces the distortion. Taking the Source of the input devices as an output, provides a Super Source-Follower (SSF) with excellent filter performance [10].

Both of them have lead to GmC filters with high FOM, as shown in Fig. 12. The Gm blocks with negative resistors and SSF's seem to be the best indeed.

For tunable filters. The best FOM values are obtained if input MOSTs are used in the linear region [11] or CMOS inverters are used with cross-coupling [12]. CMOS inverters seem thus to do better than operational amplifiers AND Gm blocks. They are discussed next (Fig. 13).

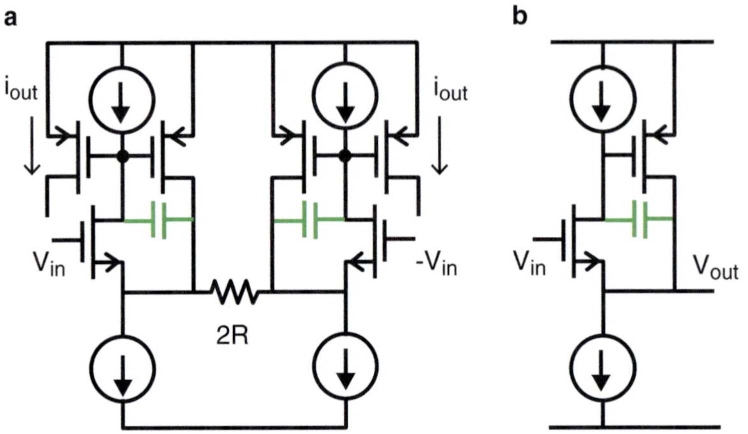

Fig. 11 Linearized Gm block (**a**) and Super Source-follower (**b**)

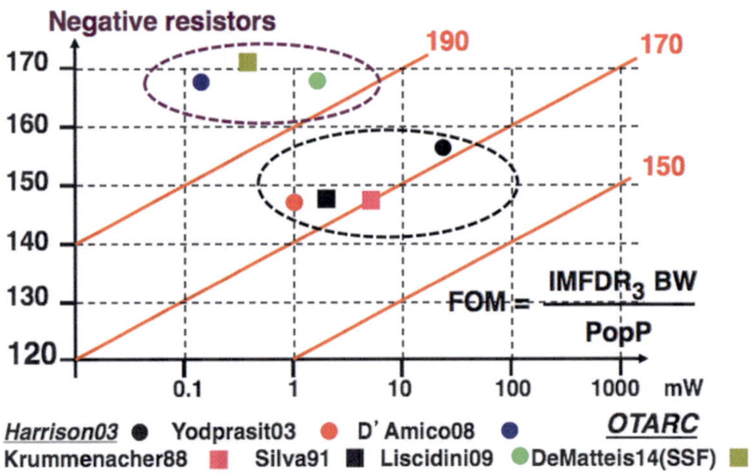

Fig. 12 Filter performance for fixed-frequency filters

5 Inverters Only

CMOS inverter amplifiers have been around for a long time. They excel in FOM, in linearity and in class-AB operation. Moreover they can operate at very low supply voltages, especially at low frequencies. In Fig. 14 the minimum supply voltage is shown versus GBW for several different channel lengths.

Their PSRR is a real drawback however. This is why they are better used with current sources, to isolate them form the supply lines as shown in Fig. 15. This turns them into class A amplifiers, which is less power efficient.

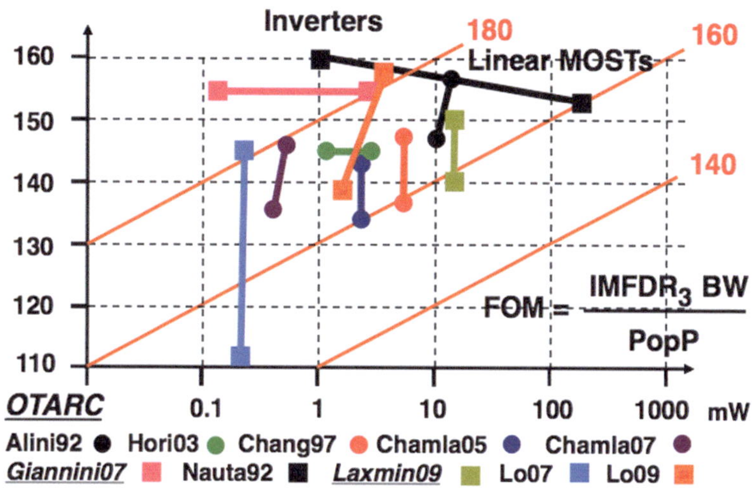

Fig. 13 Filter performance for tunable filters

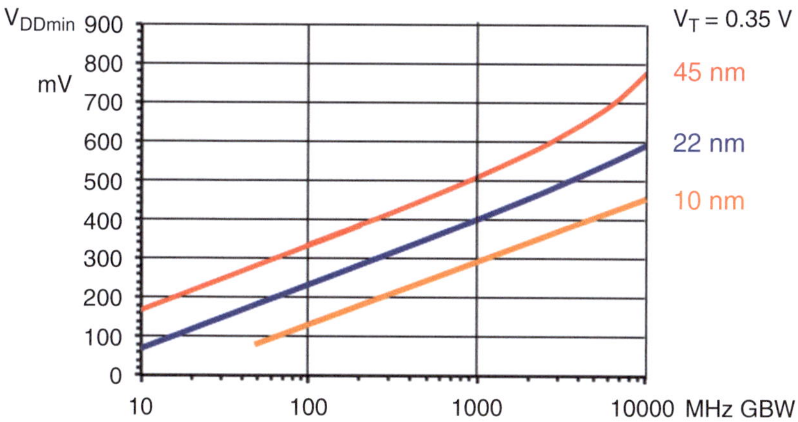

Fig. 14 Minimum supply voltage of CMOS inverter amps

Class-AB biasing is ideal for very-low-power GmC filters as shown in [12]. The tuning of the supply voltage to set the quiescent current is not an ideal solution however. It has been improved in many later publications [13].

An even better solution for power reduction is the biasing of the CMOS inverters in class C [14, 15]. This means that the supply voltage is smaller than the sum of (the absolute values of) $V_{Tn} + V_{Tp}$. For example the lowest-power audio (20 kHz) Delta-Sigma converters all use class-C inverters, as shown in Fig. 16 [15]. The input MOSTs are cascoded for higher gain. Gain boosting is adopted as well.

Another way of biasing CMOS inverters is to use dynamic biasing [16, 17]. A clock is used to provide accurate biasing of the gate inputs, in order to set the

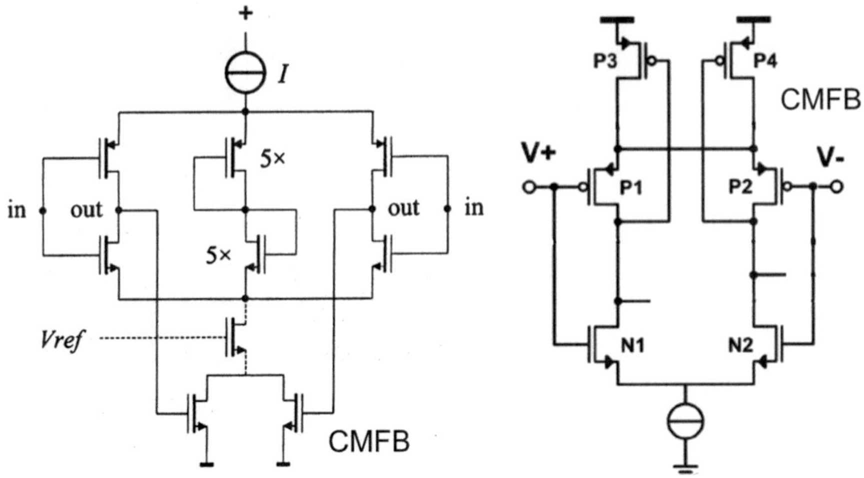

Fig. 15 Class-A CMOS inverter amps and filters

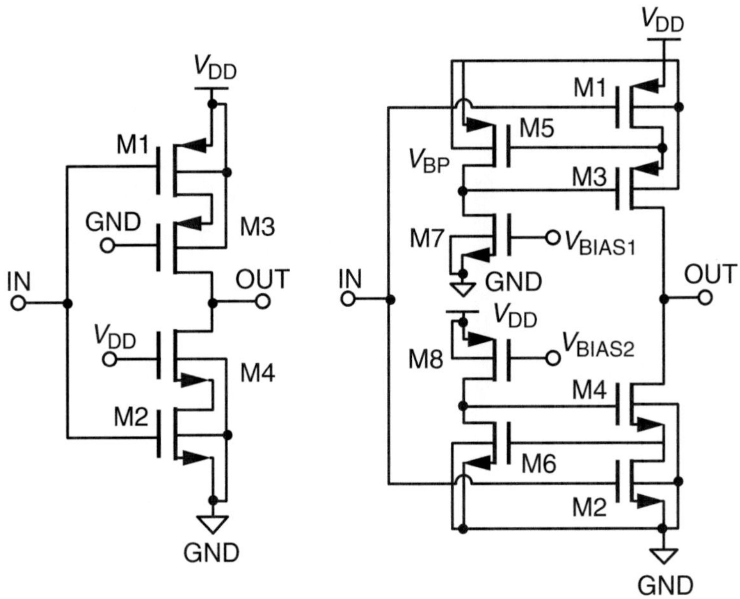

Fig. 16 Class-C cascoded CMOS inverter amplifiers [15]

quiescent current. The inverters can still operate in class AB or C for low power consumption. An example is given below (Fig. 17).

A different way to use inverters is to chain them into a ring oscillator. The input of the amplifier is then the supply line of the ring oscillator. This has been done routinely in VCO-based Delta-Sigma Converters and TDC-based PLL's.

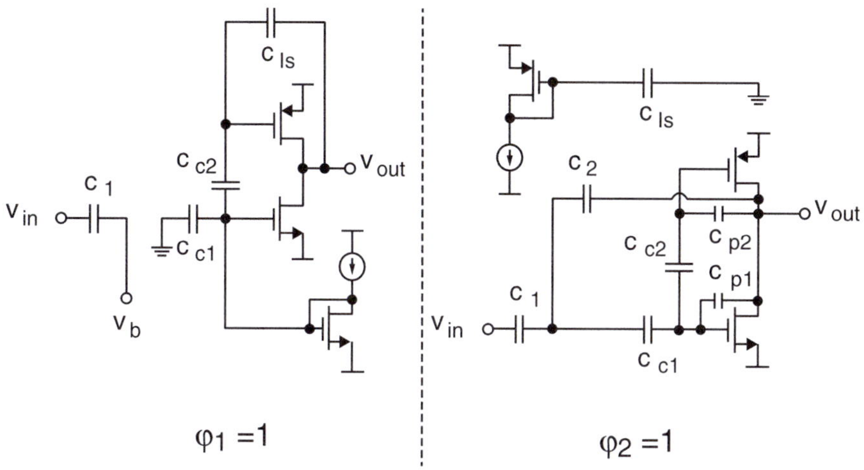

Fig. 17 Class-C cascoded CMOS inverter amplifiers [17]

Fig. 18 Ring oscillator based CMOS inverter amplifiers [18]

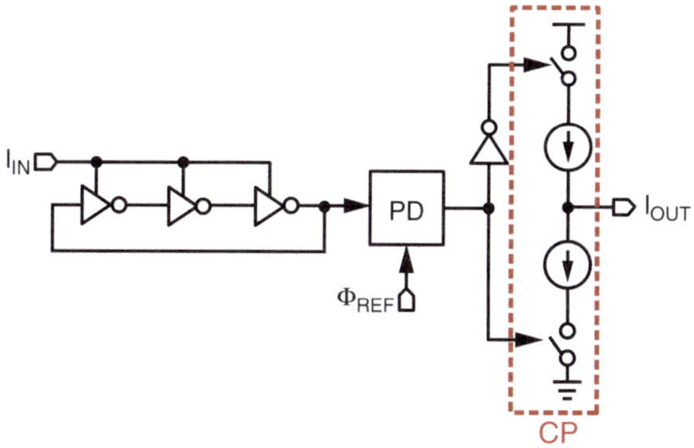

An example is shown in Fig. 18. The input current is converted into a frequency, followed by a phase detector, which them drives charge pumps to provide an analog output current.

Another example is shown in Fig. 19. The input is a differential voltage, and so is the output. Charge pumps are used again to reconstruct the analog output voltage. Several RC filters are needed however, to reduce the switching frequencies and to set the filter frequencies, 40 MHz in this case. Its IIP3 is excellent. Also its area (in 55 nm CMOS) is quite small.

Ring amplifiers on the other hand, consist of a cascade of three CMOS inverters, biased in class A, AB or C. One example is shown in Fig. 20.

The biasing of the second stage is realized by a switched-capacitor technique. The gain is quite high as three stages are involved. The output stage is biased in class C. The dead zone for its biasing is provided by the second stage. A minimum dead zone is required to ensure stability.

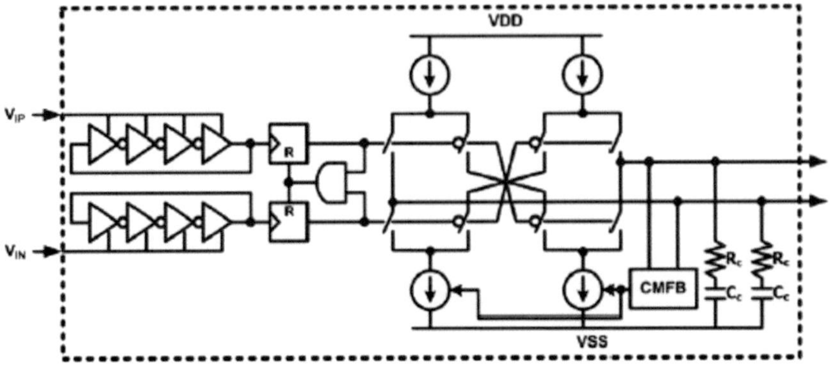

Fig. 19 VCO-based amplifier [19]

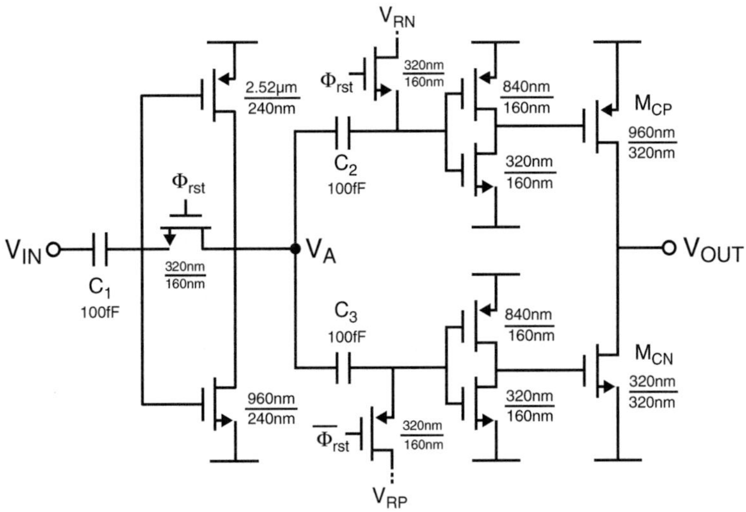

Fig. 20 Ring amplifier with three CMOS inverters [20]

A pipelined ADC was realized with such amplifier yielding only 45 fJ/conversion-step, in 018 μm CMOS.

A mere 6.9 fJ/conversion-step is achieved with a SAR-assisted pipeline ADC in which the three-stage Ring Amplifier is used, shown in Fig. 21. It is realized in 65 nm CMOS.

The first stage is switched. The resistors R_B provide the biasing in the second stage, for the class-C output stage.

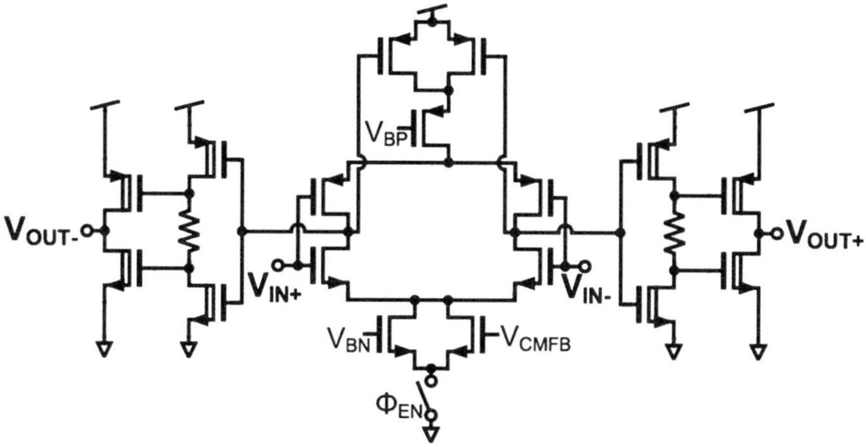

Fig. 21 Ring amplifier in SAR-assisted pipeline ADC [21]

6 Conclusions

Inverter amplifiers are taking over operational amplifiers and Gm blocks in most filter functions, including very-low-power ADCs. Circuits tricks are used in operational amplifiers to increase gain at high frequencies. Internal feedback is used in Gm blocks to improve their linearity. VCO-based and Ring amplifiers only use CMOS inverters however, with comparable and more promising performance.

References

1. Sansen W (2006) Analog design essentials. Springer, Dordrecht
2. Sansen W (2013) In: 2013 IEEE 20th international conference on electronics, circuits, and systems (ICECS), pp 337–340
3. Ohri KB, Callahan MJ (1979) Integrated PCM codec. IEEE J Solid-State Circuits SC-14(1):38–46
4. Camenzind HR, Grebene AB (1969) An outline of design techniques for linear integrated circuits. IEEE J Solid-State Circuits 4(3):110–122
5. Van de Plassche RJ (1971) A wide-band operational amplifier with a new output stage and a simple frequency compensation. IEEE J Solid-State Circuits 6(6):347–352
6. Peng X, Sansen W (2005) Transconductance with capacitances feedback compensation for multistage amplifiers. IEEE J Solid-State Circuits 40(7):1514–1520
7. Grasso AD, Palumbo G, Pennisi S (2007) Advances in reversed nested Miller compensation. IEEE Trans Circuits Syst I 54(7):1459–1470
8. Qu W, Im J-P, Kim H-S, Cho G-H (2014) 17.3 A 0.9V 6.3μW multistage amplifier driving 500pF capacitive load with 1.34MHz GBW. In: Solid-state circuits conference digest of technical papers (ISSCC), 2014 IEEE international, pp 290–291
9. Sansen W (2012) In: IEEE international solid-state circuits conference short course 2012

10. De Matteis M, Pezzotta A, D'Amico S, Baschirotto A (2014) A 33-MHz 70dB-SNR super-source-follower-based low-pass analog filter. In: ESSCIRC 2014—40th European solid state circuits conference, pp 363–366
11. Alini R, Baschirotto A, Castello R (1992) Tunable BiCMOS continuous-time filter for high-frequency applications. IEEE J Solid-State Circuits 27(12):1905–1915
12. Nauta B (1992) A CMOS transconductance-C filter technique for very high frequencies. IEEE J Solid-State Circuits 27(2):142–146
13. Christen T (2013) A 15-bit 140-μW scalable-bandwidth inverter-based ΔΣ modulator for a MEMS microphone with digital output. IEEE J Solid-State Circuits 48(7):1605–1614
14. Chae Y, Han G (2009) Low voltage, low power, inverter-based switched-capacitor delta-sigma modulator. IEEE J Solid-State Circuits 44(2):458–472
15. Luo H, Han Y, Cheung RCC, Liu X, Cao T (2013) A 0.8-V 230-uW 98-dB DR inverter-based sigma-delta modulator for audio applications. IEEE J Solid-State Circuits 48(10):2430–2441
16. Wang J, Matsuoka T, Taniguchi K (2009) A 0.5 V feedforward delta-sigma modulator with inverter-based integrator. In: Proceedings of ESSCIRC 2009, pp 328–331
17. Michel F, Steyaert M (2012) A 250mV 7.5μW 61dB SNDR SC ΔΣ modulator using near-threshold-voltage-biased inverter amplifiers in 130nm CMOS. IEEE J Solid-State Circuits 47(3):709–721
18. Drost B, Talegaonkar M, Hanumolu PK (2012) Analog filter design using ring oscillator integrators. IEEE J Solid-State Circuits 47(12):3120–3129
19. Hsu C-W, Kinget PR (2014) A 40MHz 4th-order active-UGB-RC filter using VCO-based amplifiers with zero compensation. In: ESSCIRC 2014, pp 359–362
20. Hershberg B, Weaver S, Sobue K, Takeuchi S, Hamashita K, Moon U (2012) Ring amplifiers for switched capacitor circuits. IEEE J Solid-State Circuits 47(12):2928–2942
21. Lim Y, Flynn MP (2015) In: International solid state circuits conference, pp 458–459

Linearization Techniques for Push-Pull Amplifiers

Rinaldo Castello, Claudio De Berti, and Andrea Baschirotto

Abstract Amplifiers that need to drive heavy loads (low resistances and/or large capacitances) with good efficiency generally use a push-pull output stage. This intrinsically creates large open-loop distortion components that need to be compressed through feedback to insure high closed-loop linearity. Minimizing close loop residual distortion involves three steps that will be discussed in this chapter. Eliminate all open-loop source of distortion not intrinsic to the proper operation of the push pull structure. Second, choose the amplifier topology that gives the maximum close loop compression of the open-loop distortion components for a given bandwidth. Third, maximize the open-loop gain in the signal band and/or the unity gain bandwidth of the amplifier for a given topology while insuring stability in the presence of variable loads.

1 Introduction

Push-pull amplifier (Fig. 1) closed-loop linearity can be expressed as the ratio between open-loop non-linearity component and loop-gain. As a consequence push-pull amplifier linearization techniques deal with two aspects: reducing the open-loop non-linearity and/or increase the loop-gain in the frequency band over which linearity is important. Open-loop non-linearity is mainly (but not only) limited by the output stage performance and this is typically improved by circuit design. On the other hand, loop-gain improvement refers to the amplifier frequency response and, then, it can be addressed by consideration on the amplifier topology, i.e. number of stages, compensation techniques, etc. .... The combination of these two aspects can guarantee improvements in the closed-loop linearity of push-pull amplifier.

R. Castello (✉) • C. De Berti
University of Pavia, Pavia, Italy
e-mail: rinaldo.castello@unipv.it

A. Baschirotto
University of Milano Bicocca, Milano, Italy

K.A.A. Makinwa et al. (eds.), *Efficient Sensor Interfaces, Advanced Amplifiers and Low Power RF Systems*, DOI 10.1007/978-3-319-21185-5_8

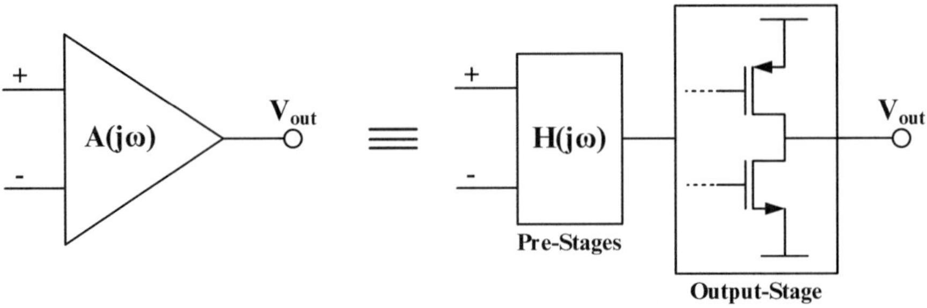

Fig. 1 Push-pull amplifier structure

2 Output Stage Distortion

Distortion in the output stage may occur in two situations depending on the signal amplitude, i.e. for signal at the peak of the voltage swing and for signal around the cross-over, as follows.

The first distortion mechanism occurs for the signal at the signal peak. Output devices driving high current into a resistive load generally enter the linear region of operation in the proximity of the signal peaks. This modulates (reduces) output impedance, and output stage gain as a function of the signal amplitude, introducing distortion. This cause of distortion can be reduced if the output transistors are designed with small saturation voltage (V_{sat} that is similar to the overdrive voltage V_{ov}) that corresponds to adopting very large W/L. This solution is bounded by the practical limit on the device size. The smaller is the resistive load, the larger the device size W/L has to be. This means that for very small resistive load (100 Ω or less) an extremely large W/L would be required.

The second situation occurs at the cross-over, e.g. when the circuit switches from sourcing to sinking and vice-versa, as shown in Fig. 2. The efficiency of these operations can be improved by dynamic turning on/off of driving devices. As a consequence, the cross-over distortions can be classified in two categories.

The first one is associated with the distortions caused by the complete shutting-off of some devices during one half of the signal cycle, while the second one concerns the distortions caused by the turning on of the output devices during the wrong phase of the signal.

The complete shut-off of an output device occurs when its V_{gs} is null, or even negative for an NMOS transistor; if that happens during one phase of the signal, the device experiences a turn-on delay during the crossing from one phase to the other. This creates a dead-time in the circuit response that introduces cross-over distortion. The problem is analyzed in Fig. 3a. When the current in transistor M_3 becomes smaller than I, node X rises close to V_{DD} and the diode device M_2 and the output device M_1 are completely shut off. Subsequently, when V_{out} approaches ground from the negative side, the current in M_3 increases to turn on M_1, however node X must fall by one threshold before M_1 turns on. This occurs after a delay

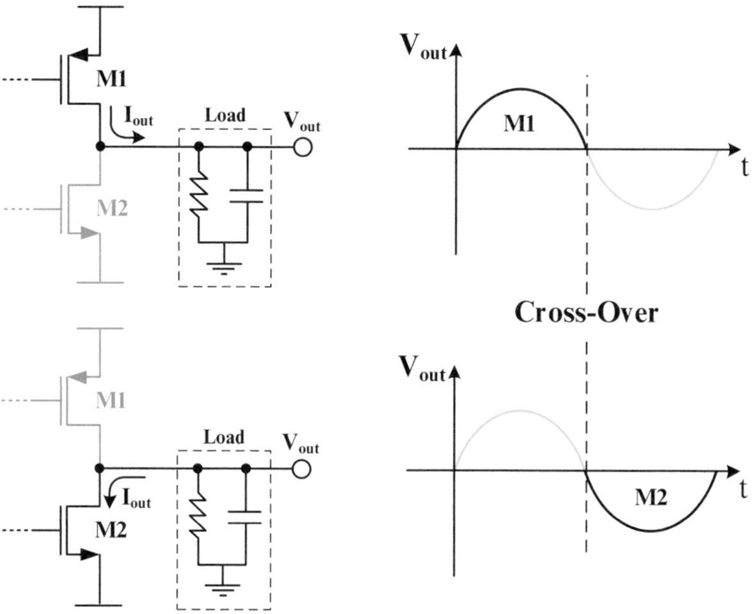

Fig. 2 Cross-over in push-pull amplifier

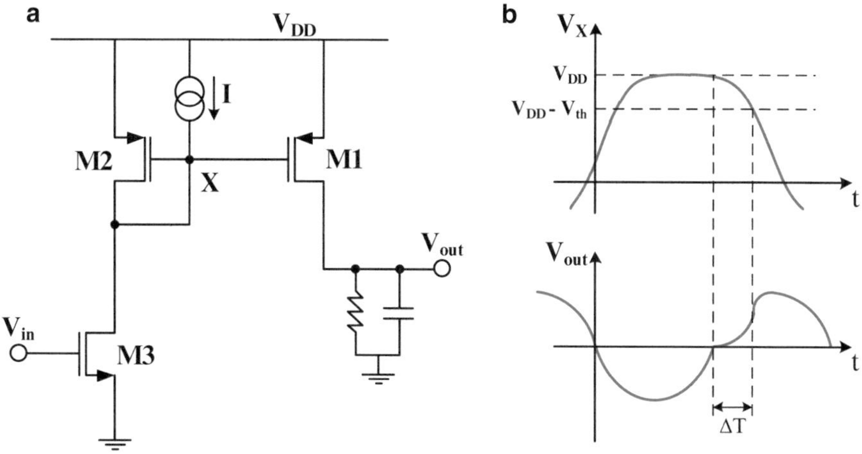

Fig. 3 (**a**) Circuit with turn-on delay of M1. (**b**) Resulting crossover distortion

equal to ΔT with respect to the crossover point, due to the large gate capacitance of M_1 and the small driving current available. The resulting cross-over distortion is depicted qualitatively in Fig. 3b.

The same effect may occur for a capacitive coupling into a high impedance node, even when no explicitly pull-up current is present, as shown in Fig. 4a. In this circuit,

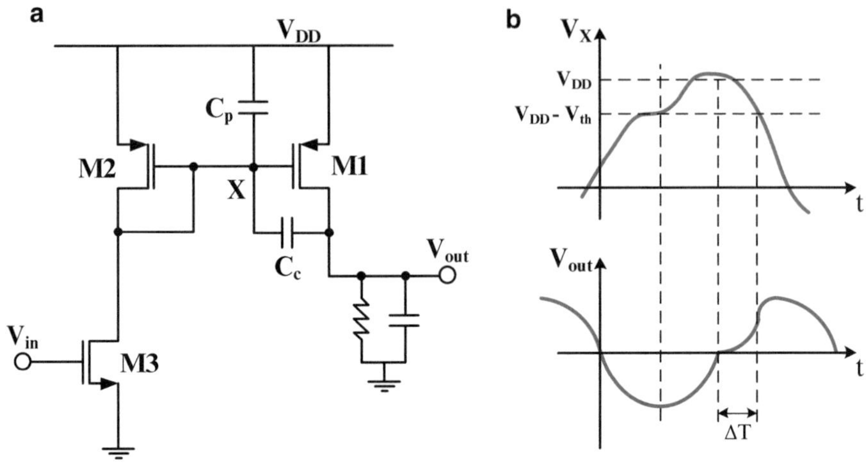

Fig. 4 (**a**) Circuit with turn-on delay of M1 caused by Cc capacitive coupling. (**b**) Resulting crossover distortion

when M_3 turns off, node X tends to rise till roughly one threshold below V_{DD}, therefore M_1 and M_2 remain very close to their conduction region. Nevertheless, when V_{out} rises from its negative peak toward ground, this voltage variation is coupled into the high impedance node X by capacitor C_c. If C_p is much smaller than C_c node X can rise even above V_{DD}. The result is a cross-over distortion similar to the one described in the previous example (Fig. 4b).

This distortion source can be reduced by reducing the turn-on delay. A possible solution is to prevent the complete shut-off of the output devices placing a clamping device at the critical high impedance node. Referring to Fig. 5a, V_{bias} is selected to keep M_C turned off when M_2 and M_1 are turned on, in a way to not interfere with the circuit during normal operation. However, while node X tends to lift up, M_C goes into the conductive region and it clamps node X at a voltage just slightly below a threshold voltage from V_{DD} (Fig. 5b). This results in a drastically decrease of cross-over distortion. Moreover, the clamping effectiveness of M_C can be adjusted changing V_{bias} tradeing-off interference in normal circuit operation with clamping voltage.

As already mentioned, the other source of cross-over distortion may happen if one of the large output devices is turned on during the wrong phase. A possible situation that leads to this problem can be again caused from an unwanted capacitive coupling at the gate of the output transistor as depicted in Fig. 6a. As V_{out} goes from zero towards the negative rail the n-MOS portion of the output push-pull stage (M_2) should be active while M_1 should be turned off. However, the output voltage variation is coupled into the high impedance node X by capacitor C_c. If the displacement current through C_c is larger than I, node X begins to fall. The transistor M_1 starts to conduct and its current is subtracted from the current that drives the load. It follows that the overall amplifier feedback tends to compensate the reduced load

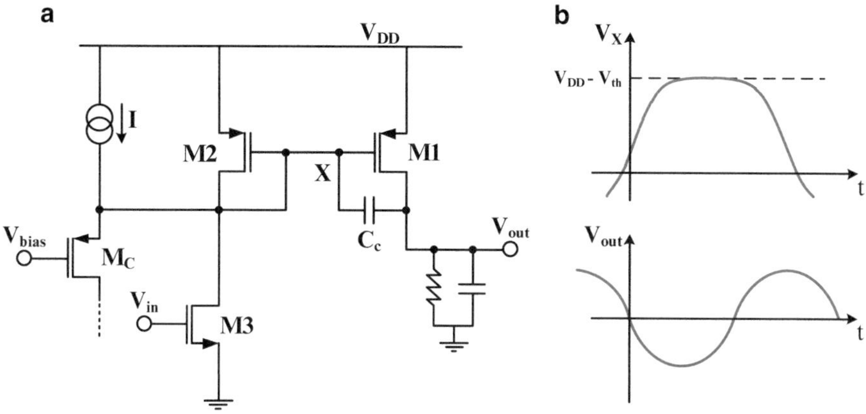

Fig. 5 (**a**) Circuit with clamping on node X. (**b**) Crossover distortion decreased

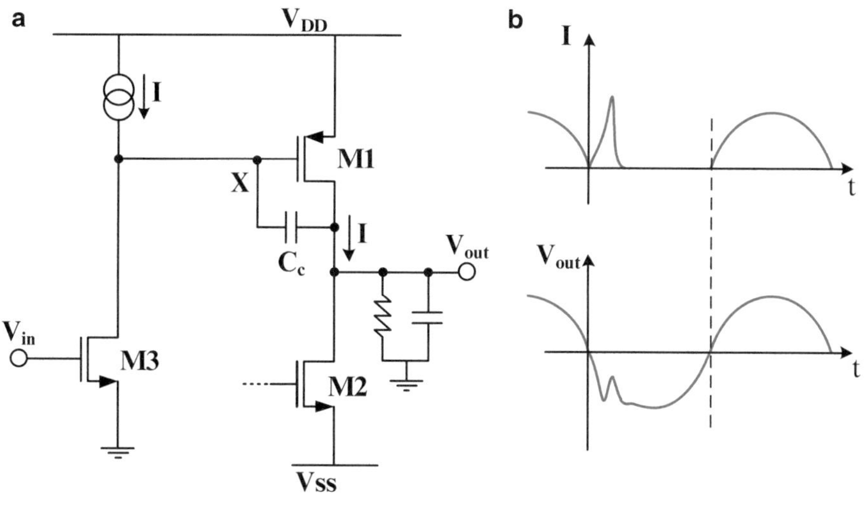

Fig. 6 (**a**) Circuit with unwanted turn-on of M1. (**b**) Resulting crossover distortion

current increasing the V_{gs} of M_2, thus resulting in a current spike between the two supplies through M_1 and M_2 (cross-bar conduction). This has the effect of distorting the output signal and can even damage the circuit Fig. 6b.

In order to prevent this effect a quasi-current mirror structures with a push-pull driver can be implemented. The operating principle of this solution, as shown in Fig. 7, is to turn off M_1 with a large current signal provided by M_3. Therefore, to create the unwanted turn on of M_1, the displacement current through C_c must be much larger than the peak current provided by M_3. Notice that this solution also intrinsically solves the previously described problem as shown in Figs. 3 and 4

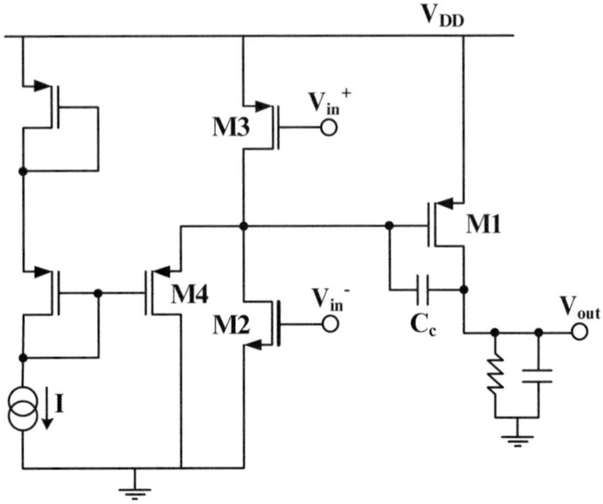

Fig. 7 Circuit to prevent unwanted turn-on of the output devices

(complete turn off of the output transistor before cross-over from off to on state). This is because transistor M_4 intrinsically creates the required clamping at the gate of M_1.

3 Effect of Circuit Topology on Closed-Loop Distortion

As a first approximation, the output stage open-loop distortion is independent from the prior gain stages but it is only a function of the load to be driven and the size/current level of the output stage. The output stage non-linearity can be modeled as a current source placed in parallel to the output push-pull transistors that inject the harmonic/intermodulation terms, as shown in Fig. 8a. While performing a closed-loop analysis, the distortion current source can be moved outside the amplifier as shown in Fig. 8b. From this model it is clear how the equivalent distortion voltage at the amplifier output and correspondingly at its virtual ground is the results of the distortion current source times the output impedance [1]. Therefore, lower output impedance at the frequency of the distortion terms means better linearization. Notice that, as opposed to a memory-less non-linear circuit, the linearity performance of a closed-loop amplifier may be significantly different when comparing intermodulation products and harmonic terms especially for closely spaced tones.

Figure 9a schematically shows a general multistage topology using a multiple feedback compensation scheme. The methodology to evaluate the output impedance is to apply a voltage source at its output and to determine the current that it absorbs. In this condition, due to the zero equivalent impedance of the voltage source, all compensation capacitors connected between the intermediate nodes and the output are effectively grounded. The resulting amplifier topology when computing the

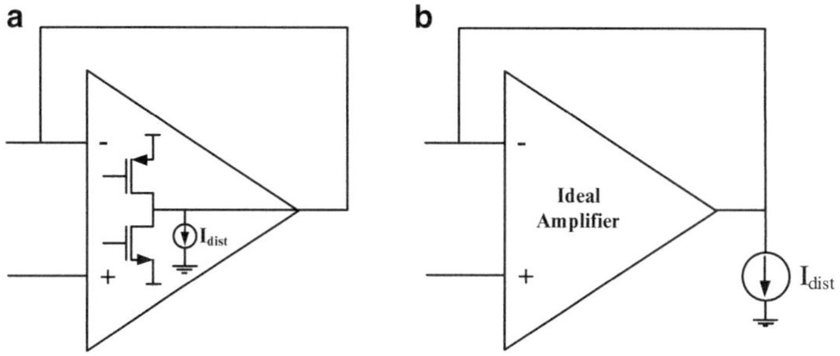

Fig. 8 Distortion modeling: (**a**) Current source placed in parallel to the output push-pull transistors. (**b**) Equivalent model for closed-loop analysis.

output impedance can be approximately substituted with the one shown in Fig. 9c with excellent accuracy for frequencies below the amplifier unity gain bandwidth. This is because the only error of such an equivalence, is having neglected the intermediate feedback paths through the compensation capacitors. However, these terms become significant, compared to the injection through the explicit feedback path only when the transconductance of each amplifier stage is comparable with the admittance of the corresponding compensation capacitance. The required condition is violated only beyond the amplifier unity gain bandwidth. Based on the circuit of Fig. 9c, it is easy to see that the output impedance versus frequency will have the shape shown in Fig. 9d which at high frequency grows with a positive slope of $(N-1) \times 20$ *dB-per-decade*, where N is the number of stages of the amplifier. A more detailed description of this behavior will be given in the example reported at the end of this section.

Another way to express this concept is to say that in closed-loop operation the open-loop distortion is reduced by the total gain between the amplifier input and the distortion source at the frequency of the distortion components. Referring to Fig. 10, the relevant gain for the distortion produced by the amplifier output stage is the one between the input node and the gate of transistor M_{out} (node A) when the amplifier output is grounded (hereinafter called Grounded-Out Pre-Gain (GOPG) or $H(j\omega)$). This situation is shown in Fig. 9e and corresponds to the same operating condition as when computing the output impedance using a test voltage source. For a multi-feedback topology the GOPG at high frequency has a negative slope of $(N-1) \times 20$ *dB-per-decade*, where N is the number of stages, as shown in Fig. 9f. In this section the relations between output admittance (in place of output impedance), GOPG and closed-loop linearity will be investigated.

As an example, three different amplifier topologies are taken in consideration: a simple two-stage Miller, a three-stage nested-Miller and a four-stage double-nested-Miller (Fig. 10). The output stage is the same for all three configurations. To make things more evident, all three amplifiers are designed to exhibit the same open-loop gain (defined as V_{out}/V_{in}) and the same unity gain bandwidth, as shown in Fig. 11a. To obtain these properties, the different transconductances and load

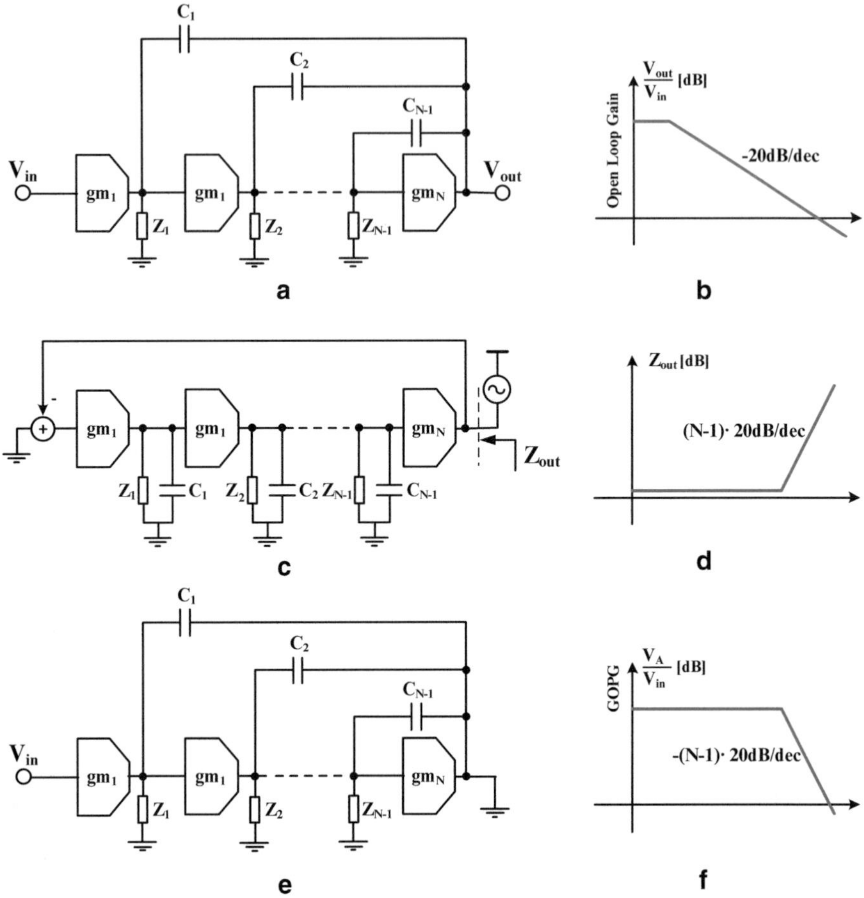

Fig. 9 (**a**) Multi-loop nested-Miller amplifier. (**b**) Resulting Open Loop Gain. (**c**) Equivalent topology for computing output impedance. (**d**) Resulting output impedance. (**e**) Equivalent topology for computing the GOPG. (**f**) Resulting GOPG.

values in the multi-stage amplifiers are chosen to match the open-loop behavior of the two-stage amplifier. The output admittance is evaluated applying a voltage source at the output and sensing the current absorbed by the circuit. By referring to Fig. 11, the output admittance can be determined as the total gain $H(j\omega)$ that exists between the input node and the gate of the last transistor (i.e. V_A/V_{in}) multiplied by the last stage transconductance. From the above discussion we have shown that $H(j\omega)$ corresponds to the GOPG, except for the second order effects introduced by the feedback paths associated with the compensation capacitors which however are negligible up to the unity gain bandwidth. Figure 11b shows the simulated and calculated (using the approximated model introduce above) output admittance for the three amplifiers. Notice from these curves that the output admittances have the

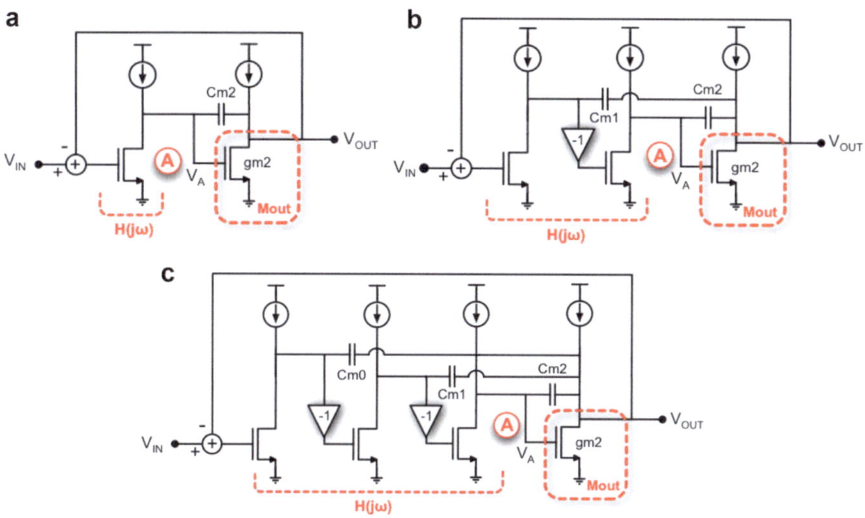

Fig. 10 (**a**) Two-stage Miller. (**b**) Three-stage nested-Miller. (**c**) Four-stage nested-Miller

same DC value because in all the three configurations the open-loop DC gains are identical. However, their slopes are proportional to the number of stages, as mentioned previously. Furthermore the simulated and calculated curves start to depart from each other only beyond unity gain bandwidth as predicted. From the linearization point of view, if the frequency of the distortion component is beyond the point where the admittance starts sloping down, higher number of stages means higher admittance and thus more linearization. When an input tone at f_{IN} is applied, the output distortion at $3 \times f_{IN}$ can be modeled with a current source parallel to the output. Figure 12 indicates that for an amplifier with several hundred MHz of bandwidth the closed loop third harmonic tone (HD3) at $3 \times f_{IN}$ is lower in the three-stage nested-Miller amplifier and even better in the four-stage nested-Miller when f_{IN} is located between approximately 100 kHz and 100 MHz.

In a more realistic situation the DC gain is also proportional to the number of stages and the poles of the GOPG are very close to each other producing the output admittance shown in Fig. 13a for the three amplifier topologies. From this it may seem that an arbitrarily small closed-loop third harmonic distortion can be obtained up to a frequency approaching unity gain bandwidth (ideally 1/3 of it) by arbitrarily increasing the number of stages. In reality a larger number of stages produces a smaller bandwidth because of stability constraints especially when a constant power consumption is assumed, as conceptually shown in Fig. 13b. It follows that in a realistic situation, for a given frequency band of interest, there is an optimum number of stages that gives the best linearization. In particular, for smaller bandwidth of interest, the resulting optimal number of stages is larger [2].

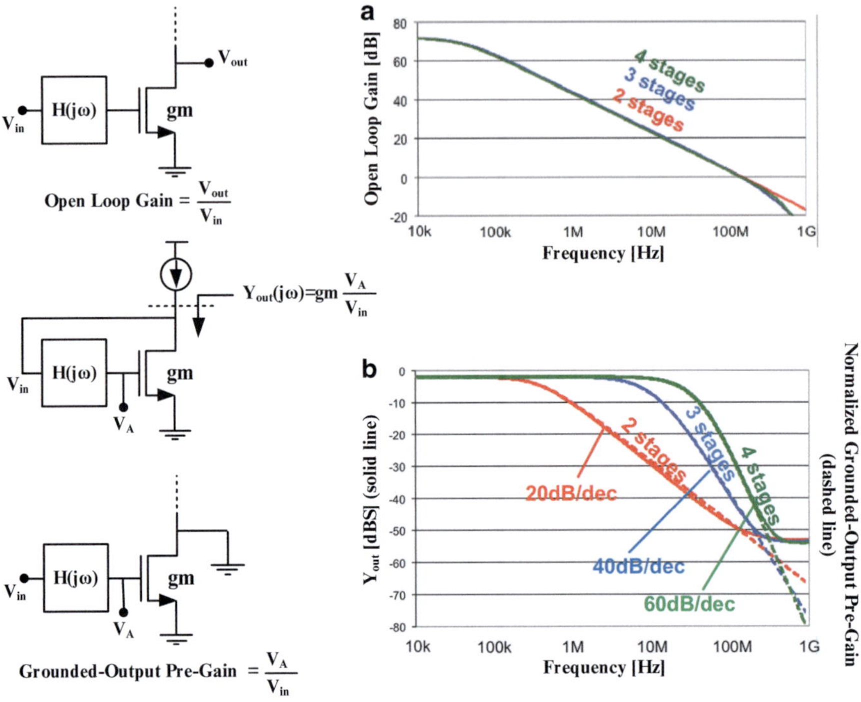

Fig. 11 (**a**) Open-loop Gain magnitude. (**b**) Closed-loop output admittance compared to normalized Grounded-Output Pre-Gain

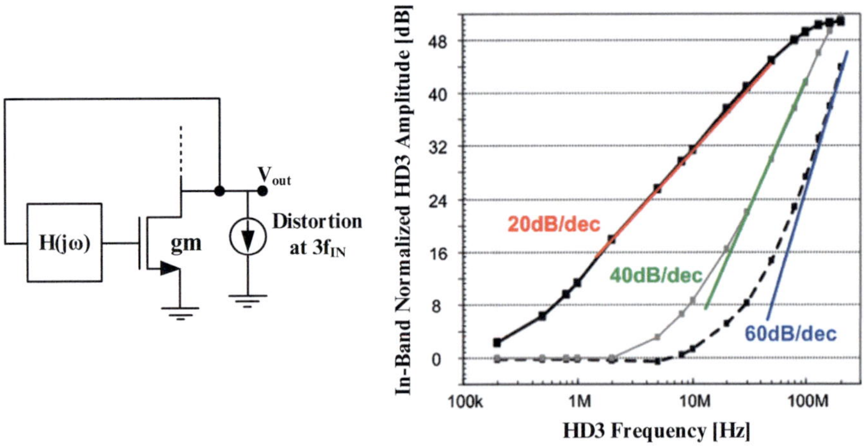

Fig. 12 Third harmonic tone (HD3) at $3 \times f_{IN}$

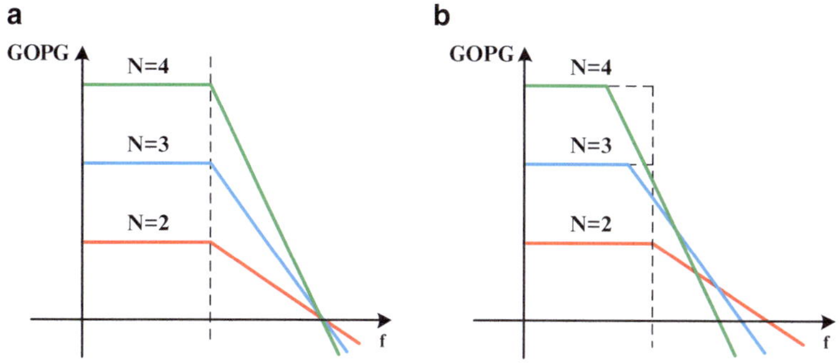

Fig. 13 GOPG for different number of stages: (**a**) Ideal case. (**b**) Real case

4 Optimization of the Open-Loop Gain

As mentioned in previous paragraph, the main drawback of a multi-stage amplifier is the reduction of unity gain bandwidth. A rule-of-thumb for multistage amplifiers with nested-Miller compensation says that to ensure the same level of stability the bandwidth needs to be halved for every additional stage. For example, a three-stage nested-Miller amplifier will need to have approximately half the bandwidth of the Miller amplifier represented by its two last stages. Notice that this does not take into consideration the additional power consumption associated with the extra stage. Therefore if a comparison with constant power consumption is made the bandwidth will become slightly less than half. Using topologies that are more complex than the conventional nested-Miller, it is possible to combine the output impedance benefits of a multi-stage amplifier with the bandwidth offered by a two-stage amplifier [3, 4].

A topology that combines nested-feedback and DC feed-forward is shown in Fig. 14a [5]. When using this configuration the outer Miller capacitor (C_1) is halved compared with the standard nested miller topology, therefore moving the dominant pole to higher (double) frequency [6, 7]. Referring to the main amplifier path made up by gm_1, gm_2 and gm_3, the second pole (located close to the unity band frequency of the inner Miller amplifier) is now falling in-band, leading to a nearly unstable behavior. To solve this problem an auxiliary path gm_f is added in parallel with the input stage. The feed-forward path is directly feeding the input of the last stage therefore bypassing the intermediate amplifying stage gm_2. Choosing the transconductance of the additional stage such that $gm_f/gm_1 = 1$, the zero produced by the feed-forward path ideally cancel out the in-band pole, and thus the unity gain bandwidth (and the phase margin) becomes the same as that of the two-stage Miller amplifier (Fig. 14b). Notice that, contrary to what typically occurs when using a parallel configuration, here an exact pole-zero cancellation can be achieved (neglecting a small mismatch error) and no in-band doublet exists. Qualitatively speaking, this amplifier can offer significantly better linearity (due to its multi-stage behavior for the output impedance) compared to a simple two-stage Miller

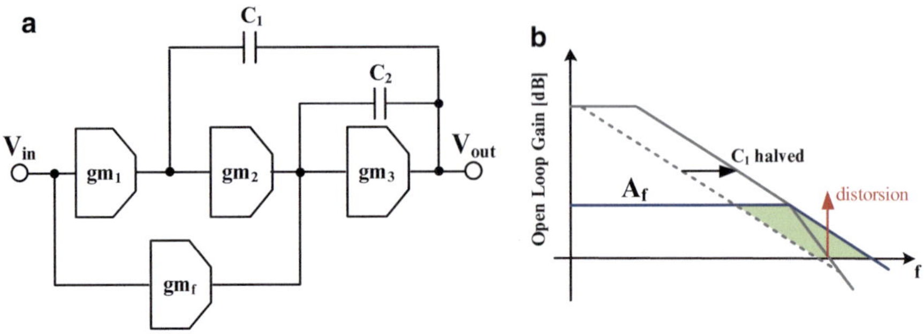

Fig. 14 (**a**) Three-stage nested-feedback with DC feedforward. (**b**) Resulting Open Loop Gain

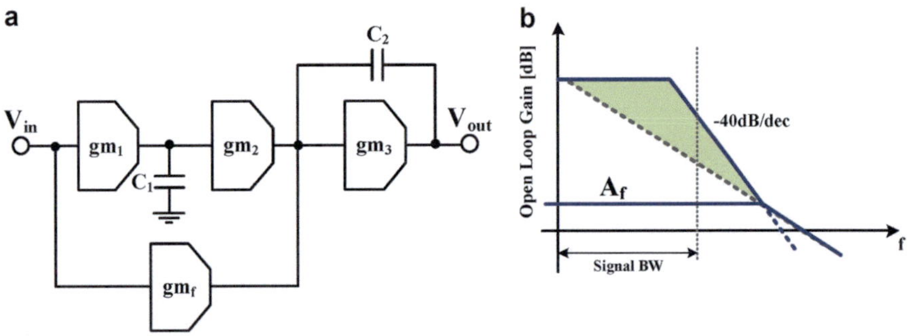

Fig. 15 (**a**) Three-stage with DC feedforward. (**b**) Resulting Open Loop Gain with -40 dB-per-decade slope

especially when the frequency of the distortion component is close to the unity gain bandwidth. The drawback is a more complex structure to be implemented.

It should be noticed that although the main open-loop distortion components are generated in the output stage these are divided by a very large gain when the overall feedback loop is closed. On the other hand, smaller distortion sources associated with the previous stages are divided by smaller gains. In particular distortion sources associated with the input stage remain unaltered when the loop is closed. Therefore if a very linear buffer must be designed, care must be taken to ensure that no slewing is produced at the input.

A possible way to decrease the input slewing distortion is to increase the gain between the virtual ground and the gate of the output transistors in the frequency range of the signal (not of the distortion) [8]. One way to achieve this goal is to implement an amplifier with an open-loop frequency response that displays a slope of −40 dB-per-decade while approaching the unity gain bandwidth, as shown in Fig. 15b.

The same multipath topology of Fig. 14a can be rearranged to produce a slope of −40 dB-per-decade in the open-loop frequency response. This can be achieved by connecting the outer Miller capacitance toward ground and re sizing the intermediate transconductor and the two capacitors (C_1 and C_2) as shown in Fig. 15a. In this way the three-stage amplifier (main path) is no longer nested-Miller compensated. As a consequence the pole at the output of the first stage is moved to a much higher frequency since C_1 no longer sees any Miller multiplication effect. Of course, the frequency response associated with the main amplifier is very close to be unstable (in practice it would be so) because at the unity gain bandwidth the slope is −40 dB-per-decade. However stability is ensured thanks to the auxiliary gain path gm_f placed in parallel with the input stage. The zero produced by the feed-forward path when placed sufficiently before the unity bandwidth (typically at a frequency between 1/3 and ½ unity gain bandwidth) restores the −20 dB-per-decade slope at the crossing point, and thus stability is guaranteed.

Comparing the two solutions assuming that they both can achieve the same unity gain bandwidth, the following points can be qualitatively concluded.

1. from the point of view slewing distortion, the second is clearly superior.
2. the same is valid with respect to the accuracy of the closed-loop frequency response over a given bandwidth (e.g. in a continuous-time filter).
3. with respect to the non-linearity of the output stage, both topologies compress the open-loop distortion terms with a gain that has a slope of −40 dB-per-decade. However, the Miller amplifier compresses a distortion term located at a given distance from unity gain bandwidth more than the uncompensated topology because of the zero in the frequency response of the latter. The ratio in the amount of distortion compression between the two topologies is approximately equal to the ratio between the position of the zero and the unity gain bandwidth.
4. the latter conclusions are valid with respect to the noise associated with the output stage.

5 Optimization of Output Stage Compensation

As already explained in Sect. 2, in push-pull amplifiers the output devices generally enter the linear region of operation in the proximity of the signal peaks, if an area efficient design is done. This changes the gain of the output stage as a function of the signal amplitude. If the output devices reach the deep linear region, the gain can become significantly smaller than 1. As a consequence of this behavior, not only significant distortion components are injected in parallel with the output, but also stability issue may result in a multi-stage amplifier. In this paragraph, the cause that leads to the risk instability is analyzed and a possible solution is presented.

For simplicity, this problem will be analyzed for a nested-Miller amplifier operated in Class A with different load resistances R_L and consequently different gains Let us focus first on the last two stages of the circuit, shown in Fig. 16. The first stage is realized by transistor M_2 (in common source configuration) with a cascoded load. The last stage is realized with transistor M_3 (also in common source configuration) whose transconductance gm_3 is assumed to be constant (class A)

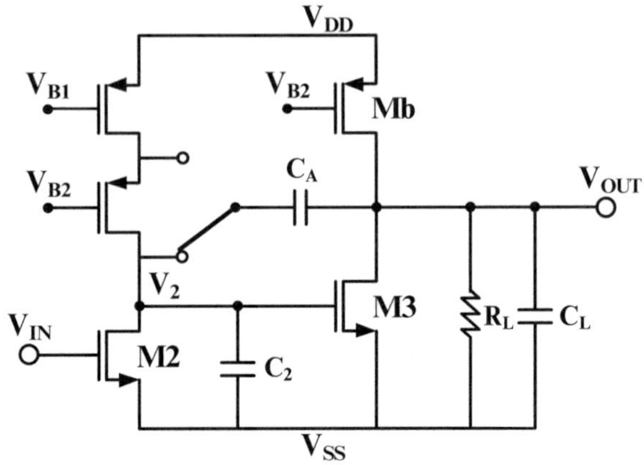

Fig. 16 Two-stage Miller-compensated topology

biased by M_B. If the resistive load R_L is sufficiently small, its gain A_3 can be expressed as gm_3 times R_L. Since gm_3 is constant, to vary A_3 it is necessary to vary the resistive load R_L. In this way it is possible to model in a simple way the gain reduction effect of push-pull amplifiers that occurs in the proximity of the signal peaks.

If the two-stage amplifier is Miller compensated, the feedback capacitor C_A is connected between the output node and the drain of the first-stage input transistor, as shown in Fig. 16. In this case, a current $C_A d(V_{OUT} - V_2)/dt$ must be provided by the first-stage, to charge and discharge C_A. It follows that the resulting Miller capacitance is $C_A(1 + A_3)$.

Assuming that the parasitic capacitance at the gate of M_3 (C_2) is much smaller than the compensation capacitor C_A and that the load capacitance C_L is also smaller than C_A, the unity gain bandwidth (*UGB*) is given by:

$$UGB \approx \frac{\mathrm{gm_2 A_3}}{C_A\,(1 + A_3)} \tag{1}$$

where gm_2 is the transconductance of M_2. This implies that, if A_3 is sufficiently larger than 1, the *UGB* is close to gm_2/C_A independently of the value of A_3. Instead, if A_3 is significantly smaller than 1 (e.g. in the proximity of the signal peaks in a push-pull amplifier), the *UGB* starts to decrease. In a two-stage amplifier, this results in an increased distortion introduced by the output stage, but it does not affect the stability. Instead, in a multi-stage topology, a lower *UGB* in the last two stages causes instability if it becomes smaller than the nominal (i.e. A_3 greater than 1) *UGB* of the last three stages. These aspects are demonstrated through the simulations reported in the fallowing section.

Assuming to have $gm_3 = 0.1\ S$, two different values of R_L are considered, $50\ \Omega$ and $0.5\ \Omega$. The corresponding output stage gains are, respectively, $A_3 = 5$ and $A_3{'} = 0.05$. The other design values of the two-stage amplifier are: $C_L = 10\ pF$, $C_A = 12\ pf$, $C_2 = 250\ fF$, $gm_2 = 4.8\ mS$ and the equivalent impedance at the drain

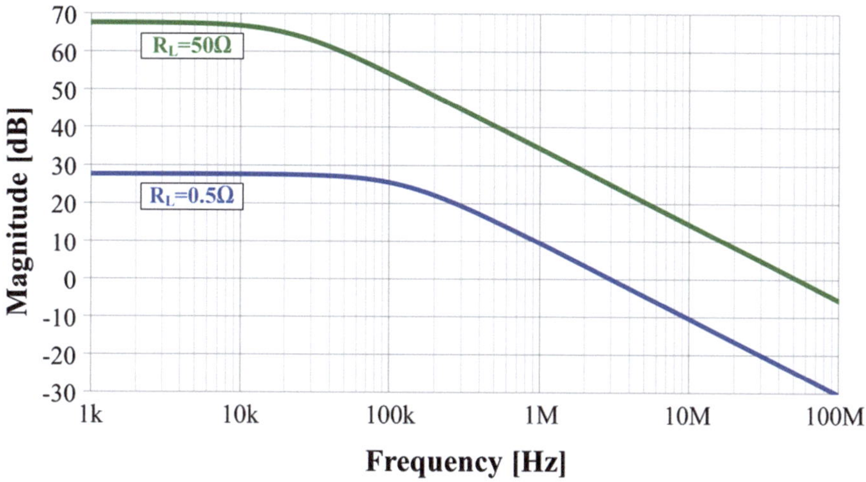

Fig. 17 AC magnitude for the two-stage Miller-compensated

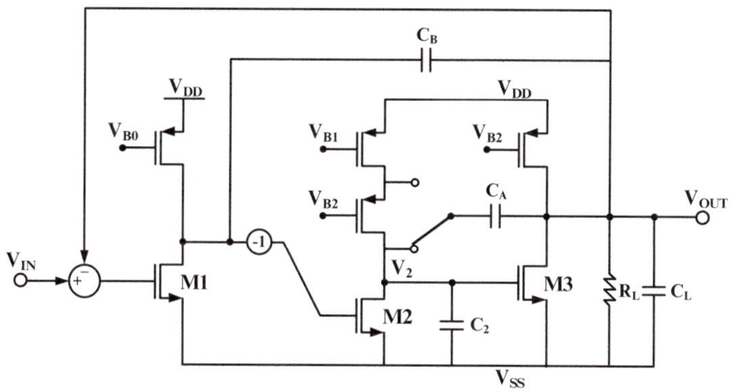

Fig. 18 Two-stage Miller-compensated amplifier

of the first-stage driver is $R_2 = 100\ k\Omega$. Notice that the value of the above design parameters (e.g. the ratio between C_A and C_2) have been chosen to strongly emphasize the phenomenon that will be discussed in the following sections. In a real situation things may be less exacerbated, although other effects that have been neglected for simplicity (e.g. the presence of a cascade in the output stage) may make things even more critical.

Figure 17 shows the simulated *UGB* of the two-stage amplifier for the two possible gains. Notice that the *UGB* is decreased from 53 to 3 MHz when the gain A_3 is reduced by a factor of 100. To show the effect of this behavior on a multi-stage amplifier, a three-stage nested-Miller topology is simulated. The circuit is shown in Fig. 18. In this case, not only the open-loop unity gain bandwidth is decreased when A_3 is reduced by a factor of 100, but, more importantly, the phase margin is reduced

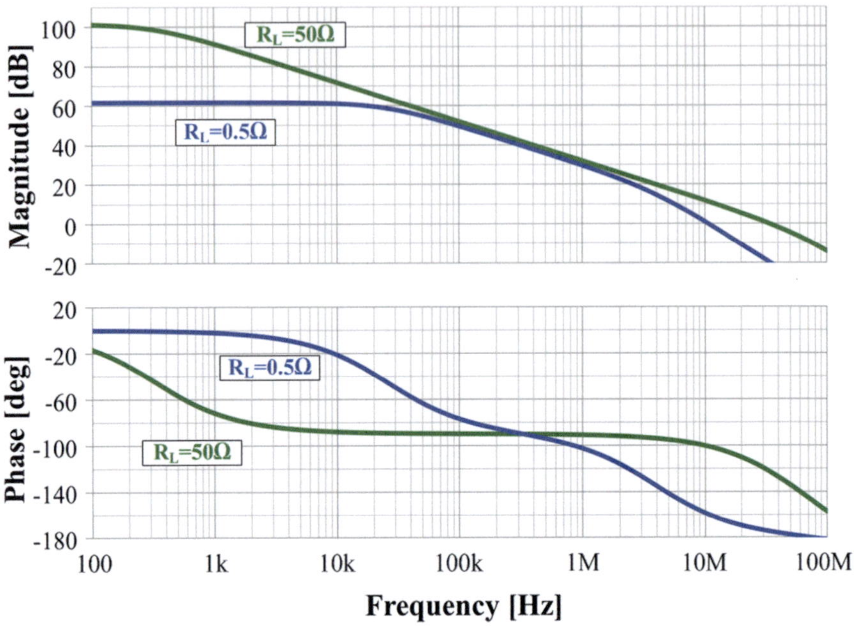

Fig. 19 Bode plots for the three-stage Miller-compensated

from 60° to 20° (Fig. 19). This is caused by the non-dominant pole generated by the amplifier last two stages that operate in unity gain configuration (due to the presence of C_B). Such a pole is located near the unity gain bandwidth of the inner Miller amplifier, which is drastically reduced when the load resistance is reduced, as shown above. The step response of this circuit in closed-loop configuration is shown in Fig. 20. When $R_L = 0.5\ \Omega$, the 1 V step has an overshot greater than 0.5 V compared to the step response when $R_L = 50\ \Omega$ and, consequently, the settling time is ten times longer.

A possible solution to guarantee enough phase margin when $A_3 = 0.05$ is to further compensate the three-stage amplifier increasing C_B. The drawback is the reduction in the *UGB* of the three-stage amplifier and the associated lower linearization factor. A more efficient technique would be to limit the correlation between the gain A_3 and the *UGB* of the last two stages.

Returning to the case of a simple Class-A two-stage amplifier, the same *UGB* can be obtained using the cascade Miller compensated topology shown in Fig. 21 i.e. removing the feed-forward path from V_2 to V_{OUT} while still producing a dominant pole thanks to the Miller effect [9]. In the new circuit the feedback capacitor C_A is connected between the output node and the source of the load cascade transistor. Therefore, a current $C_A d(V_{OUT})/dt$ must be provided by the first-stage to charge and discharge C_A, as opposed to $C_A d(V_{OUT} - V_2)/dt$ for the simple Miller case i.e. the feed-forward term associated with V_2 has been removed. It follows that the

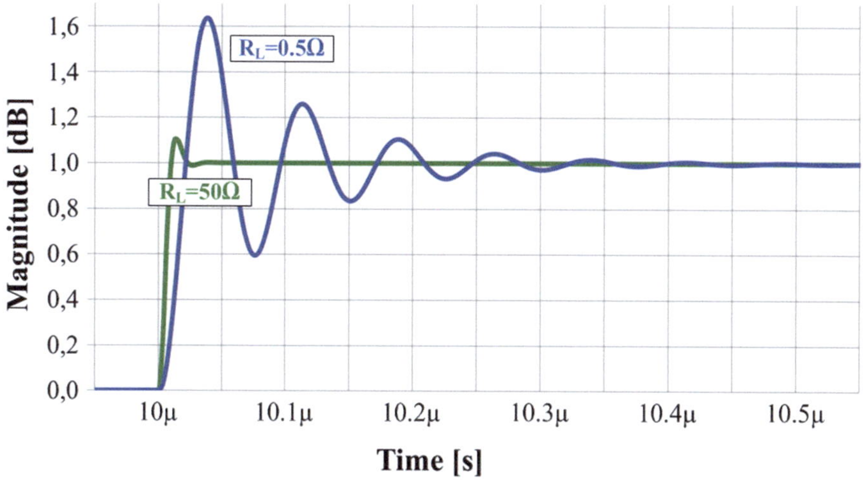

Fig. 20 Step response for the three-stage stage Miller-compensated

Fig. 21 Two-stage cascode Miller-compensated

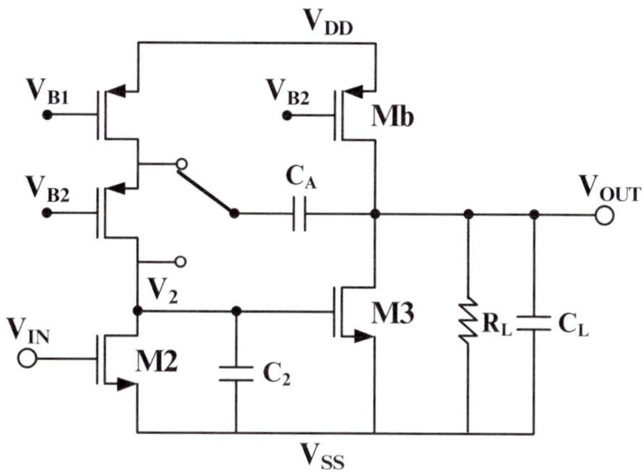

equivalent Miller capacitance is now $C_A A_3$ rather than $C_A(1 + A_3)$. Thus, using the same assumptions used to derive (1), the unity gain bandwidth (*UGB*) is given by:

$$UGB \approx \frac{gm_2 A_3}{C_A A_3} = \frac{gm_2}{C_A} \tag{2}$$

The *UGB* is now (approximately) independent from the gain of the output stage. As it was done for the two-stage Miller-compensated amplifier, the behavior of the cascode compensated topology when used within a multi-stage amplifier is shown (using simulations) in the following section.

Figure 22 shows the simulated *UGB* of the modified two-stage amplifier for the two possible gain levels. That thanks to the use of cascade Miller compensation, the

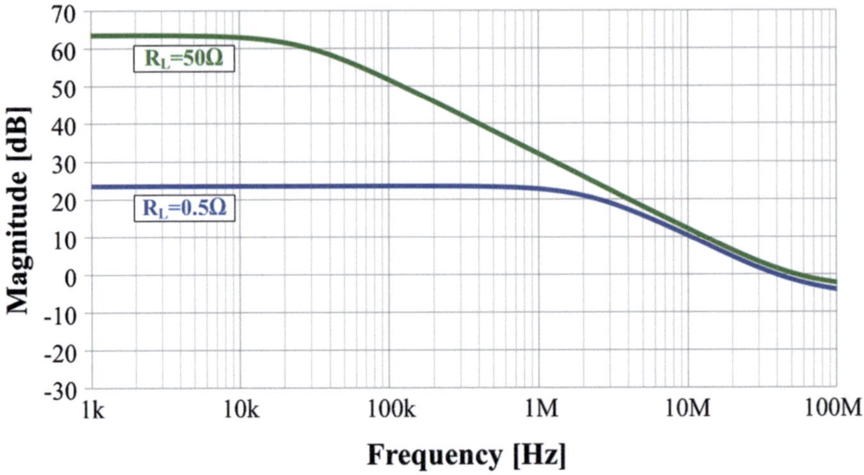

Fig. 22 AC magnitude for the two-stage cascode Miller-compensated

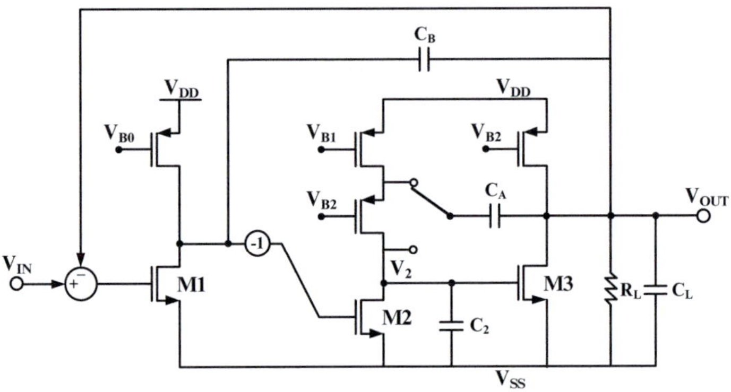

Fig. 23 Three-stage cascode Miller-compensated

UGB is only slightly decreased (from 53 to 49 MHz) when the gain A_3 is reduced by a factor of 100. This means that the distortion terms introduced by the output stage are compressed by a much larger factor. Moreover, in a multi-stage topology, a quasi-constant *UGB* in the last two stages guarantee stability for all possible load resistances. Using the new compensation within the nested-Miller amplifier of Fig. 23, the simulated AC response, shown in Fig. 24, exhibits a variation of only 1° of phase margin when the gain A_3 is changed by a factor of 100. It follows that the closed-loop step response remains almost the same in both cases, as shown in Fig. 25. Therefore, in the proximity of the signal peaks, a multi-stage push-pull amplifier that uses a cascade Miller compensation in the inner loop can remain stable even when the output devices enter into the deep linear region. Moreover, such a topology is significantly superior also for its driving capability and linearity. In fact

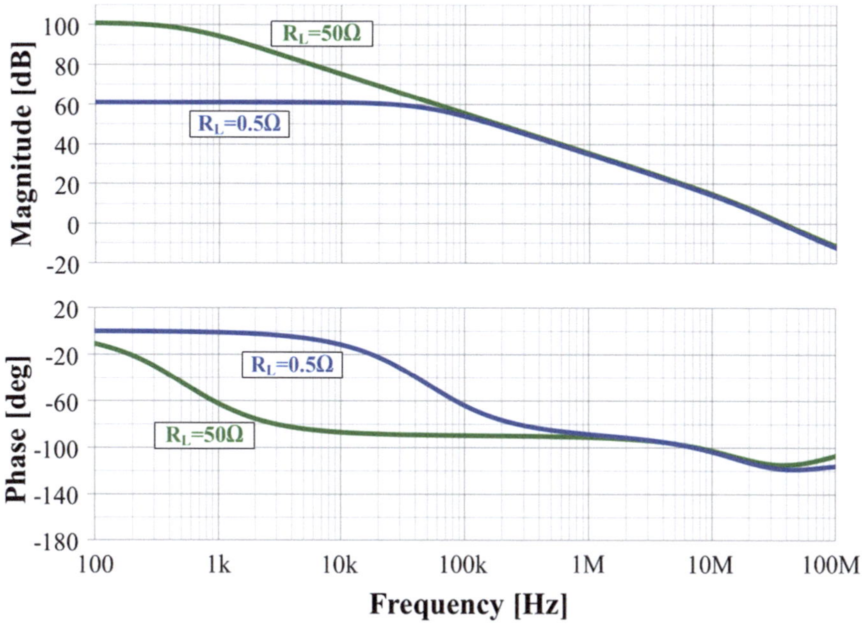

Fig. 24 AC magnitude for the three-stage cascode Miller-compensated

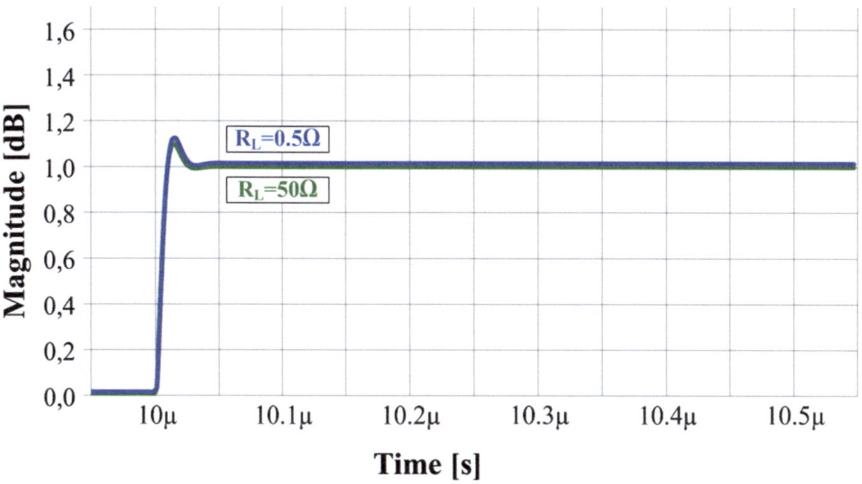

Fig. 25 Step response for the three-stage cascode Miller-compensated

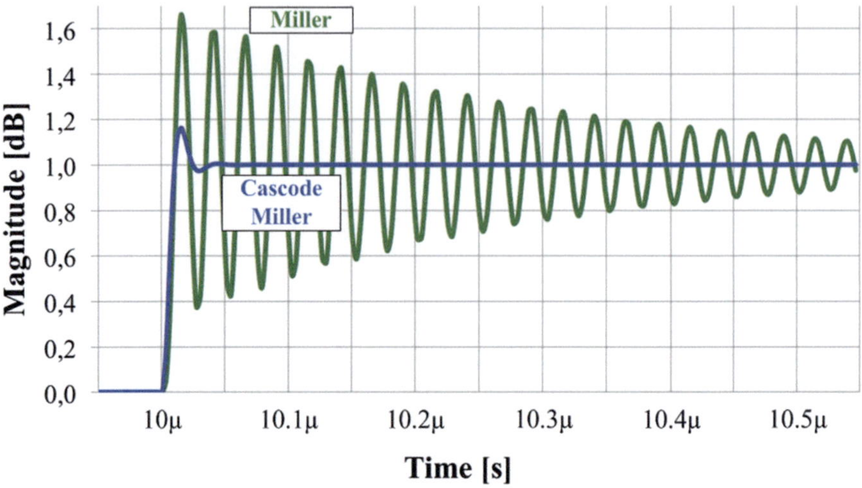

Fig. 26 Step response when $C_L = 400$ pF

it can drive a capacitive load more than an order of magnitude greater compared with an amplifier that uses a simple Miller inner loop before instability occurs. Such a behavior is evident from Fig. 26 that shows the step response for the two topology considered (for the design parameter used in all previous simulations) with a capacitive load of 400 pF and $R_L = 50\ \Omega$. In addition any distortion due to the output stage (e.g. in a push pull topology when the output transistor enter the deep linear region at the peaks of the swing) is compressed by a much larger gain compared with the simple Miller case. This is due to the fact that the closed-loop output impedance for the cascade Miller amplifier is smaller than that for the simple Miller one by a factor equal to the ratio between the compensation capacitance C_A and the parasitic capacitance at the input of the output transistor C_2.

References

1. van der Zee R (2009) Output impedance shaping for frequency compensation of MOS audio power amplifiers. IEEE J Solid-State Circuits 44(3):928–934
2. Pernici S, Nicollini G, Castello R (1993) A CMOS low-distortion fully differential power amplifier with double nested Miller compensation. IEEE J Solid-State Circuits 28(7):758–763
3. Monticelli DM (1986) A quad CMOS single-supply op amp with rail-to-rail output swing. IEEE J Solid-State Circuits 21(6):1026–1034
4. Eschauzier RGH, Ruud GH, Hogervorst R, Huijsing JH (1994) A programmable 1.5 V CMOS class-AB operational amplifier with hybrid nested Miller compensation for 120 dB gain and 6 MHz UGF. IEEE J Solid-State Circuits 29(12):1497–1504
5. Eschauzier RGH, Kerklaan LPT, Huijsing JH (1992) A 100-MHz 100-dB operational amplifier with multipath nested Miller compensation structure. IEEE J Solid-State Circuits 27(12):1709–1717
6. Gray PR, Meyer RG (1982) MOS operational amplifier design: a tutorial overview. IEEE J Solid-State Circuits SC-17:969–982

7. Solomon JE (1974) The monolithic op amp: a tutorial study. IEEE J Solid-State Circuits SC-9:314–332
8. Eschauzier RGH, Huising J (1995) Frequency compensation techniques for low-power operational amplifiers. Kluwer Academic Publishers, Dordrecht, pp 166–173
9. Ahuja BK (1983) An improved frequency compensation technique for CMOS operational amplifiers. IEEE J Solid-State Circuits 18(6):629–633

Ultra Low Power Low Voltage Capacitive Preamplifier for Audio Application

Olivier Nys, Daniel Aebischer, Stéphane Villier, Yves Kunz, and Dequn Sun

Abstract This paper discusses the design strategy of an acquisition chain for microphone input signal. First, target specifications are given. The overall strategy for the acquisition of the signal, pre-amplification and A/D conversion is then discussed, together with the main constraints, and a new architecture for the loop is proposed. The most critical analog blocks, such as first amplification stage, loop filter and biasing circuitry, are then investigated in more details, followed by measurement results.

1 Introduction

While designing low power low voltage audio amplifiers, the most severe constraint to take into account is thermal noise, as it is generally the dominant noise source and its noise power is inversely proportional to the spent current consumption, at least for a given circuit topology. Hence, starting from a given circuit implementation, the noise can be reduced by two in straightforward way by doubling all the currents and at the same time the width of all transistors, resistors and capacitors while keeping constant the current densities and the voltage levels. However, doing so, the current consumption is also doubled. For a given current consumption budget, the noise can only be reduced by optimizing the circuit topology. The strategy consists into minimizing the number of transistors significantly contributing to noise, and spending most of current in the devices most critical with respect to noise. A classical implication of this is that a sufficient gain should be performed by the input stage, in order to reduce the noise contributions of the next stages, and most of the current should be spent in the first stage in order to minimize its noise. Another implication of this is that the simplest topologies are generally the most efficient ones. As an example, compared to a differential input pair with one input grounded or at a fixed voltage, a single common source transistors will achieve the same transconductance with half the current consumption and half the noise. Hence single ended topologies are generally more efficient than fully differential ones.

O. Nys (✉) • D. Aebischer • S. Villier • Y. Kunz • D. Sun
SEMTECH SA, Gouttes d'Or, 40, CH 2000, Neuchâtel, Switzerland
e-mail: onys@semtech.com

K.A.A. Makinwa et al. (eds.), *Efficient Sensor Interfaces, Advanced Amplifiers and Low Power RF Systems*, DOI 10.1007/978-3-319-21185-5_9

First of all the architecture must be selected in such a way as to minimize the number of components significantly contributing to thermal noise and hence requiring more power. Noise optimization also implies biasing all transistors acting as transconductors in the sub-threshold or weak (or at least moderate) inversion region, while those acting as current sources or current mirrors are classically biased in strong inversion or even degenerated with resistors in order to reduce their current noise contribution. At very low supply voltage however, strong inversion biasing or degeneration with resistors is hardly feasible, due to the increased voltage drop required to keep the transistors saturated. By this fact, the noise contribution of current load transistors and current mirrors is unfortunately not negligible with respect to that of active transconductors in the signal path. The biasing circuitry, generation of voltage and current reference, also needs to be designed with care not only for minimizing noise, but also to ensure a high PSRR.

2 Target Specifications

In order to fix the ideas, some target specifications are defined hereafter. Operation should be guaranteed for supply voltage as low as 0.8 V. A dynamic range of 98 dB is required, corresponding to an effective resolution of 16 bits, which, assuming a peak to peak input signal from microphone of 0.6–0.75 V, leads to an lsb of 9 μV and an input referred noise level around 3 μV RMS. Notice that the low noise floor is mainly required at low signal power, and higher noise can be tolerated at large signal level. The conversion rate should be selected between 16 and 32 kHz, corresponding to a bandwidth between 8 and 16 kHz. The noise must be weighted by a curve representative of the sensitivity of the ear, such as typically the A-weighting curve. This means that more importance is given to noise components in the middle of the audio band, between 1 and 6 kHz, while less weight is given to components below 1 kHz and above 6 kHz. In particular, at 100 Hz, the sensitivity is already reduced by 20 dB, and even more for lower frequencies. For this reason, flicker noise is not such a critical issue as thermal noise. The target current consumption is of the order of 100 μA.

3 Classical Approach and Design Challenges

The classical architecture (Fig. 1) is made of a first amplification stage followed by a sigma delta ADC. Sigma delta is a natural choice for audio applications, not simply because low bandwidth makes oversampling very easy, but mainly because it lowers every noise sources by averaging, not only quantization noise but also all analog noise sources such as thermal noise. Moreover it strongly reduces the noise aliasing into the baseband. The preamplifier reduces the input referred noise contribution of the sigma delta ADC. However, the preamplifier can only achieve gain when signal

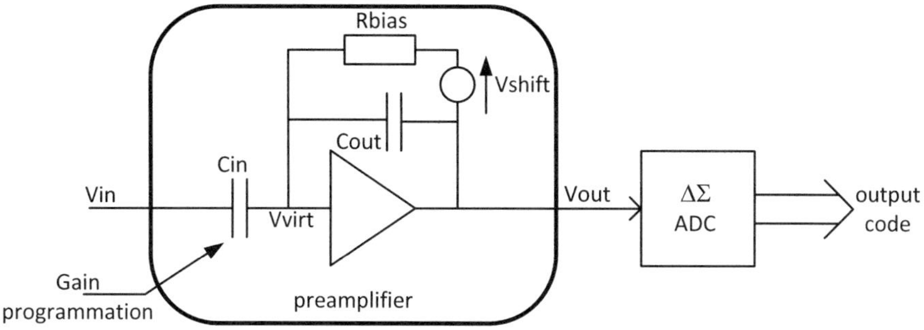

Fig. 1 Classical architecture with capacitive preamplifier and sigma delta A/D converter

level is low, otherwise the signal would be clamped by the supply or saturated within the ADC. The gain achieved by the preamplifier must thus be adapted according to the signal level in order to minimize the noise without saturating the acquisition chain. This is not a problem as long as this gain setting remains static or quasi static. However for signals such as speech with a strongly varying envelope, the gain would need to be changed dynamically, typically by changing the ratio of capacitors or resistors in the feedback path of the amplifier. The problem is then that some perturbations on the acquisition chain are hardly avoidable when the gain is switched, and these perturbations appear as spikes, thus high frequency noise components which may easily be audible, especially when gain is increased at low signal level.

Theoretically the gain of the preamplifier could be implemented as a ratio of either resistors or capacitors in the feedback path. Achieving this gain as a ratio of resistors [1, 2] however requires passing the signal through resistors of very low value, a few kiloohms, in order to limit the additional thermal noise 4kTR, implying a high current consumption. Moreover, the voltage on the virtual ground has to be made independent on the DC level of the input signal. Hence achieving the gain as a ratio of capacitors is preferred [3], as shown in Fig. 1. The advantage is that the DC component of the input signal is directly eliminated by capacitive coupling. However, some resistive feedback path, represented by resistor Rbias, must be created between output and input of the amplifier in order to have the virtual ground properly biased while having the DC output level centered around a target value. Though apparently simple, this solution presents some design issues.

First the DC output level of the amplifier does not necessarily correspond to the voltage on virtual ground, especially at low supply voltage. Typically, assuming common source input transistor, virtual ground voltage should be one VGS above negative supply or below positive supply, while the output of the amplifier should better be centered around middle of supply in order to maximize output swing. Hence some DC voltage shift must be realized somehow in the feedback path.

Second, the value of the resistor Rbias must be extremely high. Namely, it defines together with the feedback capacitor Cout a high pass filter with cut off frequency equal to $1/(2\pi \text{ Rbias Cout})$. Assuming for instance Cout $=4$ pF and a high pass cut off frequency of 100 Hz, a minimum resistor Rbias = 400 MΩ is already derived. However a much more severe constraint is set by thermal noise current 4kT/Rbias of this resistor, which, input referred, corresponds to a very high noise spectral density at low frequency, given by

$$S(f) = \frac{4kT}{Rbias(2\pi f\ Cin)^2} \tag{1}$$

Assuming the noise should be integrated down to 200 Hz, and with 20 pF input capacitance, a resistance of 5 GΩ would be required in order to have 1 μV RMS noise contribution, making classical resistors such as high resistive poly completely unrealistic. Other solutions are nevertheless possible to realize this resistive path, such as cross coupled diodes. The problem with this solution is that the path is strongly nonlinear, the conduction of the path occurring mainly in the signal peaks, such that high frequency audio components can be demodulated into low frequency. Another solution consists into achieving the resistive path with a switched capacitor branch. For instance switching a small capacitor of 10 fF at the output rate of 20 kHz, a very linear resistance of 5 GΩ is realized. Higher resistance value can be achieved by reducing the switching frequency. The drawback of this solution is that the components around the multiples of the switching frequency are then aliased in the baseband. The output of the amplifier must then be filtered out in front of the feedback resistor in order to reduce aliasing. However achieving an analog antialiasing filter with low cut off frequency is extremely expensive in silicon area.

4 New Proposed Architecture

A solution to achieve a high gain in the first stage independently on the input signal level consists into including the capacitive preamplifier already in a feedback loop [4]. If the feedback signal precisely tracks the input signal, then the difference is small and can be amplified with a high gain. Moreover, digitalization can also be included within the loop, the digital output signal being reconverted back into analog and subtracted from the input signal.

4.1 Delta Modulation

The solution illustrated in Fig. 2 thus basically corresponds to delta modulation. The input amplifier, called residue amplifier, amplifies the difference between input signal Vin and feedback signal Vdac. After some analog filtering, the output Vresidue of this amplifier is quantized by an ADC and the result numerically accumulated. The output of the accumulator is then converted back into a voltage

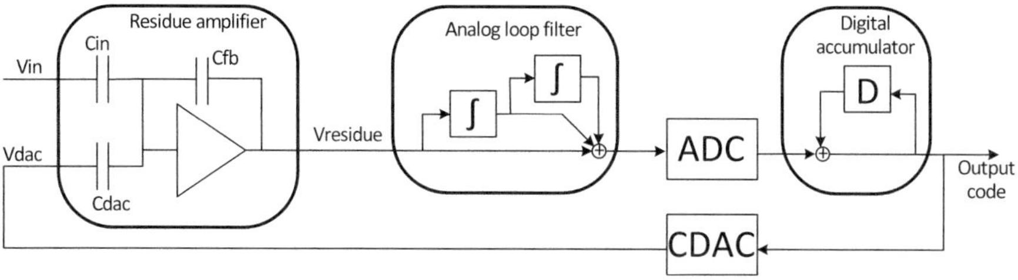

Fig. 2 New proposed architecture with the residue amplifier within a delta loop

Vdac by the capacitive DAC CDAC in the feedback path, which drives the feedback capacitor. Thus the output of the accumulator corresponds to the output code of the loop, representative of the input signal.

A proper tracking by the feedback loop is only possible with a high oversampling factor (typically 128) and a high resolution DAC (typically 8–10 bits) in the feedback path. With proper tracking the error on the virtual ground of the residue amplifier remains small (a few lsb's of the CDAC), which reduces non linearity errors in the forward path and sensitivity to clock jitter, and the residual error can be processed in continuous time.

The required bandwidth for the residue amplifier is close to the oversampling frequency, the criteria being that a compensation of a given number of lsb's at the output of the CDAC must be detected at the next sampling of the ADC with almost the corresponding number of lsb's of the ADC, in order to have a close to unity gain in the correction applied by the delta loop and guarantee the stability of the loop.

Notice that the requirements on ADC and CDAC in the loop are quite different. The ADC only needs to roughly quantify the residue, so 3 or 4 bits resolution are sufficient, while the CDAC needs a high resolution and low DNL for closely tracking of the input signal.

4.2 *Noise Shaping*

However, as such, pure delta modulation would not be sufficient in order to achieve a high resolution such as 16 bits or more, because the tracking cannot be precise at that resolution. Thanks to oversampling, overall resolution is somewhat improved with respect to that of the feedback DAC, as the components outside the baseband of the quantization noise, which can be monitored at the output of the residue amplifier, may be filtered out.

Unfortunately, this improvement is very depending on the signal type, being related to the noise spectrum. In particular for small or slowly varying input signals, most of the quantization noise components may be located at low frequencies, in the audio baseband. In such a case the resolution would not be significantly improved by oversampling and averaging.

Fortunately, this loop can easily be transformed into a sigma delta loop by adding integrators in the forward path of the loop, within the analog loop filter after the residue amplifier, in order to push the quantization noise outside the baseband. The difference with a classical sigma delta loop is that, with the digital accumulator after the ADC, one can combine a high resolution DAC in the feedback path with a very rough ADC in the forward path, while for classical sigma delta loops the resolution of ADC and feedback DAC are the same. Combining 9-bits DAC with a second order integration at an oversampling of 128, the quantization noise can easily be reduced below 120 dB and made negligible with respect to other noise sources such as thermal noise.

4.3 DC Biasing

Two further problems need however to be solved with the solution of Fig. 2. The virtual ground of residue amplifier may not be left floating, thus a resistive path must be created towards it in order to properly bias the amplifier. Moreover the digital output must have its DC level centered around the middle of the range of CDAC in the feedback path, in order to maximize the tracking range. Fortunately, these two issues can be solved together by adding a second feedback path, called resistive path (Fig. 3).

First the value corresponding to the middle of the range of CDAC is subtracted from the unfiltered output code. The difference is then low-pass filtered by the digital decimation low pass filter producing the filtered output code. The filtered output code is then accumulated, and proportional and integral components summed up with appropriate Kp and Ki gain coefficients. The sum is then truncated within

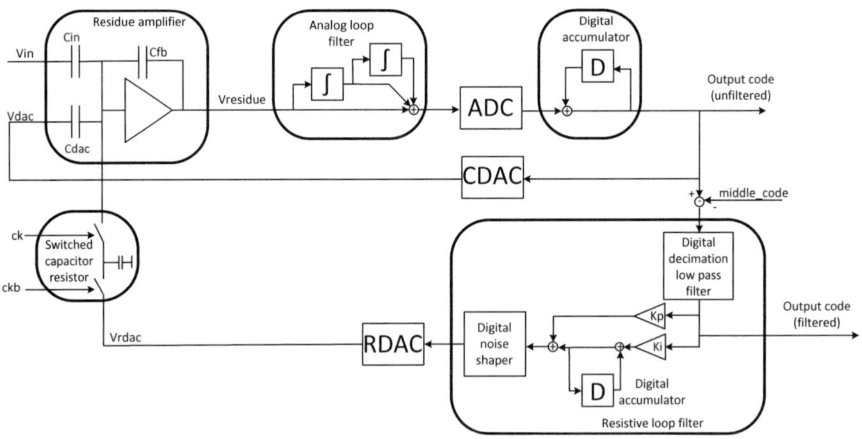

Fig. 3 New proposed architecture with the residue amplifier within a delta loop, with additional loop for resistive feedback

a digital noise shaper in order to match the word length of the second DAC called RDAC (DAC of resistive path) while rejecting the truncation noise at higher frequencies. Finally a resistive path made of a small switched capacitor is inserted between the output of the RDAC and the virtual ground of residue amplifier.

Hence charge will be injected into the virtual ground by adjusting the RDAC input code until the digital output code is centered around the middle range of the main CDAC. This second loop also allows compensation of any leakage current on the virtual ground node.

The advantage of this solution is that the time constant for the resistive feedback path can be adjusted digitally by controlling the gains Kp and Ki. Thus a very high time constant can be achieved without requiring huge resistors and capacitors as with the analog solution. Only very low frequency components (a few Hz or tens of Hz) may be sent back through this path to the residue amplifier, solving the aliasing issue with the switched capacitor resistor. In fact, this resistive path mainly tracks leakages on virtual ground of residue amplifier.

5 Practical Implementation

The loop was implemented in a 65 nm technology. In order to achieve accurate signal tracking and high gain in the first stage, a CDAC resolution of 9 bits and an oversampling factor of 128 were selected, leading to a typical clock rate for the loop of 2.56 MHz for 20 kHz Nyquist rate. The ADC resolution of 4 bits (15 levels, corresponding to an error between −7 and +7 lsb) was selected in order to allow proper tracking of signals up to full scale at the upper limit of the bandwidth, typically at 10 kHz.

In order to properly control the gain of the residue amplification stage, all the capacitors are realized based on the same elementary MOM capacitor cell of 40 fF (Fig. 4). Both Cin and Cdac are made of 512 elementary cells, corresponding to capacitors of 20.48 pF, while the feedback capacitor Cfb is made of 24 elementary cells, thus 0.96 pF, leading to a gain of 21.33 in this stage. Cfb can optionally be increased by 2 by connecting cfb_sup in parallel with it in order to reduce the gain of the stage by 2.

In order to allow adjustment of the overall gain of the front end, the input capacitor Cin can be programmed, by connecting some of the capacitors to ground instead of to the input signal.

The CDAC is directly included within the residue amplifier, being implemented by connecting to the reference voltage a number N of elementary capacitors corresponding to the input code, the 512-N other elementary capacitors being tied to ground. The selection of the cells is performed by thermometric coding in order to guarantee the monotonicity and the DNL, though this implementation is more expensive in area.

The core of the residue amplifier is shown in Fig. 5. It is a two-stages amplifier. In order to limit the transistors contributing to noise, the first stage has been realized

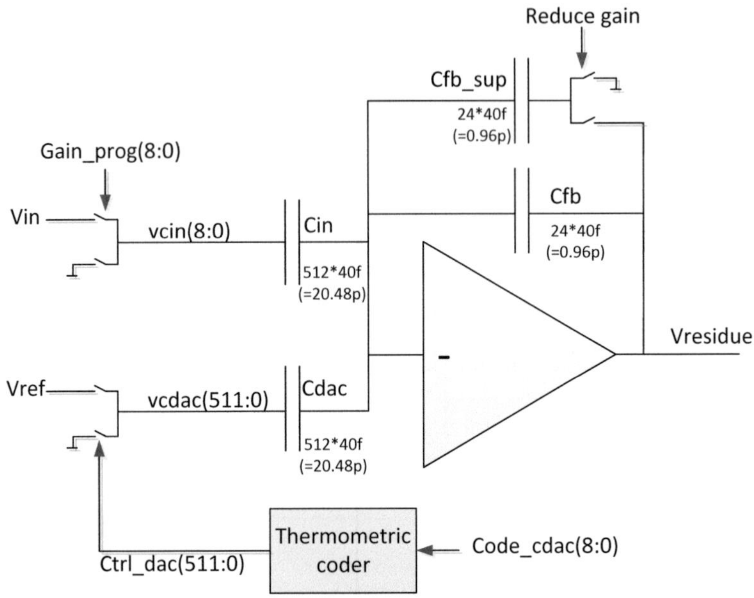

Fig. 4 Residue amplification stage, including the capacitive DAC

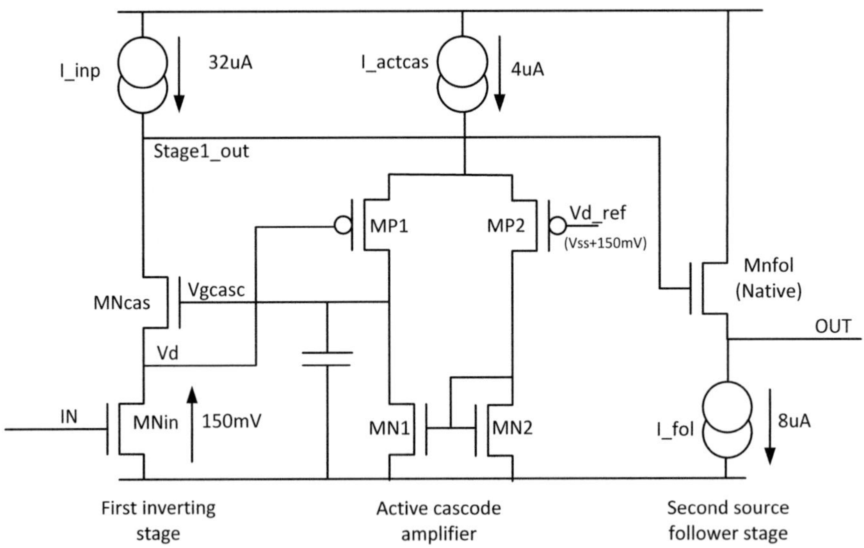

Fig. 5 Core amplifier of the residue amplification stage

by a single common source transistor MNin loaded by current source I_inp. This stage is biased at a relatively large current, 32 μA, because it is the most critical with respect to thermal noise. Moreover, MNin gate has a large gate area in order to limit flicker noise and is realized with thick oxide for minimization of leakages. In order to achieve high DC gain and minimize coupling due to parasitic gate drain capacitance, drain voltage of MNin is regulated at a voltage close to 150 mV by cascode transistor MNcas and active cascode amplifier MP1-MP2-MN1-MN2. As the first stage is inverting, the second stage must be non-inverting and is realized with source follower transistor MNfol, of native type in order to have very low gate-source voltage and optimize the output swing.

Beside MNin, the second main noise contributor is current source I_inp, because biasing the corresponding transistors in very strong inversion is not compatible with low supply voltage.

The principle of the bias current generation is shown in Fig. 6. The reference current is generated by applying a reference voltage (0.6 V) across a reference resistor Rref through a loop with error amplifier and a PMOS current mirror MPS1–MPS2– … –MPSn, such that several copies of the current through the reference resistor are generated. In order to achieve a high PSRR the drain voltages of the PMOS current mirrors are regulated at 200 mV below positive supply through active cascode loops.

As the current source I_inp is the most critical one and biases the input stage of the residue amplifier, it has its own active cascade amplifier aux_ota2 in order to boost the DC gain of the residue amplifier. The other less critical current sources have simple cascode transistors with their gates tied to that one of MPC1 transistor in the input branch.

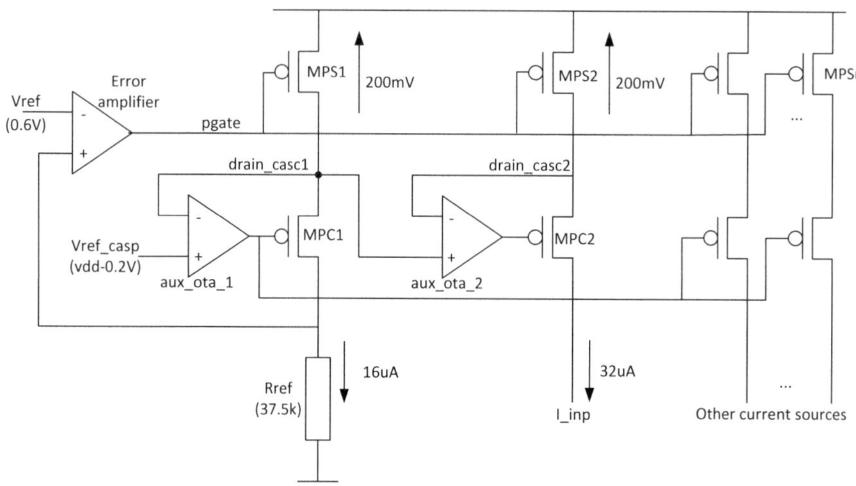

Fig. 6 Principle of bias current generation

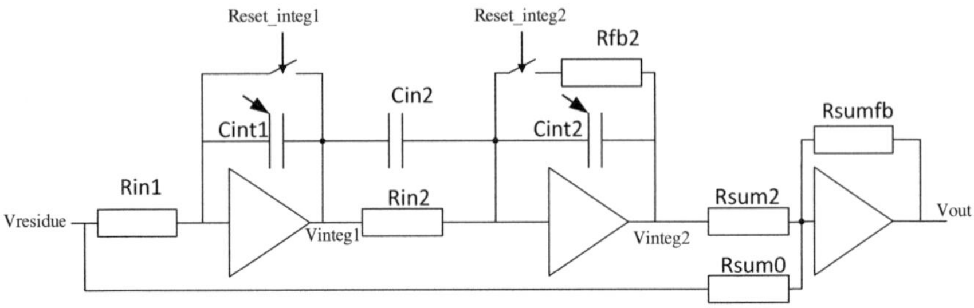

Fig. 7 Principle schematic of the analog loop filter

As the PMOS current mirror cannot be biased really in strong inversion for low voltage operation, its thermal noise contribution is important, especially that one of MPS1 and MPS2. For this reason, half of the bias current I_inp, thus 16 μA, is spent in the current branch with MPS1-MPC1 and reference resistor, as a trade-off when optimizing current consumption and thermal noise. Moreover, even with 200 mV voltage drop, the transistors of the current mirror are not deeply saturated and the current remains depending on the drain voltage, such that the noise of the active cascade transistors must also be taken into account. The noise contribution of the first active cascode amplifier aux_ota_1 can however be compensated by tying the positive input of second active cascode amplifier aux_ota_2 directly to drain_casc1 instead of vrefp_casc, improving the current mirror.

The analog loop filter (Fig. 7) is realized with RC integrators and performs a second order integration for noise shaping of the quantization noise. A third stage is required in order to sum up the residue with the output vinteg2 of the second integrator. The output vinteg1 of the first integrator does not need to be summed up, at is passed with inversion directly to the second integrator through capacitor Cin2 and Cint2. All the amplifiers used within the analog loop filter have the same structure as that one within the residue amplification stage, but with much less current, being less critical with respect to thermal noise. The integrating capacitors Cint1 and Cint2 are adjustable in order to tune the gain of the integrators as a function of the oversampling frequency. In situation of large signals overloading the delta loop, detected by saturation of the ADC during a few consecutive cycles, the integrators may be reset automatically in order to re-stabilize the loop. In such a case, first the second integrator is reset, and then, if still necessary, also the first integrator. A resistor Rfb2 has been connected serially with the reset switch of the second integrator, such that this stage then behaves as a simple inverting stage in order to still pass the output of the first integrator through the chain. The reset is automatically released when tracking is detected.

The ADC is a coarse flash with 14 comparators, producing an output code between −7 and +7 corresponding to the tracking error in lsb's. Power is optimized by enabling dynamically only the comparators near the detected transition.

Concerning the resistive feedback path, the RDAC is a classical 9 bits resistive ladder spanned between regulated supply voltage and ground in order to guarantee that leakages can always be compensated, and the switched capacitor resistive branch it drives is programmable with up to seven capacitors of 18 fF each. However, proper operation of the circuit was still achieved while de-activating all these capacitors, the parasitic capacitance of the node and related switches estimated to around 10 fF being sufficient to compensate the leakages.

6 Measurement Results

The acquisition chain has been implemented in a 65 nm technology. The area, without the reference and biasing circuitry, is around 0.3 mm^2, with two third of it being occupied by the CDAC and the related thermometric decoding. The current consumption per channel is around 80 µA, with close to 50 µA for the residue amplifier, 18 µA for the analog loop filter and 12 µA for the coarse ADC, and less than 1 µA for the resistive DAC. To this, one should however add the consumption of the buffer driving the CDAC (around 12 µA), and the generation of reference voltages and currents.

Figures 8 and 9 respectively show the signal to noise ratio and the signal over noise + harmonic distortion ratio, for a 1 kHz input signal. The full range corresponds to −12 dBV (700 mV peak to peak) and the dynamic range to 98 dB.

Figure 9 shows that the distortion becomes dominant over noise above −40 dBV. However, even when approaching the full scale, the distortion remains more than 70 dB below the signal. No significant performance degradation was observed for input signal up to Nyquist rate (10 kHz), even close to full scale, which means that the feedback path still tracks the input signal under these conditions.

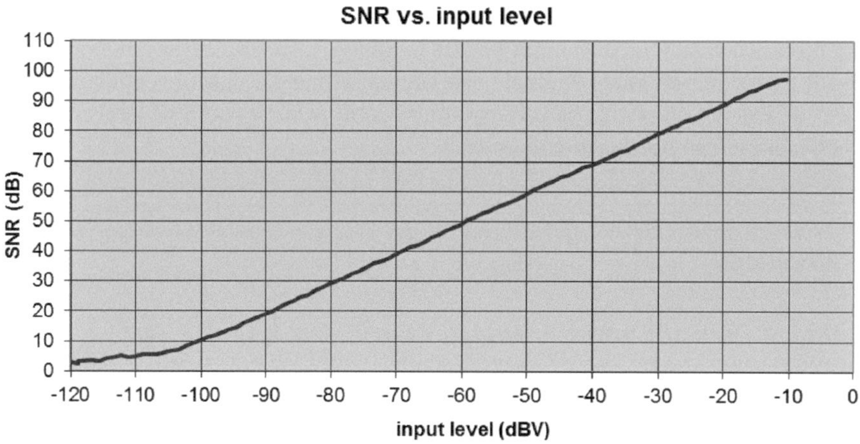

Fig. 8 Measured signal to noise ratio

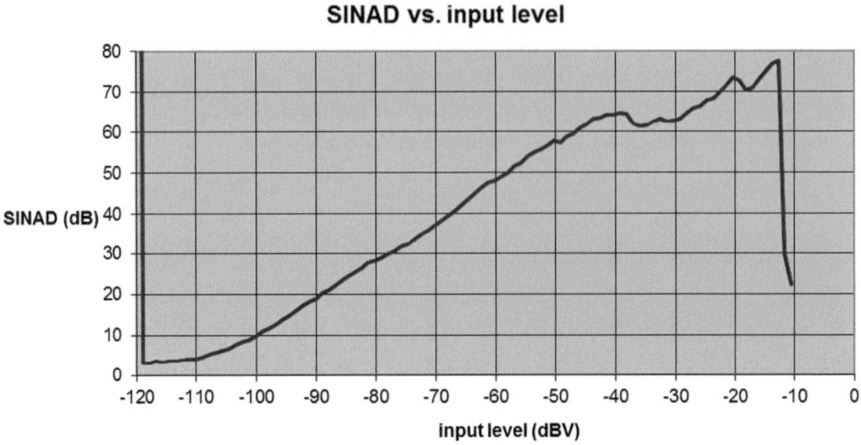

Fig. 9 Measured signal over noise plus harmonic distortion ratio

7 Conclusions

When realizing ultra-low power low voltage high resolution front ends, the main issue to fight against is thermal noise, which is inversely proportional to current consumption. It is thus of prime importance to select an architecture which minimizes the number of devices significantly contributing to noise, and hence requiring more power. A single ended implementation based on delta modulation was selected, the high gain in the first stage measuring the error being made possible by a precise signal tracking by the feedback path. Spreading the available current consumption between the most critical current branches requires a careful optimization. Low voltage sets additional constraints, as unfortunately transistors acting as current sources cannot be put in strong inversion or degenerated, and hence their contribution to thermal noise cannot be neglected. More complex active cascode biasing must then be used in order to still achieve high DC gain and high PSRR, though with reduced voltage drop.

References

1. Dörrer L, Kuttner F, Santner A, Kropf C, Hartig T, Torta P, Greco P (2006) A 2.2mW, continuous-time sigma-delta ADC for voice coding with 95dB dynamic range in a 65nm CMOS process. In: Proceedings of the 32d European solid-state circuits conference, pp 195–198
2. Jiang X, Kim MG, Cheung F, Lin F, Zheng H, Chen J, Chen A, Cheung D, Abdelfattah K, Lee S, Huang H, Kasichainula K, Cong Y, Wu J, Lee CH, Chih G, Tu Y, Brooks TL, Jiang E, Kong H, Zhao C, Keskin M (2012) A 40 nm CMOS analog front end with enhanced audio for HSPA/EDGE multimedia applications. In: Proceedings of the European solid-state circuits conference, pp 414–417

3. Klootsema R, Nys O, Vandel E, Aebischer D, Vaucher P, Hautier O, Bratschi P, Bauduin F, Van Oerle G, Jakob A, Menzl S (2000) Battery supplied low power analog-digital front-end for audio applications. In: Proceedings of the 26th European solid-state circuits conference, pp 156–159
4. Nys O. (2007) Electronic circuit for the analog-to-digital conversion of an analog input signal. European Patent EP1869771 B1

Design and Technology for Very High-Voltage Opamps

Giulio Ricotti, Dario Bianchi, Fabio Quaglia, and Sandro Rossi

Abstract This paper address two main aspects of the design of High Voltage Operational Amplifiers, both oriented to obtain high linearity. The first one is related to extend the bandwidth working on the transconductance as functions of the strongly nonlinear parasitic capacitance. The second describes a technique to draw the integrated feedback network based on a resistive voltage divider with particular focus to low power dissipation and high linearity.

1 Introduction

The description of HV-OpAmp and High Linearity feedback is going through two example of possible applications, one is for linear pulsers in ultrasound echography medical tools and the other is in MEMS actuators driving. Both systems requires HV driving in the range between 100 and 200 V peak-to-peak.

2 HV OP-Amp

The feedback amplifier uses a high gm transconductor, employing thin oxide BiCMOS devices, and a trans-impedance stage using long length, high-voltage transistors [3]. The solution is aimed at maximizing the operating frequency with minimum quiescent power and harmonic distortions meeting requirements for high performance ultrasound imaging systems. The amplifier has been designed in BCD6-SOI technology provided by STMicroelectronics. The technology has two poly and four metal layers and embeds 5 V npn bipolar devices and 0.35 μm CMOS transistors with a nominal supply of 3.5 V [4]. Employed power DMOSFets have 1 μm minimum channel length and support a maximum V_{DS} of 100 V. The maximum cut-off frequencies, at the overdrive voltage of ~2.5 V, are 6 GHz and

G. Ricotti (✉) • D. Bianchi • F. Quaglia • S. Rossi
Smart Power Designers in STMicrolectronics, Cornaredo 20010, Italy
e-mail: giulio.ricotti@st.com

K.A.A. Makinwa et al. (eds.), *Efficient Sensor Interfaces, Advanced Amplifiers and Low Power RF Systems*, DOI 10.1007/978-3-319-21185-5_10

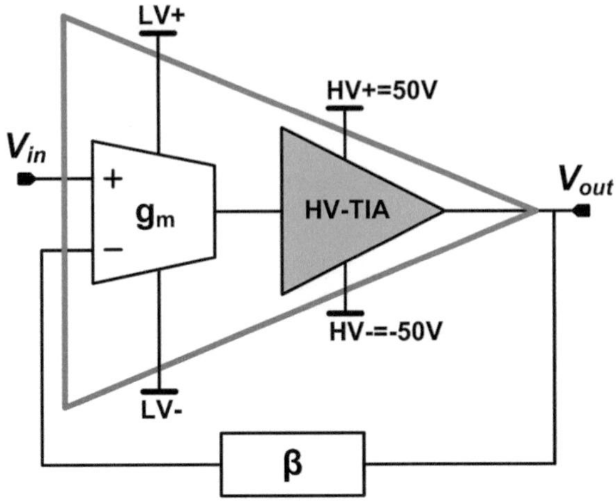

Fig. 1 Block diagram of the proposed high-voltage feedback linear amplifier

2.2 GHz for nDMOS and pDMOS, respectively. Looking at Fig. 1, the amplifier uses two different supplies. The high-voltage trans-impedance stage operates under the maximum supply voltage of ± 50 V but for minimum quiescent power consumption, devices are all biased in sub-threshold (class-B).

The transconductor [1], uses the ± 3 V low voltage supply in order to save power consumption while using large biasing current for maximum g_m. The transconductor is in class-AB, enabling peak output currents larger than biasing currents to faster charge and discharge of large parasitic capacitors at the high impedance internal node of the trans-impedance stage where the current signal develops high voltage swing. The details of the trans-impedance stage are reported in Fig. 2. All devices are high-voltage DMOSFets, except M_1 and M_2 which are thin-oxide devices for minimum input impedance. Node A bridges the two circuit sections: half positive and negative sinusoids of current injected from the transconductor are absorbed by common-gate devices M_1 and M_2, respectively. M_3 to M_6 shield the drain of M_1 and M_2 sustaining the large voltage drop. Transistors M_4 and M_5 (M_7 and M_8) mirror the half signal current with unity gain so as to develop the high voltage swing at node X. A complementary source-follower M_{10}–M_{12} with transistors biased at the threshold voltage by diode connected M_9–M_{11}, drives the off-chip load made of resistor R_L and capacitor C_L, which emulate the ultrasound transducer impedance.

2.1 *Large-Signal Frequency Response and Circuit Design*

Small-signal approximation is a common analysis technique used to model non-linear devices. The linearization is carried out at the DC bias points and can be accurate for small excursions about this point. The trans-impedance of the proposed amplifier works in class B and the assumption of small signal falls

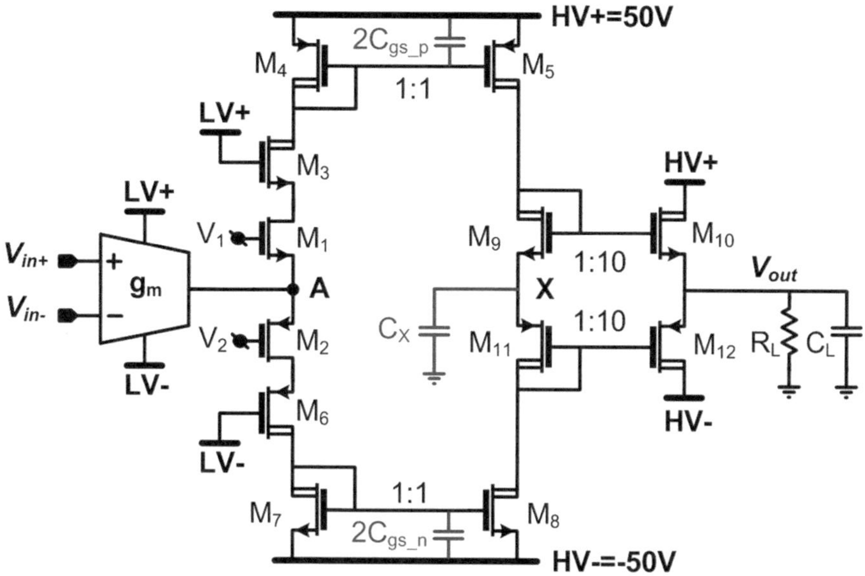

Fig. 2 Schematic of the high-voltage trans-impedance stage

because device currents and voltages significantly change depending on the input signal amplitude. Circuit analysis to estimate gain and frequency response requires inspecting device characteristics under the effect of a large signal i.e. voltage and current magnitudes in excess of biasing standing value. To this purpose we can employ the describing function approach [2], i.e. model the input–output characteristic of each device disregarding all the output harmonic components except the one equal to the input signal frequency. The device describing function depends on the input amplitude whereas information about harmonic components generated by non-linear mechanisms are set aside. The purpose of the analysis is to predict the open-loop gain-bandwidth product of the amplifier and its dependency from the signal amplitude in order to derive useful guidelines for circuit design. Looking at Fig. 2, the input transconductor is assumed to have a constant (signal independent) transconductance, gm. The output current develops the voltage swing (V_X) at the high impedance node X, driving the output buffer. The load capacitance (C_L) introduces a pole at high frequency, beyond the amplifier gain-bandwidth product and the latter is therefore determined by the capacitance of the internal high impedance node, C_X:

$$GBW \approx \frac{g_m}{2\pi C_X} \tag{1}$$

C_X is made of parasitic (C_p) of devices M_5, M_8, M_9, M_{11} plus the loading effect (C_{buf}) of the complementary source follower M_{10}–M_{12}. The latter changes significantly when the signal is applied because the gate to source capacitance is bootstrapped while entirely loading the node without applied signal, i.e. when devices work in sub-threshold. Referring to Fig. 2, the capacitance seen at source follower input is derived as:

$$C_{buf} = \frac{C_{gsN} + C_{gsP}}{1 + G_M R_L} \tag{2}$$

C_{gsN} and C_{gsP} are the gate to source capacitance of $N(M_{10})$ and $P(M_{12})$ output stage devices respectively, and $G_M = (I_{out}/V_{od})|_{\omega 0}$ is the trans-conductance describing function, derived as follows. Assuming M_{10} and M_{12} are biased at the threshold voltage by the diode-connected M_9–M_{11}, with a sinusoidal overdrive voltage, $V_{od}(t) = V_{od}\sin(\omega t)$, the output current is

$$I_{out}(t) = \begin{cases} \frac{\beta_n}{2} V_{od}^2 \sin^2(\omega_0 t) & \text{for } 0 \le \omega_0 t < \pi \\ \frac{\beta_p}{2} V_{od}^2 \sin^2(\omega_0 t) & \text{for } \pi \le \omega_0 t < 2\pi \end{cases}$$

Assuming $\beta_n = \beta_p = \beta$, the fundamental Fourier component of the output current, at ω_0, is defined by:

$$I_{out}|_{\omega 0} = \frac{4\beta V_{od}^2 \sin(\omega_0 t)}{3\pi} \tag{3}$$

The expression for the transconductance describing function follows as

$$G_M = \frac{4\beta\,|V_{od}|}{3\pi} \tag{4}$$

revealing a direct dependence on the developed V_{od} signal, as intuitively expected. By circuit inspection, V_{od} can be easily expressed as a function of the output voltage amplitude and replacing (4) in (2), the higher the signal amplitude the lesser the input referred capacitance because bootstrap of C_{gs} is more effective. Figure 3 shows calculated and simulated $C_X = C_{buf} + C_p$ versus the output amplitude. A capacitance C_p of 6.5 pF representing the parasitic of devices M_5, M_8, M_9, M_{11} in the high impedance node is taken into account. A very good agreement is evident and proves that the describing function approach is able to capture the effect of large signal operation.

The target closed loop gain of the amplifier is 40 dB with a 3 dB bandwidth larger than 5 MHz. This sets a minimum required gain-bandwidth product, GBW > 500 MHz. From Fig. 3, the lower is the output voltage amplitude and the larger is the high impedance node capacitance. Assuming a minimum delivered output voltage of $2V_{pk\text{-}pk}$, $C_X = 13.5$ pF. From (1) the minimum g_m required for the low voltage transconductor to meet the required GBW is 42 mS. A value of 60 mS has been selected to keep some margin considering also the impact of other non-idealities, such as slew-rate and secondary poles, on the closed loop

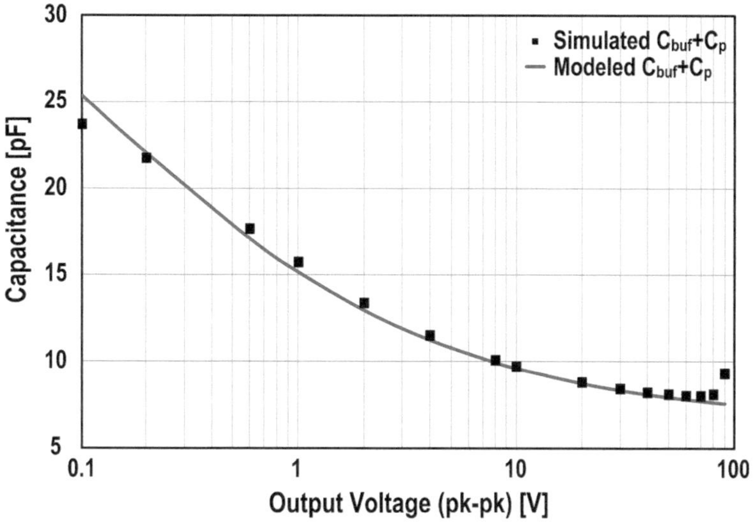

Fig. 3 Calculated and simulated equivalent capacitance at high impedance node versus the sinusoidal output voltage amplitude

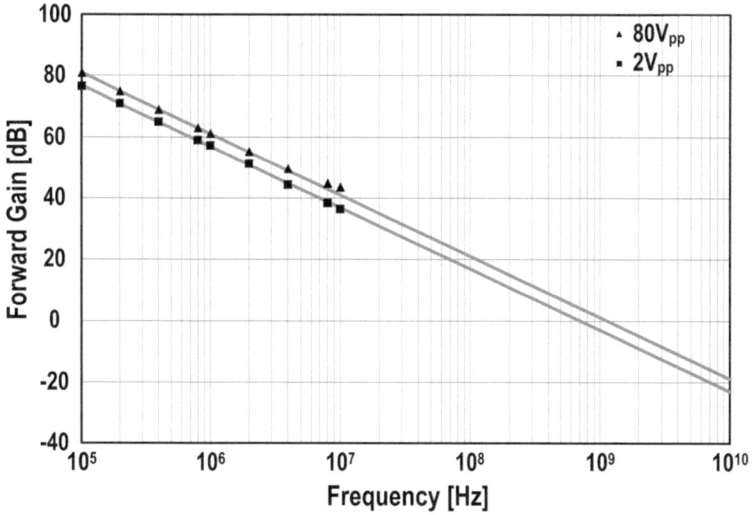

Fig. 4 Simulated amplifier gain—bandwidth product at 2 and 80 V output amplitude

bandwidth. Figure 4 shows the simulated open-loop gain versus frequency for output voltage swings of 2 and 80 V assuming a transconductance stage of 60 mS with no bandwidth limitation and the complete trans-impedance stage of Fig. 2. The extrapolated GBW are 713 MHz and 1.2 GHz for 2 V and 80 V respectively. The corresponding capacitances at the high impedance nodes are 13.4 pF and 8.1 pF respectively, and the results of Figs. 3 and 4 are in good agreement. Notice that up to 10 MHz no secondary poles emerge. The describing function can be applied to estimate the frequency location of the secondary poles introduced by the common-gate and current-mirror devices in Fig. 2 under large signal operation. It can be verified that also at moderately low output voltages, the branch current amplitude is high enough to push the secondary poles beyond 10 MHz.

The selected circuit topology for the low-voltage transconductor introduces an additional pole at relatively low frequency which could affect the closed loop stability. Its effect is compensated using lead technique through an off-chip capacitor feedback capacitor. The capacitor determines a pole at ∼8 MHz in the closed-loop transfer function.

2.2 Experimental Results

The proposed operational amplifier has been fabricated by STMicroelectronics and the chip photograph is shown in Fig. 5. Realized prototypes have been tested in standard ceramic dual-in-line packages without any heat sink. To avoid excessive self heating, measurements of the frequency response and output power have been carried out with a pulsed input signal having a duty cycle less than 10 %. The feedback network is off-chip and defines a closed loop gain of 40.9 dB. A load impedance, comprising a 100 Ω resistor in parallel with a 150 pF capacitor, emulating the transducer's impedance, has been used in all experiments. The high

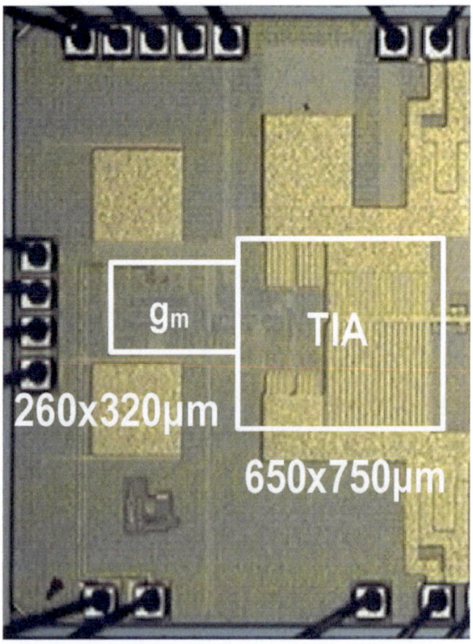

Fig. 5 Chip Micrograph

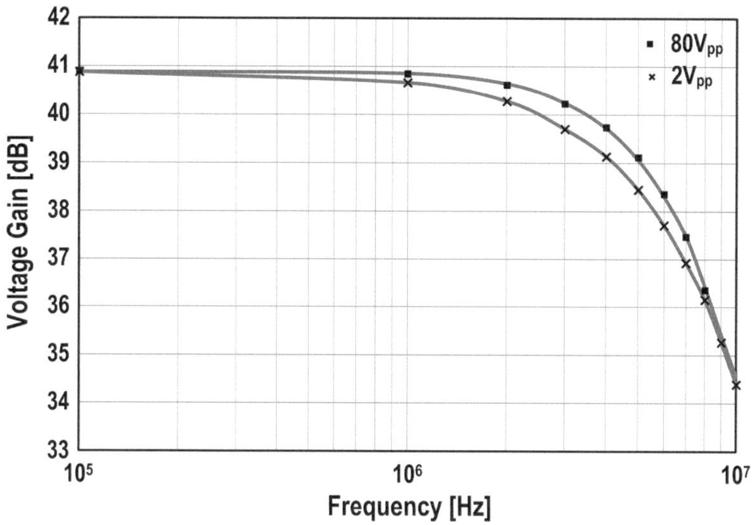

Fig. 6 Measured (*dots*) and simulated (*continuous line*) closed loop frequency response for output voltage swings of 2 and 80 V

and low voltage supplies, set to ±50 V and ±3 V respectively, deliver static currents of 100 μA and 4.5 mA leading to a quiescent power dissipation of 37 mW.

Figure 6 shows the amplifier closed loop frequency response for output voltage swings of 2 V and 80 V, respectively. Simulations are also reported for comparison showing a very good agreement. The −3 dB bandwidths are 5.5 MHz and 6.5 MHz for 2 V and 80 V output swings, respectively. Considering the frequency location of the pole introduced by the off-chip compensation capacitor, and on the gain-bandwidth products of Fig. 4, the closed loop bandwidth estimates are 5 and 6.5 MHz i.e. in good agreement with measured results. The maximum measured output signal amplitude is 90 V. The positive and negative slew rate, measured at maximum output voltage, are +2 kV/μs and −2.2 kV/μs respectively. This reflects into a full swing maximum sinusoidal frequency of less than 8 MHz not to be subject to slew rate limitations.

Linearity performances have been analyzed through second harmonic distortion measurements. Gaussian envelops, typical of ultra-sound systems, have been used for testing. Screenshots of the oscilloscope, showing the output signal and the Fourier Transform for a 2 MHz sinusoid having Gaussian envelope for various pk-to-pk amplitudes, are shown in Fig. 7.

Linearity measurements vs. frequency are reported in Fig. 8 for an $80V_{pk\text{-}pk}$ output voltage and two different closed loop gains of 40 dB and 46 dB, respectively. Being the output voltage constant, the loop gain reduces correspondingly by 6 dB.

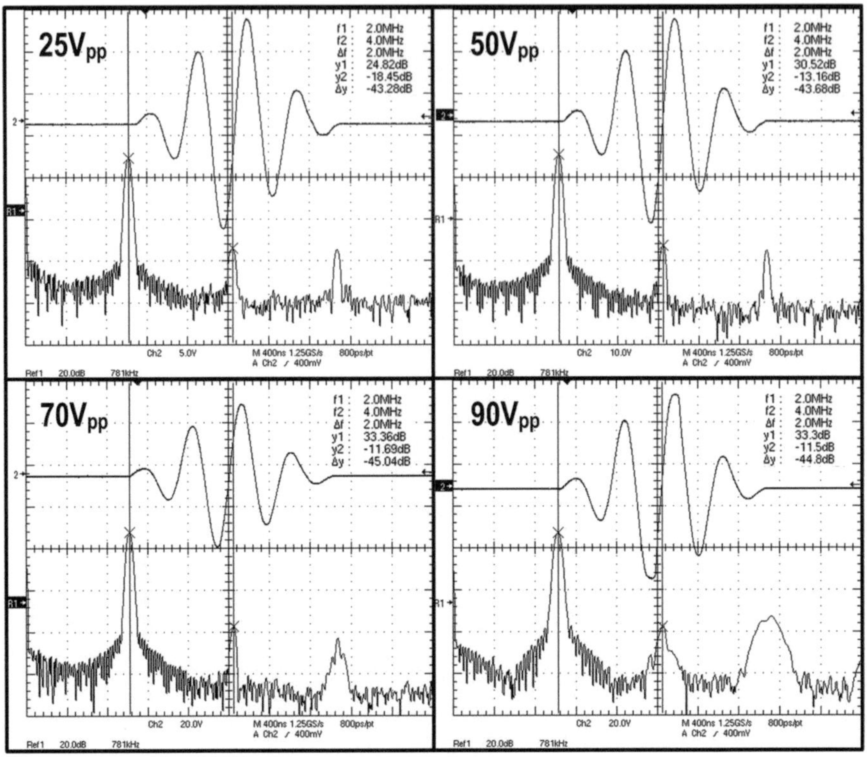

Fig. 7 Time-domain response and DFT of various peak to peak sine output signals at 2 MHz with Gaussian envelope

3 The Integrated Feedback Network in HV Operational Amplifier

Many actual projects are oriented to drive MEMS (Micro Electro Mechanical System) actuators by electrostatic force developed between stator and rotor electrodes, for example like comb fingers. To generate enough actuation force the driving voltage is high, in the range of 100–250 V.

The required driving voltage, generally is and arbitrary waveform and the linearity response of the HV-OpAmp is a key parameter.

This kind of ASICs generally are entering in portable devices, it means they are supplied by a battery with typical voltage range of 2.5–4.2 V. To generate the HV-Supply there is a DC/DC converter on the same chip that inject switching noise in the common substrate.

Another key parameter is the low noise, in terms of thermal noise, 1/f noise and also in terms of noise immunity from the power supply or from the substrate.

The load of these HV-drivers is a strongly not linear capacitor in the range of few pico-farad.

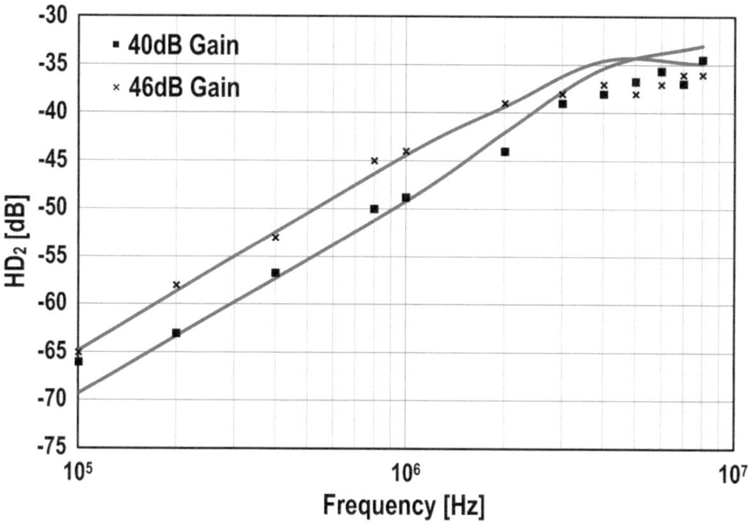

Fig. 8 Measured (*dots*) and simulated (*continuous line*) and HD_2 versus frequency at $80V_{pk\text{-}pk}$ output voltage for two different closed loop gains of 40 and 46 dB

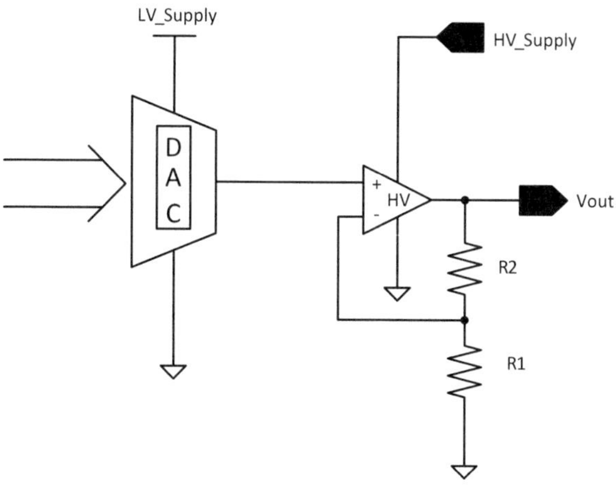

Fig. 9 General example of HV driver

The current consumption of the OpAmp must be very low, typically 2–4 μA including the load power but excluding the resistive feedback network.

To satisfy all these specifications, the important element is the feedback network, the voltage divider done by R1 and R2 as reported in Fig. 9.

The signal quality is strongly related to the quality of the voltage divider.

First of all the divider has to dissipate as low as possible power but it's a tradeoff between resistor value, silicon area and performances.

For example 20 μA at Vout = 200 V having R1 = 100 K and R2 = 9.9 M and the DAC voltage is multiplied by 100.

There are many issues to keep into account and to manage.

The matching must be in the range of 1 % at 3σ, with so large area of the divider the parasitic capacitor between the body of the resistor modules and bottom layers risks to assume a large value that strongly influence the HV Op-Amp stability and bandwidth. The linearity remain the key issue, because it's simple to analyze that across R1 the maximum voltage is the DAC output, due to the virtual short in closed loop, that is maximum 2 or 2.5 V, but across R2 the voltage is strongly variable from 0 to 200–250 V.

In the integrated resistors there is a strong modulation effect due to the voltage contrast between the resistor body and the bottom layer or the adjacent resistor.

In the following is reported a real example of the implementation of all the technique to adopt to target high quality feedback voltage divider.

The resistor structure that has been chosen for a single module is represented in Fig. 10 and exploits the poly-poly cap structure as a resistance. The poly//poly_cap layer, which has the same doping concentration of HIPO (High Poly Resistor), is the resistor body while the field oxide under the bottom plate is used to sustain high voltage fields.

To increase the attenuation of noise coming from substrate a pwell is included but it should be connected to ground and not left floating in order to avoid the breaking of pwell-nwell junction when a switching signal is applied.

Pwell, nwell and isolation trench are common to all resistors so the parasitic model should be the one represented in Fig. 11.

R1 and R2 have been considered as done by 25 KΩ modules and in particular R1 is made up by 16 modules (4 + 4 in series connected in parallel with other 4 + 4 in series) while R1 is made up by 396 modules connected in series.

In order to have the best matching configuration all those resistors are connected in a matrix of 4 modules per raw for a total of 103 rows so divided: 12 rows for R2, 1 row for R1, 25 rows for R2, 1 row for R1, 25 rows for R2, 1 row for R1, 25 rows for R2, 1 row for R1, 12 rows for R2.

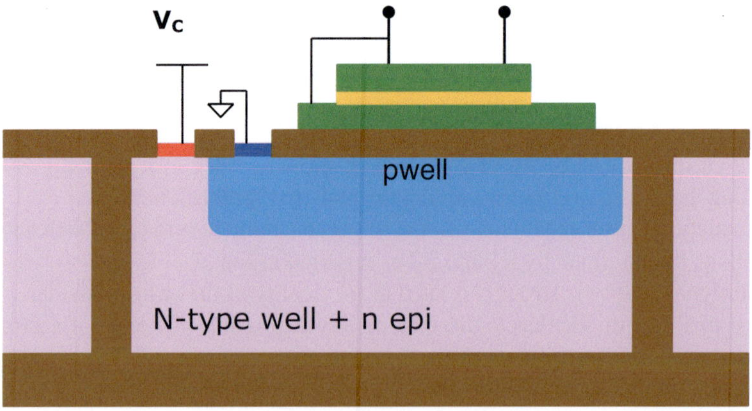

Fig. 10 Single module of the voltage divider, in a SOI technology, with Nwell biased at LV clean supply and a Pwell biased at GND to shield against substrate noise

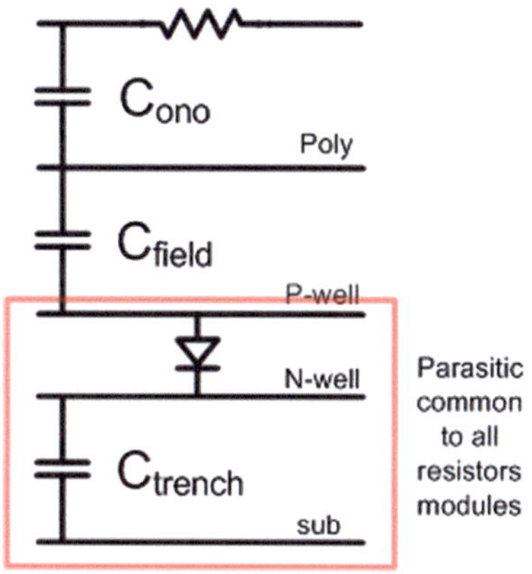

Fig. 11 Equivalent circuit associated to the single resistor module including the parasitic components

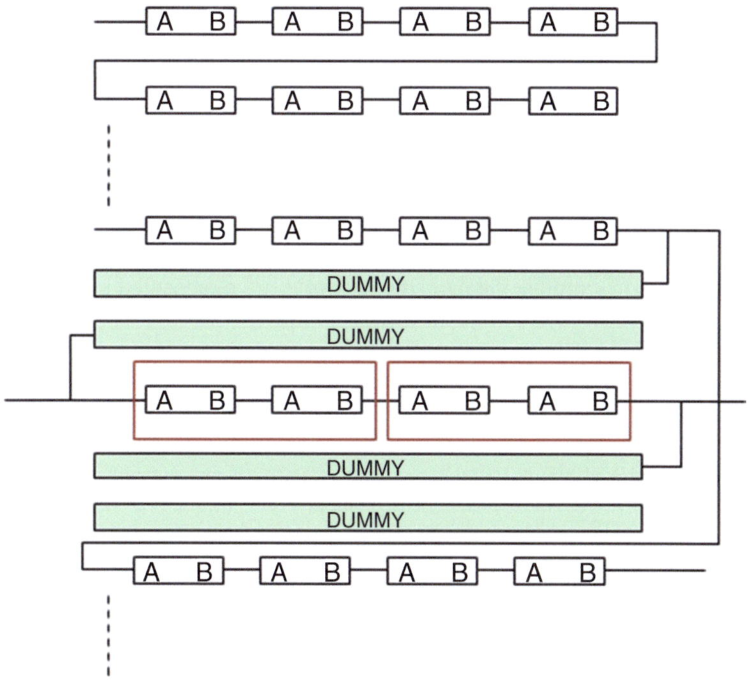

Fig. 12 Singularity in the modules between R2 modules and R1 modules where a huge voltage contrast appears

Where there is an interface between R1 and R2 modules a dummy structure is required (Fig. 12) in order to avoid the depletion of real resistor modules due to high field seen by resistors themselves. So dummies are used only to sustain the depletion effect of high fields.

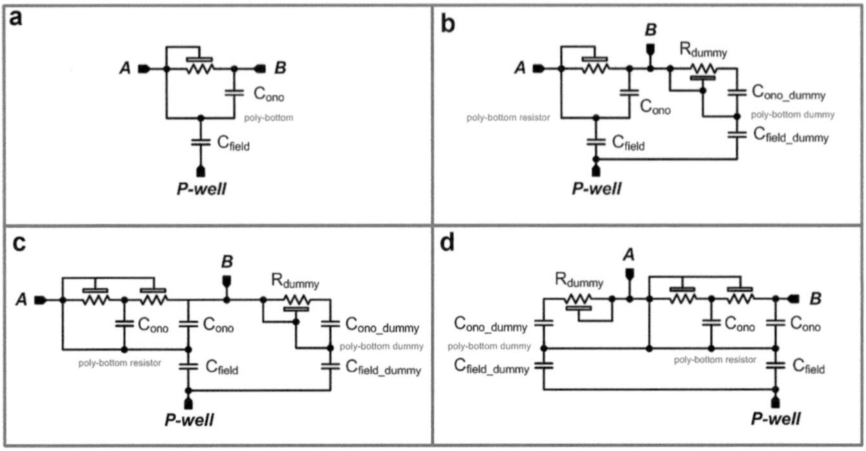

Fig. 13 Resistor module for R2, for R2 connected to dummy, for R1 with the two different dummies connections

In addition, resistors modules making up R1 have a slight different structure compared to those used for R2 since they see nearly half the voltage drop across their terminals. To avoid different voltage modulation effects they have been designed as modules of two resistors of 25 kΩ sharing the same poly bottom layer. The schematics used to simulate the parasitic networks are represented in Fig. 13.

References

1. Su D, McFarland W (1997) A 2.5V, 1W monolithic CMOS RF power amplifier. In: Proceedings of the custom integrated circuits conference, IEEE, pp 189–192
2. Sokal N, Sokal A (1975) Class E–A new class of high-efficiency tuned single-ended switching power amplifiers. IEEE J Solid-State Circuits 10(3):168–176
3. Sen S, Leung B (1996) A class-AB high-speed low-power operational amplifier in BiCMOS technology. IEEE J of Solid State Circuits 31(9):1325–1330
4. Bianchi D, Quaglia F, Mazzanti A, Svelto F (2014) Analysis and design of a high voltage integrated class-B amplifier for ultra-sound transducers. IEEE Trans Circuits Syst I 61(7): 1942–1951

Advances in Low-Offset Opamps

Qinwen Fan, Johan H. Huising, and Kofi A.A. Makinwa

Abstract This paper focuses on the design of amplifiers that achieve micro-volt offset by employing dynamic offset reduction techniques, which include chopping, auto-zeroing and chopper stabilization. The working principles and non-idealities of these techniques will be described. The up-modulated offset associated with chopping causes ripple, which can be a significant source of error if not filtered effectively. Thus, various ripple reduction techniques are introduced to suppress this ripple to the micro-volt level. Also discussed is the ping-pong architecture, which enables the realization of auto-zeroed amplifiers with continuous-time behavior. Examples of chopped amplifiers, auto-zeroed amplifiers and chopper stabilized amplifiers are presented, as well as designs in which multiple techniques are combined.

1 Introduction

Low-offset amplifiers are widely required in precision applications such as current sensing, precision sensor readout such as strain gauge, thermocouple, and hall sensor. In these applications, micro-volt offset is often required, as well as low $1/f$ noise since the signals of interest in these applications are not only at DC, but also are at low frequencies. A traditional way to reduce offset is trimming, which usually involves measuring the offset of the opamp at certain temperatures and then cancelling it via programmable devices or settings built inside the opamp [1]. Trimming thus incurs significant extra costs when used in high-volume production. To minimize these costs, digitally-assisted trimming can be employed [2], which involves sampling the opamp's offset during startup and then compensating for it with the help of on-chip analog-to-digital and digital-to-analog converters. However, both types of trimming do not compensate for offset drift due to temperature

Q. Fan (✉)
Delft Lab, Mellanox, Delft, The Netherlands
e-mail: qinwenf@mellanox.com

J.H. Huising • K.A.A. Makinwa
Electronic Instrumentation Lab, Delft University of Technology, Delft, The Netherlands

K.A.A. Makinwa et al. (eds.), *Efficient Sensor Interfaces, Advanced Amplifiers and Low Power RF Systems*, DOI 10.1007/978-3-319-21185-5_11

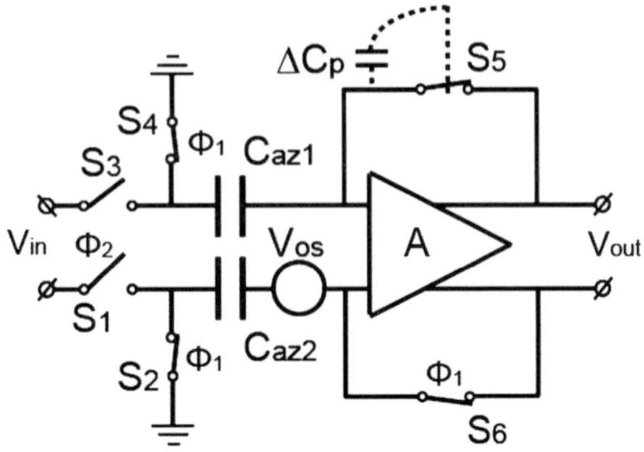

Fig. 1 Basic block diagram of an auto-zeroed amplifier

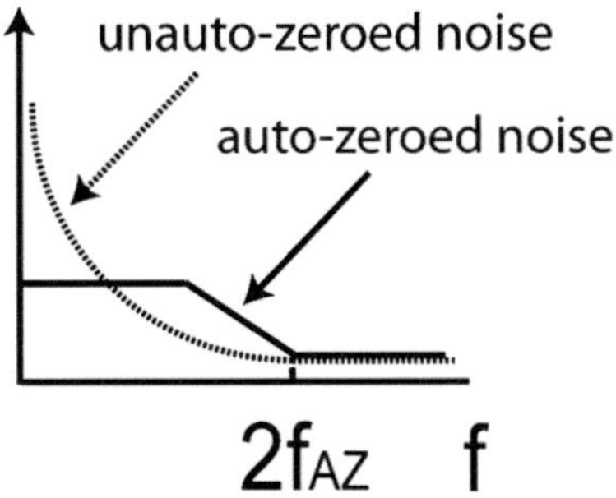

Fig. 2 Noise folding caused by auto-zeroing

variations. Furthermore, they do not reduce 1/*f* noise, which will still be a significant source of error in low-frequency readout systems.

To avoid these drawbacks, dynamic offset reduction techniques can be employed. These include auto-zeroing, chopping and chopper stabilization and will be briefly introduced in the following. These techniques function as long as the circuit is powered, and thus cancel offset and suppress 1/*f* noise at all temperatures.

The basic principle of auto-zeroing is illustrated in Fig. 1. A SC network driven by a digital clock is built around an opamp *A*. In the first clock phase Φ_1, *A* is connected in unity-gain configuration, and its offset is thus sampled on C_{az2} and meanwhile appears at its output. In the next clock phase Φ_2, *A* amplifies the input signal, and its offset V_{os} is cancelled by the voltage stored on C_{az2} in Φ_1. Ideally, the voltage stored on C_{az2} should be equal to V_{os}, and thus, *A* appears to be offset free. The low frequency 1/*f* noise components are also stored on $C_{az1,2}$ and so are canceled. However, the higher frequency 1/*f* noise components are less correlated and so cannot be effectively canceled [3]. Auto-zeroing has a major drawback: increased baseband noise. The sample-and-hold (S&H) action of $C_{az1,2}$ in Fig. 1 will result in noise folding, which increases the noise level at low frequencies [1, 3–5]. This effect is illustrated in Fig. 2. It can be seen that without auto-zeroing, the low frequency noise is dominated by the 1/*f* noise, while with auto-zeroing, the low frequency noise is dominated by the white noise that has been folded back from high frequencies. For the complete (and rather complicated) theory of noise

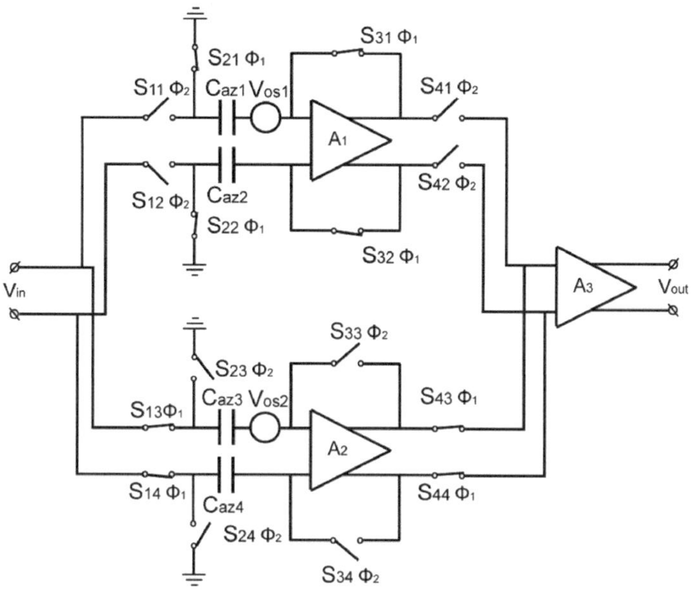

Fig. 3 A ping-pong auto-zeroing topology

Fig. 4 Voltages in the frequency domain in a chopper amplifier

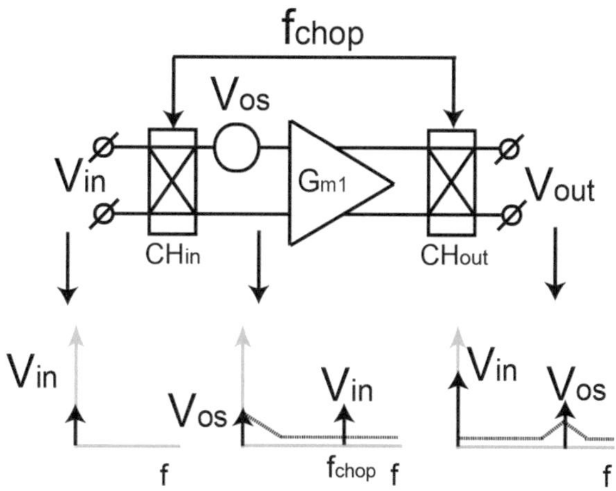

folding, readers are suggested to refer to [3, 5]. Another drawback of auto-zeroing is that it is not continuous time. To obtain a continuous-time operation, a ping-pong structure [6] can be employed which is shown in Fig. 3. In this case, two duplicate input stages are employed. While one stage is amplifying the signal, the other stage is being auto-zeroed. In this manner, a continuous-time operation is obtained.

Chopping involves the use of two synchronized polarity-reversing choppers [1, 3, 4, 7] for precise modulation and demodulation. Each chopper consists of four switches driven by clock signals with two complementary phases at a certain chopping frequency (f_{chop}). When chopping is applied to an opamp, as

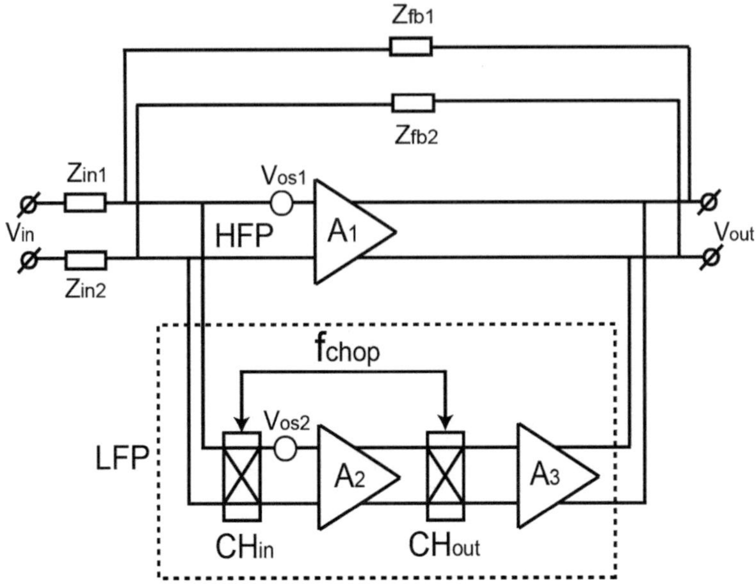

Fig. 5 Block diagram of a chopper-stabilized amplifier with a global negative feedback network

shown in Fig. 4, the input signal is first moved to the odd harmonics of f_{chop} by CH_{in}, then amplified and finally moved back to DC. Meanwhile, the offset and 1/*f* noise of G_{m1} are up-modulated by CH_{out} to the odd harmonics of f_{chop}. Thus ideally, an offset and 1/*f* noise free opamp is obtained. The drawback of chopping, however, is that the up-modulated offset and 1/*f* noise appear as ripple at the output of the amplifier. When the amplifier is followed by a sampling system, the ripple can cause significant errors. Thus, the ripple must be minimized. This will be discussed in Sect. 3.2.

The basic topology of a chopper stabilized amplifier is shown in Fig. 5 [1, 3, 8]. The amplifier consists of two signal paths: a main signal path consisting of A_1, and an auxiliary signal path consisting of A_2, A_3. This topology has been used in several state-of-the-art designs [8–10]. The main signal path provides wide signal bandwidth and is thus often referred to as the high-frequency path (HFP); while the auxiliary path provides low offset, high DC gain and usually has limited bandwidth, and thus is often called as the low-frequency path (LFP). The LFP determines the offset and the low frequency noise of the opamp, while the HFP determines the bandwidth and the high frequency noise of the opamp. To achieve low offset, the offset of the HFP must be taken care of. In the presence of a global negative feedback as shown in Fig. 5, the offset of A_1 (V_{os1}) will be amplified and then fed back to the input of the LFP. Thus, it will be corrected by the high-gain LFP. The residual offset due to V_{os1} can be expressed as [1, 3]:

$$V_{error} = \frac{V_{os1} \times A_1}{A_2 \times A_3}. \tag{1}$$

Thus, as long as there is sufficient gain in the LFP, the residual offset is negligible. It is worth mentioning that the low-frequency 1/*f* noise of the HFP is also suppressed by the LFP in the same manner. The offset of the LFP, however, is removed by chopping. The up-modulated offset and 1/*f* noise of the LFP are not suppressed in this case, and thus will create ripple. The solution to this problem will be presented in Sect. 3.2.

Although auto-zeroing, chopping and chopper stabilization can ideally reduce the offset to zero, they all have their own non-idealities which give rise to residual offset and ripple. This will be discussed in Sect. 2. Design examples employing these techniques will be presented in Sect. 3, followed by a discussion section in regard to the pros and cons of each design in Sect. 4. Finally, a conclusion will be drawn in Sect. 5.

2 Non-idealities of Auto-Zeroing, Chopping and Chopper Stabilization

Both auto-zeroing and chopping have non-idealities which can give rise to residual offset. The main cause of this is the mismatched charge injection and clock feedthrough errors of the switches.

In the case of auto-zeroing, as shown in Fig. 1, an extra parasitic capacitance between the clock line and the input terminal of $S_5 \Delta C_p$, which can be introduced by layout asymmetry, will contribute an error charge on C_{az1} as soon as S_5 is opened. This error charge directly contributes to an offset error, which cannot be eliminated by auto-zeroing itself. With a clock voltage of V_{clk}, the error voltage stored on C_{az1} will roughly be equal to $V_{clk} \times \Delta C_p / C_{az}$. For instance, with $V_{clk} = 1$ V, $\Delta C_p = 1$ fF and $C_{az1} = 1$ pF, the error voltage will be 1 mV. Moreover, when S_5 and S_6 are not totally matched, the channel charge released by S_5 and S_6 when they become open is not equal. The net result is again an error charge stored on $C_{az1,2}$. A more accurate auto-zeroing implementation can reduce such errors and will be presented in Sect. 3.1.

In the case of chopping, a mismatched parasitic capacitance ΔC_{po1} (Fig. 6) from the clock line to one of the inputs of CH_{out} results in an AC current. This current can be modeled as an AC voltage at the input of G_{m1}, which in turn, can be modeled as a residual offset V_{off1} at the input of CH_{in}. This can be roughly estimated as [3]:

$$V_{off1} = \frac{V_{clk} \times \Delta C_{po1} \times 2 f_{chop}}{G_{m1}}. \tag{2}$$

For instance, with $V_{clk} = 3$ V, $\Delta C_{po1} = 1$ fF, $f_{chop} = 30$ kHz and $G_{m1} = 100$ µS, V_{off1} is then 1.8 mV. Similarly, a mismatched parasitic capacitance ΔC_{pi1} from the clock

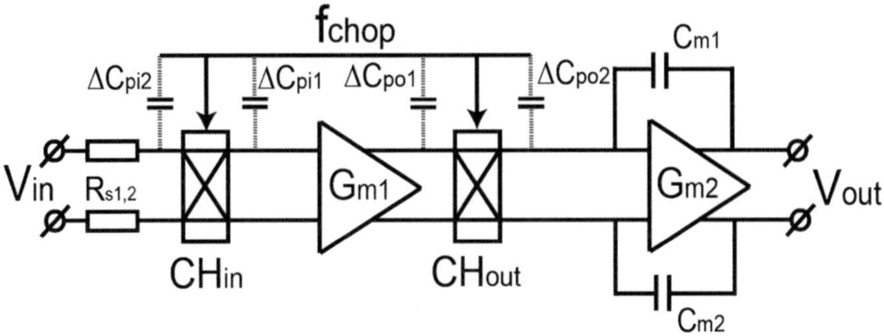

Fig. 6 A chopper opamp with mismatched parasitic capacitors

line to one of the outputs of CH_{in} again results in an AC current, which is then demodulated by CH_{in} and converted into a voltage by the source resistance R_s. Thus, a second residual input offset V_{off2} is obtained, which can be estimated by [3]:

$$V_{off2} = \frac{V_{clk} \times \Delta C_{pi1} \times 2f_{chop}}{R_{on}}. \tag{3}$$

Furthermore, a mismatched parasitic capacitance ΔC_{po2} (Fig. 6) introduces an AC clock feed-through spike, which is then filtered by the integrator built around G_{m2} and appears as an output ripple. The amplitude of this ripple V_{rip1} can be estimated by:

$$V_{rip1} = \frac{V_{clk} \times \Delta C_{po2}}{C_{m1,2}}. \tag{4}$$

Similarly, the mismatched parasitic capacitance ΔC_{pi2} again introduces an AC clock feed-through current spike at the input of the amplifier, which is then converted into a voltage by the source impedance R_s, and then filtered by the whole amplifier and appears as another ripple component V_{rip2} at the output:

$$V_{rip2} = \frac{V_{clk} \times \Delta C_{pi2} \times R_s \times G_{m1}}{C_{m1,2}}. \tag{5}$$

Finally, the mismatch between the chopper switches will result in mismatched charge injection errors, which, as explained above, have the same effects as mismatched clock feed-through.

In the case of chopper stabilization, all the non-idealities associated with chopping will apply. An extra non-ideality is that the DC gain of the LFP is not infinite, thus the offset of the HFP will not be completely eliminated, which will reveal itself as a residual offset, and can be calculated by (1).

To reduce these non-idealities, the clock voltage should be as small as possible and the clock frequency should be as low as possible. The switches layout should be as symmetrical as possible to reduce all the mismatched parasitic capacitances; and the size of the switches should also be minimized to for less charge injection errors.

3 Design Examples of Low-Offset Opamps with Dynamic Offset Reduction Techniques

In this section, two amplifier designs that employ chopping and auto-zeroing will be discussed. Later a design employing only chopping is presented. To eliminate the chopping ripple, a ripple reduction technique is presented. Finally a chopper-stabilized opamp is introduced.

3.1 Two Micro-volt Opamps Featuring Chopping and Auto-Zeroing

From the introduction, it is clear that chopping reduces offset and $1/f$ noise but creates high frequency ripple, while auto-zeroing ideally does not introduce ripple, but suffers from increased base-band noise. When auto-zeroing is combined with chopping, however, the increased base-band noise can be up-modulated to high

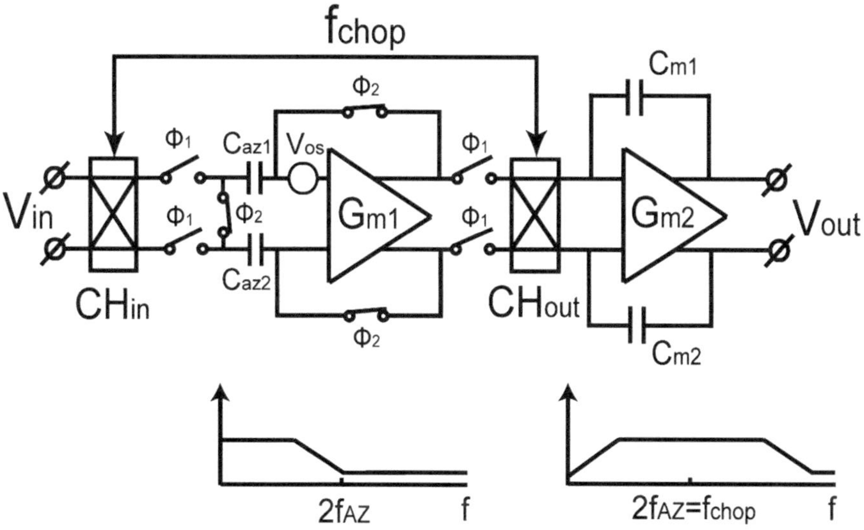

Fig. 7 Block diagram of a two-stage amplifier that employs both chopping and auto-zeroing

frequencies. Thus, a low base-band noise floor can still be obtained. Also the offset is canceled by auto-zeroing, thus the ripple caused by the up-modulated offset is also eliminated. As a result, an opamp with both low offset, low noise and low ripple can be expected when both techniques are employed. One design example is shown in Fig. 7 [11]. It consists of a two-stage miller-compensated opamp. G_{m1} is auto-zeroed and the offset of G_{m2} is suppressed by the gain of G_{m1} when referred to the input of the opamp. Two choppers are added around G_{m1}. The increased base band noise bandwidth in this case is about $2\times$ auto-zeroing frequency. Thus, the chopping frequency is chosen to be two times the auto-zeroing frequency. As a result, low noise is obtained at low frequencies, as shown in Fig. 7.

A second example is shown in Fig. 8 [1, 12], where an offset compensation loop is implemented around the input transconductor G_{min}. During Φ_1, G_{min} is disconnected from the signal source. Its input is shorted so that its offset voltage is converted into an offset current and then integrated on the integrator built around G_{mAZ}. The output voltage of G_{mAZ} is then converted into a current by G_{mc}, which will cancel the offset current of G_{min} completely. In Φ_2, G_{min} is connected to the input signal, and the input of G_{mAZ} is disconnected from the output of G_{min}. The integrator built around G_{mAZ}, however, holds the compensation voltage stored in Φ_2, so that the offset of G_{min} is also compensated in Φ_2. The advantage of using an auto-zeroing loop rather than its simplified counterpart (Fig. 7) is that the errors of auto-zeroing (such as the charge injection and clock feed-through errors of S_{5-8} in Fig. 8) are better suppressed by its high loop gain. In the case of Fig. 7, the charge injection errors will be directly stored on $C_{az1,2}$, thus directly contributing to residual offset errors. Thus, the auto-zeroing loop is often more accurate. The noise bandwidth of the auto-zeroing loop in [12] is determined by the time constant of the loop. By making the settling time of the loop longer, the bandwidth of the loop is reduced. As a result, it is not necessary to choose a chopping frequency equal to twice the auto-zeroing frequency. In this design [12], a lower chopping frequency is chosen, which results in less charge injection and clock feed-through errors.

3.2 A Chopper Opamp with Ripple Reduction Loop

This design is presented in Fig. 9. It consists of a two-stage Miller-compensated opamp G_{m1} and G_{m3}. The offset of G_{m1} is up-modulated and appear as a ripple at the output of the opamp. When the opamp is followed by a sampling ADC, for instance, the ADC will sample the ripple together with the signal. Depending on the relationship between the ADC's sampling frequency and the chopping frequency, the sampled ripple will either appear as an offset or as an intermodulation tone. To be negligible, the ripple amplitude should be suppressed to below the level of the offset. In this design, a ripple reduction loop (RRL) is employed [13]. It consists of two sense capacitors C_{s1} and C_{s2} to sense the ripple at the output of G_{m3}, and convert the AC ripple voltage to an AC current, which is then demodulated to a DC current and then integrated on C_{int} through a current buffer (CB). The voltage on C_{int} is then converted to a current by G_{m4}, which is used to compensate the offset

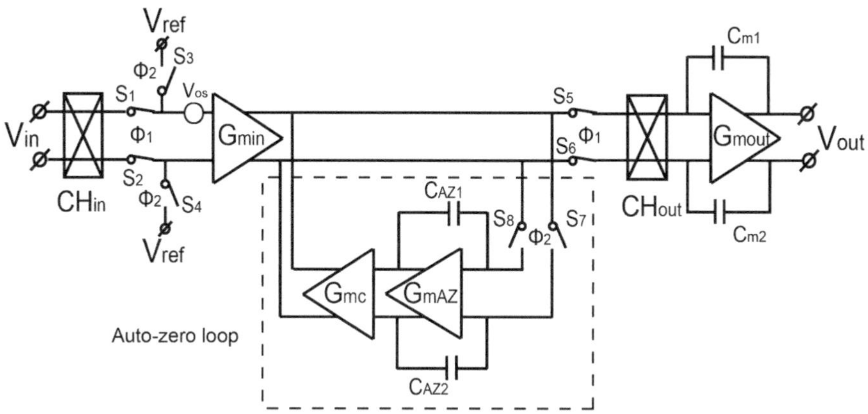

Fig. 8 Block diagram of an amplifier that employs both chopping and auto-zeroing

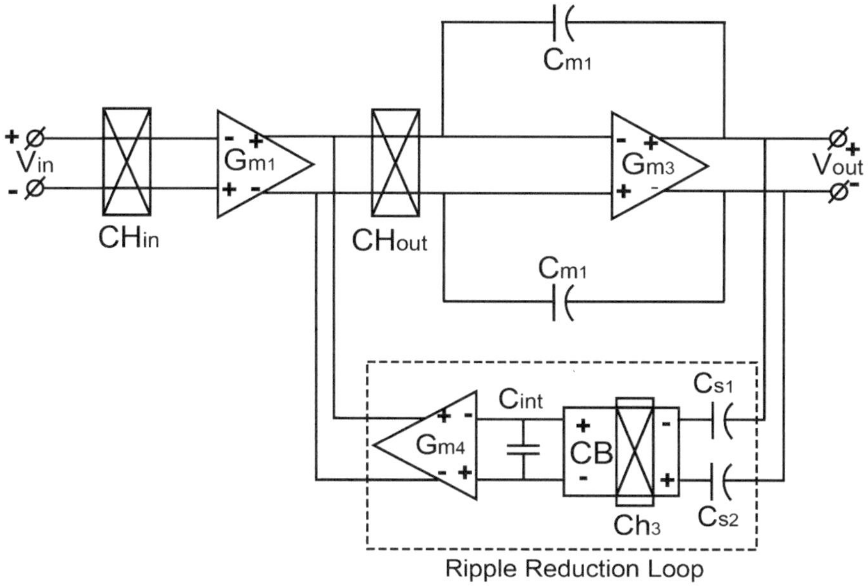

Fig. 9 Block diagram of a chopper opamp with a ripple reduction loop

current of G_{m1}. Such a ripple reduction loop does not necessarily increase the noise of the opamp, as long as G_{m4} is minimized. This choice ensures that all the noise and error sources from the RRL are negligible.

It is interesting to point out that the RRL shares some similarities with the auto-zeroing loop shown in Fig. 8. They both compensate the offset of the input stage by injecting a correction current through a transconductor. The difference, however, is that the auto-zeroing loop sense the offset of the input stage by sampling its offset directly, while the RRL senses the up-modulated offset of the input stage and

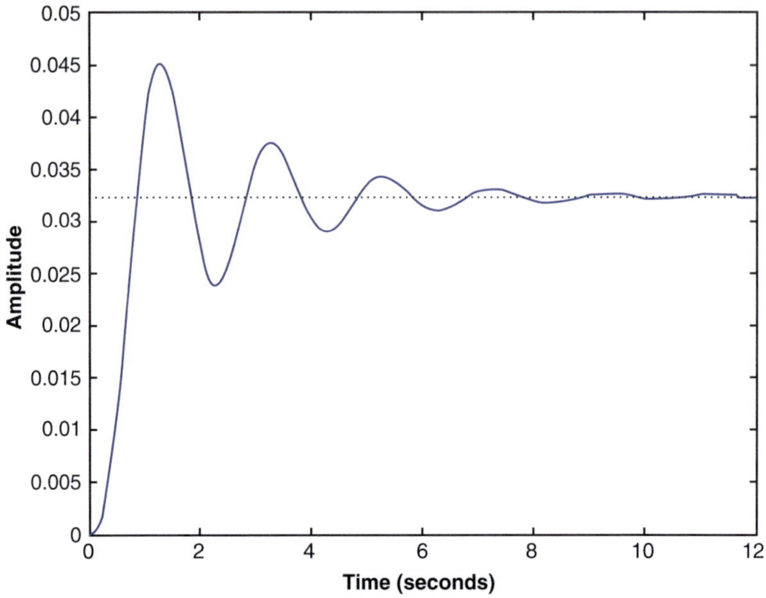

Fig. 10 The step response of a notch filter

then de-modulate it. The advantage of the RRL, however, is that it does not involve sampling, and thus does not suffer from noise-folding effects.

One concern about this simple, but effective, design is that the RRL serves as a notch filter, which does not only cancels the offset of G_{m1}, but also the input signal that is at the chopping frequency. One common issue associated with such a notch filter is that its step response exhibits ringing as shown in Fig. 10 and requires a long settling time. The settling time of the ringing is determined by the relative position of the poles and zeros of the notch filter, as explained in [14]. This is undesirable in applications where fast settling is required. In the following, a chopper-stabilized opamp is introduced to solve this problem.

3.3 A Chopper-Stabilized Opamp

A chopper stabilized opamp is shown in Fig. 11 [9]. As previously introduced, a chopper stabilized opamp consists of two paths: a HFP and a LFP. The HFP consists of a two-stage miller compensated opamp G_{m11} and G_{m3}. The LFP consists of four-stage opamp: G_{m21}, G_{m4}, G_{m5} and G_{m3}. To achieve stability, hybrid miller compensation is employed, which requires G_{m11}/C_{m11} to be equal to G_{m21}/C_{m31} [1, 3] for a 20 dB/dec roll off. As previously introduced, the offset reduction factor of a chopper stabilized opamp relies on the high gain of the LFP, thus, the DC gain

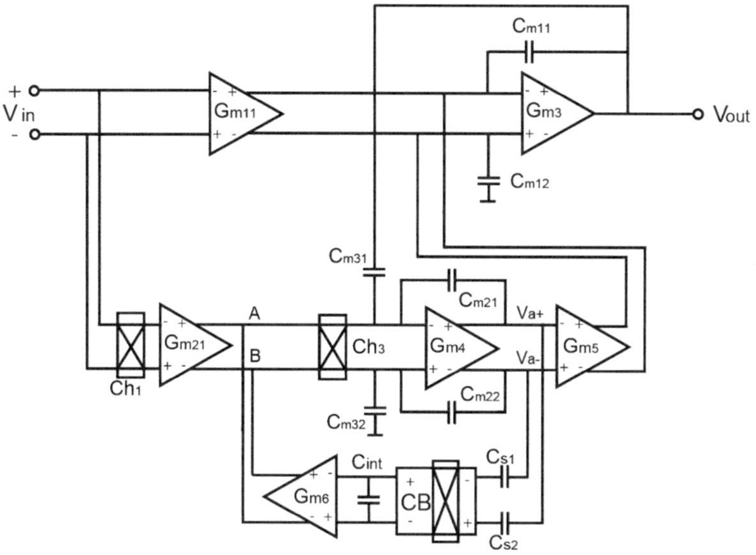

Fig. 11 Block diagram of a chopper stabilized opamp

of G_{m4} should be optimized. Since the offset of G_{m21} is up-modulated by Ch_3, the integrator built around G_{m4} serves as a low-pass filter to reduce the ripple. The ripple is even further reduced by the integrator built around G_{m3}. However, the amplitude of the ripple may still be too large in many applications. As a result, a RRL is again employed in this design to ensure the ripple is suppressed to the minimum. However, since the opamp consists of two signal path, the notch that is created by the RRL can be effectively overcome by the HFP [15]. In this way, a much smoother transfer function is obtained, thus a much faster step response can be achieved.

4 Discussions

The presented designs all feature micro-volt offset and low $1/f$ noise. They also have their own pros and cons.

The design presented in Fig. 7 is the most power efficient. However, as mentioned earlier, the charge injection and clock feed through errors during Φ_2 are stored on $C_{az1,2}$, thus limiting the accuracy of auto-zeroing. Another concern is that due to noise folding, the chopping frequency needs to be equal or above 2× of the auto-zeroing frequency. A higher chopping frequency often results in more charge injection and clock feedthrough errors, which in turn give rise to more residual offset and ripple. The design presented in Fig. 8 can reach a better accuracy since the charge injection errors associated with $S_{5\text{–}8}$ can be suppressed by the loop gain of the auto-zeroing loop. The noise bandwidth of the auto-zeroing loop can also be adjusted according to design specifications. Thus, a lower chopping frequency can be chosen for lower charge injection and clock feedthrough errors. The cost, however, is the more power consumption and a relatively slow settling auto-zeroing loop, which may be a problem for applications where fast settling is preferred.

Both designs presented in Figs. 7 and 8 are not continuous-time designs, which means that the opamps are only available for 50 % of the time. This can be a problem for applications where the signals must be continuously monitored such as electrocardiogram. To make these two designs continuous-time, a ping-pong topology has been employed in [11, 12], which involves the use of a duplicate input stage as discussed in Sect. 1. However, in precision applications, to achieve low noise, the current consumption of the input stage can often be dominating, thus, employing two duplicate input stages will likely increase the power consumption significantly.

The design presented in Fig. 9 can be a very effective and power efficient for applications as long as fast step response is not a concern. When fast step response is indeed a concern, chopper-stabilized opamp is a good choice. Another concern associated with the design in Fig. 9 may be that since the RRL is sensing directly at the output of the opamp, it will also sense other components around the chopping frequency and its harmonics coming from the outside. Thus, the RRL will be disturbed by the presence of external noise and interference.

The chopper-stabilized opamp in Fig. 11 has a better step response, but appears not to be very power efficient since it employs two input stages G_{m11} and G_{m21}. G_{m11} and G_{m21} are often chosen to be equal and this is mainly because hybrid miller compensation requires G_{m11}/C_{m11} to be equal to G_{m21}/C_{m31}. Thus, for optimal matching purposes, G_{m11} is often designed to be equal to G_{m21} [9, 10]. However, it is also possible to choose a smaller G_{m11} and accordingly a smaller $C_{m11,12}$ as long as G_{m11}/C_{m11} is equal to G_{m21}/C_{m31}. In this case, the matching should be taken care of with better layout plan. For instance, if G_{m21} is $4 \times G_{m11}$, G_{m21} should be broken into at least four pieces equal to G_{m11} and G_{m11} should be placed in the center of G_{m21} to optimize matching. The same goes for the layout of C_{m11} and C_{m31}. In this way, the power efficiency of a chopper stabilized opamp can be optimized. The consequence of choosing a smaller G_{m11} is that the high frequency noise will increase. However, in many precision applications, high frequency noise may not be a concern.

5 Conclusions

In this paper, various techniques and design examples of low-offset opamps have been presented. The non-idealities associated with auto-zeroing, chopping and chopper stabilization have been discussed, as well as the pros and cons of each of the design examples. For applications where power efficiency and continuous-time operations are most important, chopped amplifiers are the best choice. For applications where continuous-time operation is not needed, auto-zeroed amplifiers can offer great results. And for applications where wide bandwidth and fast step response is required, a chopper stabilized amplifier can be employed.

References

1. Huijsing JH (2011) Operational amplifiers: theory and design, 2nd edn. Springer, New York
2. Xu J, Yazicioglu RF, Grundlehner B, Harpe P, Makinwa KAA, Van Hoof C (2011) A 160 μW 8-channel active electrode system for EEG monitoring. IEEE Trans Biomed Circuits Syst 5(6):555–567
3. Enz CC, Temes GC (1996) Circuit techniques for reducing the effects of op-amp imperfections: autozeroing, correlated double sampling, and chopper stabilization. Proc IEEE 84(11): 1584–1614
4. Witte F, Makinwa KAA, Huijsing JH (2009) Dynamic offset compensated CMOS amplifiers. Springer, Dordrecht
5. Kundert K. Simulating switched-capacitor filters with spectre RF. http://www.designers-guide.org/Analysis/sc-filters.pdf
6. Opris IE, Kovacs GTAA (1996) A rail-to-rail ping-pong op-amp. IEEE J Solid State Circuits 31(9):1320–1324
7. Menolfi C, Huang Q (1999) A fully integrated, untrimmed CMOS instrumentation amplifier with submicrovolt offset. IEEE J Solid State Circuits 34(3):415–420
8. Burt R, Zhang J (2006) Micropower chopper-stabilized operational amplifier using a SC notch filter with synchronous integration inside the continuous-time signal path. IEEE J Solid State Circuits 41(12):2729–2736
9. Fan Q, Huijsing JH, Makinwa KAA (2012) A 21 nV/$\sqrt{}$Hz chopper-stabilized multipath current-feedback instrumentation amplifier with 2 μV offset. IEEE J Solid State Circuits 47(2):464–475
10. Witte JF, Huijsing JH, Makinwa KAA (2009) A chopper and auto-zero offset-stabilized CMOS instrumentation amplifier. In: VLIS circuits, pp 210–211
11. Tang ATK (2002) A 3μV-offset operational amplifier with 20 nV/$\sqrt{}$Hz input noise PSD at DC employing both chopping and auto zeroing. In: IEEE ISSCC digest of technical papers, pp 386–387
12. Michel F, Steyaert M (2012) On-chip gain reconfigurable 1.2 V 24μW chopping instrumentation amplifier with automatic resistor matching in 0.13 μm CMOS. In: IEEE ISSCC digest of technical papers, pp 372–374
13. Wu R, Makinwa KAA, Huijsing JH (2009) A chopper current-feedback instrumentation amplifier with a 1 mHz $1/f$ noise corner and an AC-coupled ripple-reduction loop. IEEE J Solid State Circuits 44(12):3232–3243
14. Analog Device. Chapter 8 Analog filters. http://www.analog.com/library/analogdialogue/archives/43-09/EDCh%208%20filter.pdf
15. Fan Q, Huijsing JH, Makinwa KAA (2013) A multi-path chopper-stabilized capacitively coupled operational amplifier with 20 V-input-common-mode range and 3 μV offset. In: IEEE ISSCC digest of technical papers, pp. 176–177

Amplifier Design for the Higgs Boson Search

Jan Kaplon and Walter Snoeys

Abstract Integrated circuits and devices have been a cornerstone in the recent discovery of the Higgs boson by the ATLAS and CMS experiments at the Large Hadron Collider (LHC) at the European Laboratory for High Energy Physics (CERN) in Geneva. Particles are accelerated and brought into collision at well-defined interaction points. Detectors, giant cameras of about 40 m long by 20 m in diameter, constructed around these interaction points take pictures of the collision products as they fly away from the interaction point. They contain millions of channels often generating a small ($\sim$1 fC) electric charge upon particle traversals. Integrated circuits provide the readout in a very aggressive radiation environment and accept collision rates of about 40 MHz with on-line selection of potentially interesting events before data storage. Power consumption directly impacts the measurement quality as it governs the amount of material present in the detector, and often the fraction of the power consumed by the front end amplifiers is significant if not predominant. We present basic architectures and a selection of front end amplifiers we hope as a representative overview for various types of particle detectors operated at LHC.

1 Introduction and General Specifications

During the past century physics research on the nature of matter has culminated in the standard model stating that all matter in the universe is constructed from a small number of elementary particles interacting via four elementary forces. The recent discovery of the Higgs Boson independently by the CMS [1] and ATLAS [2] experiments at CERN validated the Englert-Brout-Higgs mechanism, explaining the origin of mass of the subatomic particles in the standard model, and lead to the award of the Nobel Prize in physics to François Englert and Peter Higgs for their theoretical prediction of this mechanism. Many results including the Higgs discovery have been obtained in physics experiments where particles

J. Kaplon (✉) • W. Snoeys
CERN, Geneva, Switzerland
e-mail: Jan.Kaplon@cern.ch; Walter.Snoeys@cern.ch

K.A.A. Makinwa et al. (eds.), *Efficient Sensor Interfaces, Advanced Amplifiers and Low Power RF Systems*, DOI 10.1007/978-3-319-21185-5_12

are first accelerated to relativistic energies, and then brought into collision at well-determined interaction points, producing thousands of fragments flying away from the collision point. These fragments or their products—many decay after having travelled only for a few millimetres—are detected and visualized by detectors about 40 m long and 20 m in diameter constructed in underground caverns around these interaction points. Collisions have to be produced at large rates (40 MHz at LHC) as the interest lies in extremely rare events: for instance the first LHC run produced a candidate Higgs event only every one trillion collisions. It is this low probability to produce interesting events which pushes for a high collision rate and is the cause for important system requirements in terms of radiation tolerance, data reduction and timing resolution.

These large detectors consist of many layers (see the schematic overview of the ATLAS detector in Fig. 1). The inner layers are part of the tracker which reconstructs the particles' tracks in the presence of a magnetic field (4 T in CMS), which resolves the momentum of the particle via the track curvature and also indicates whether a particle originated from the collision itself or from a particle decay immediately thereafter. Tracking layers need to detect particle traversals with a high position resolution and a time resolution good enough to associate the traversal with the correct collision (every 25 ns), but the dynamic range of the input charge is not very high. Almost all of the inner tracking layers are based on reversed biased silicon p-n diode arrays, two-dimensional for pixel detectors and one-dimensional for silicon strip detectors. The calorimeter is constructed around the tracker and absorbs most of the particle fragments and has to measure their energy over a very wide range. Calorimeters can be homogenous, using a single material to stop and detect the particles, or heterogeneous, when absorber and

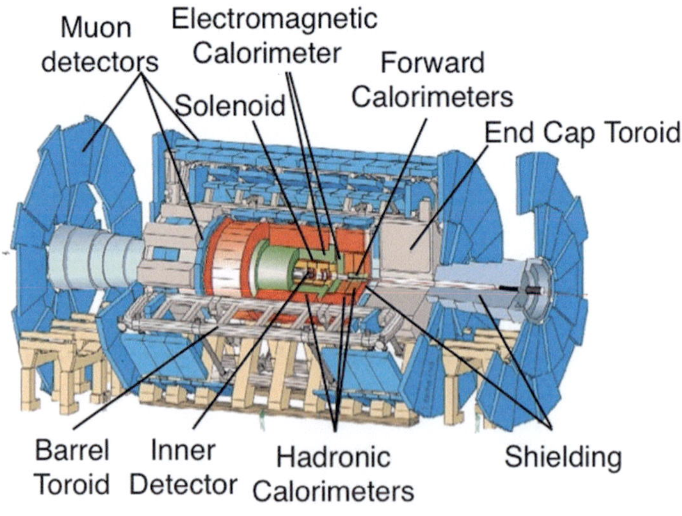

Fig. 1 Schematic overview of the ATLAS detector at CERN. ATLAS Experiment © 2015 CERN

detector layers are alternated, and often use scintillators as detecting material. Muons are extremely penetrating particles and are typically not absorbed by the calorimeter but they are often the signature of interesting events, and for that reason the outer layers of CMS and ATLAS are muon detectors.

ATLAS and CMS are the two main experiments designed for the Higgs boson search at the LHC. Apart from typical requirements concerning power, noise and dynamic range, the front end electronics in these experiments must be able to deal with very high signal rates and withstand the severe radiation environment. In the paper we focus on prototypes developed for the high energy upgrade finalized in 2014 and for the high luminosity upgrade to be completed around 2022.

The location of the given system inside the detector volume will specify requirements concerning radiation hardness as well as granularity of the sensors and timing response of the readout electronics, since the system has to distinguish different events both in space as well as in time. Systems close to the interaction point, like the silicon pixel vertex detector need to tolerate up to 500 Mrad Total Ionizing Dose (TID) and of the order of 10^{16} N/cm^2 NIEL fluencies. For the detectors installed at larger radii radiation tolerance requirements are less demanding, but silicon strip detectors used in inner tracker detectors must stand radiation doses of about 50 Mrad and 1×10^{15} N/cm^2.

After initial developments in specific radiation tolerant CMOS and BiCMOS technologies [3–6], most presently installed detectors adopted a commercial 250 nm CMOS technology for the readout circuits. This technology provided a sufficiently thin gate oxide and confirmed the early predictions in the early 80s of which the gate oxide was sufficiently thin to guarantee sufficient radiation tolerance [7, 8], when combined with special layout techniques like annular NMOS transistors [9, 10]. Present designs for pixel and strip detector front ends are implemented mainly in 65 and 130 nm CMOS technology nodes, both providing the necessary radiation tolerance even without special layout [11]. Systems installed farther away from the interaction point, where radiation is not a primary concern, often still use the cost effective 250 nm CMOS node, but sometimes SiGe BiCMOS is used if performance is the driving key and not the power (e.g., AMS 350 nm SiGe for MAROC or IBM8WL 130 nm SiGe for ATLAS Liquid Argon Calorimeter).

2 Front End Electronics from the Particle Detectors' Standpoint

Regardless of the big variety of the fundamentals, construction and physical properties, almost all particle detectors can be represented by a current source, loaded with parasitic capacitances, delivering fast signal pulses which have to be amplified and filtered for noise optimization by the front end amplifier. Figure 2 shows two basic examples of the front end amplifier connected to the particle detector. In the first case the front end electronics connected to the highly segmented sensor must provide

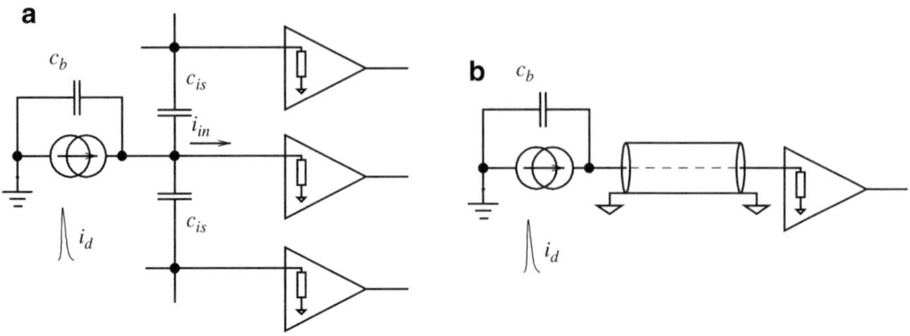

Fig. 2 Two examples of the front end amplifiers connected to the particle detectors. (**a**) Front end electronics connected to highly segmented particle detector with parasitic fringing capacitance to the neighbors and (**b**) front end amplifier connected to the distant detector using transmission line

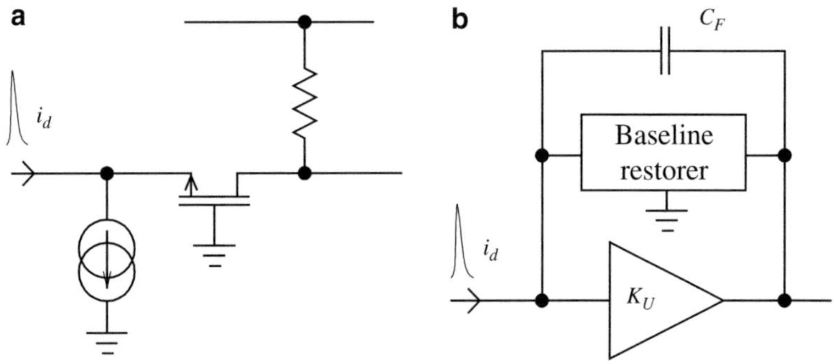

Fig. 3 Basic low input impedance architectures. (**a**) Common gate amplifier and (**b**) high gain amplifier with shunt-shunt feedback

efficient charge collection avoiding crosstalk signals to neighboring channels via parasitic fringing capacitances. The second example shows a front end connected to a distant detector using a transmission line. The input impedance has to be low for the first case and well controlled for the second in order to match the line impedance.

In general two architectures of amplifiers can provide low input impedance (see Fig. 3), namely the one based on common gate (base) configuration and the one employing high gain amplifiers with shunt-shunt feedback with integration capacitor and passive or active baseline restoration circuit. Although both solutions can meet requirements for low or controllable input impedance they will differ considerably in the achievable speed, power and noise performance, which will be the basis for the choice of the configuration for a given application.

While particle detectors can be represented by a current source delivering very fast (pico-seconds to nano-seconds range) pulses, in most cases (except for example ATLAS Liquid Argon Calorimeter discussed later) the quantity which has to be measured is the charge generated in the detector volume. For that reason, the noise performance of the given front end amplifier is described by the Equivalent Noise Charge (ENC), which is the noise of the amplifier expressed in Coulombs or number of electron charges at its input.

In most cases the very small signals delivered from the detectors impose improvement of the SNR by band-pass filtering performed after the preamplifier stage. For a charge sensitive amplifier (CSA) followed by the popular CR-RC band-pass filter (shaper), the ENC follows the formula [12]:

$$ENC^2 = \left(F_V \overline{v_n} c_d / \sqrt{\tau}\right)^2 + \left(F_i \overline{i_n} \sqrt{\tau}\right)^2 \qquad (1)$$

where v_n and i_n are power densities of equivalent series and parallel noise sources (for MOS transistor respectively $\sqrt{\frac{4kT\gamma n}{g_m}}$ and $\sqrt{4kT\gamma n g_m}$), F_V and F_i are the coefficients dependent on filter order and τ is the peaking time of the CSA response. As can be seen from (1), the contribution of parallel and series equivalent noise sources to the ENC for a given detector capacitance can be optimized by varying the time constant of the shaper. For the electronics at the LHC this optimization is constrained by time resolution requirements as the collisions take place every 25 ns and particle traversals have to be associated with the correct collision. In a well-designed amplifier (sufficient gain of the first stage, active loads degenerated) the dominant noise source should be the input device(s). Consequently, for the input stages where the dominant noise source is the series noise of the input device(s) (true for feedback amplifiers) the ENC can be optimized by varying the bias current affecting directly the noise spectral densities of the input transistor(s), keeping at the same time the parallel noise contribution (detector leakage, feedback circuit) at an acceptable level. This way a compromise is made between ENC performance and power consumption within the required boundaries. On the other hand, the ENC performance of the common gate (base) amplifier with the main parallel noise contribution from the input transistor, does not depend directly on the input capacitance. A short review of the basic architectures discussing basic analog parameters together with examples of the prototyped designs is presented in the following sections. All calculations for the presented formulas for the input impedances and noise contributions from the input devices can be found in [12] unless otherwise noted.

3 Front End Amplifiers Based on the Common Gate Architecture

Figure 4 shows three basic configurations of common gate (base) amplifiers used for the input stages of front end amplifiers for particle detectors. Figure 4a shows a simple common gate amplifier biased with a current source and loaded with a resistor. The input impedance of the amplifier depends directly on the transconductance of the transistor M_1 ($r_{in} = 1/g_m$) and since the circuit works in

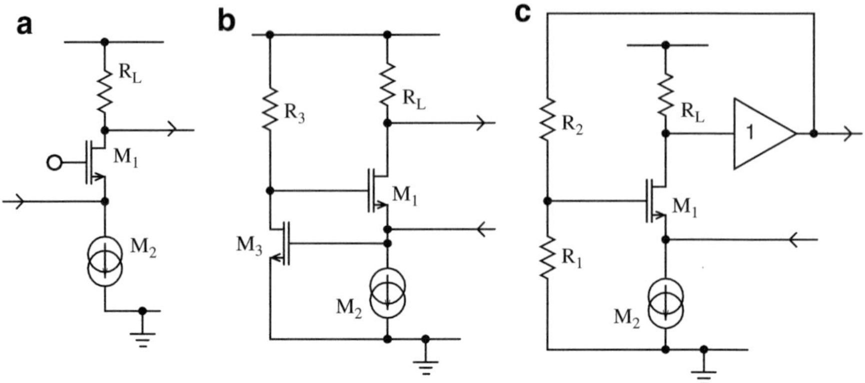

Fig. 4 Basic configuration of the common gate input stages. (**a**) Common gate amplifier, (**b**) super common-gate amplifier, and (**c**) common gate amplifier with series-shunt feedback

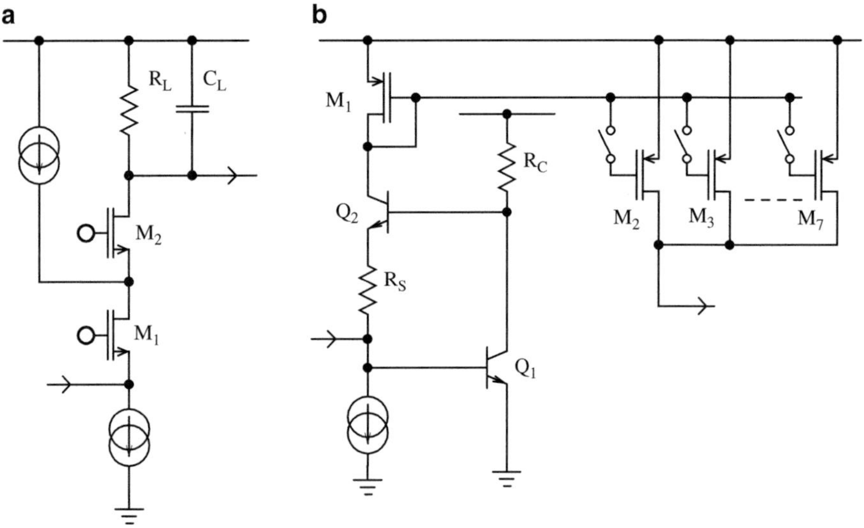

Fig. 5 (**a**) Simplified schematic of the NINO preamplifier and (**b**) schematic of the MAROC preamplifier

an open loop configuration, it is constant over whole frequency bandwidth of the preamplifier. The drawback of this circuit is the high level of parallel noise depending directly on the g_m of the input transistor, which defines at the same time the input impedance of the preamplifier. Despite this disadvantage, this architecture can be used for readout of photodetectors delivering relatively high signal charges. Figure 5a shows the simplified schematic of one branch of the pseudo differential pair of the preamplifier of the NINO chip implemented in CMOS 250 nm process [13]. Originally designed for ALICE Time-Of-Flight (TOF) detector employing

Multigap Resistive Plate Chamber (MRPC) it has been applied in many other places like the ATLAS beam condition monitoring system.

The input impedance of the preamplifier is controlled by the bias circuit, stabilizing the current in the first transistor and providing matching to the impedance of the transmission line used for connection to the MRPC (40–75 Ω). By adding a second common gate amplifier built with M_2 and employing an extra current source biasing M_1 to achieve required input impedance it is possible to increase the load resistor R_L and improve the gain of this stage.

Due to the pseudo-differential structure of the input stage, a relatively good stabilization of the operating point at the preamplifier output is achieved. Since the dominant noise contribution is parallel noise due to g_m of the input transistor M_1, one can expect that this circuit can work with very large detector capacitances. In practice, the extra time constant formed by the input impedance and detector capacitance affects the response of the amplifier, degrading the achievable time resolution as well as noise performance (τ in formula (1)). The intrinsic peaking time of the NINO response is of the order of 1 ns. This limits the input capacitance range to about 20 pF maximum. For the nominal bias and input load condition (5 pF detector capacitance) an ENC of about 2500e$^-$ has been measured. For nominal signals from an MRPC of more than 100 fC, 25 ps rms time jitter has been reported. The power consumed by one channel is about 30 mW.

The circuit which partially solves the trade-off between noise performance and input impedance is shown in Fig. 4b. In the super common gate configuration, the input impedance depending originally for common gate amplifier on the transconductance of M_1 only, is improved by the gain of the booster amplifier built with M_3 and R_3 i.e., $r_{in} = 1/(g_{m1}\, g_{m3}\, R_3)$. The signal gain of the super common gate is equivalent to the simple configuration i.e., it depends only on the load resistor R_L and the parameters of the filter following the preamp. Instead of one input device, now we have two, both contributing to the ENC of this stage. The M_1 transistor contributes with its g_m as a parallel noise source and M_3 with its g_m as a series noise source. Running the bias current of M_1 at a relatively low level to minimize the parallel noise contribution, it is still possible to achieve the desired input impedance by increasing the gain of the booster amplifier i.e., increasing the transconductance of M_3 which, at the same time, minimizes the contribution of the series noise. Although the super common gate amplifier offers much more freedom to optimize noise and to achieve desired input impedance, there is still some correlation between these variables. An additional drawback of running M_1 with low bias current is related to the fact that high input signals can modulate its g_m i.e., the input impedance of the preamplifier, what can result in reflections for very large signals if the circuit is used for terminating a transmission line. One can also notice that the super common gate amplifier has in fact the Gyrator-C active inductor topology. Care must be taken in order to prevent oscillations due to inductive characteristic of the input impedance at high frequency by minimizing the parasitic capacitances and the introduction e.g., of a series resistor at the source of M_1. Despite these drawbacks, the circuit is used often for readout of photomultipliers for photon counting applications, where the connection in-between detector and readout electronics is relatively short. An example of the usage of the super common gate is the preamplifier of MAROC chip [14, 15] shown in Fig. 5b. The MAROC chip is used in ATLAS luminosity detector for readout of multi-anode photomultipliers. The 64-channel chip has been implemented in AMS 350 nm SiGe process. The set of switchable current mirrors in the load of the input stage allows for compensation of the PMT gain variation which can be as large as a factor of three. The full channel of MAROC consists of slow (160 ns peaking time for energy measurement) and fast (10 ns peaking time for auto trigger function) shapers, discriminator and sample and hold buffer allowing for energy measurement with per-channel 12-bit Wilkinson ADC. The overall power consumption is around 5.5 mW for the complete channel (no data for preamp only). For the nominal biases, the input impedance of the preamplifier is about 100 Ω and the ENC for the 160 ns shaping is around 5000e$^-$ [15]. The reported figure showing the ENC as a function of shaping time clearly shows two competing contributions from parallel and series noise sources (Fig. 13 in [15]).

A very interesting approach for a low noise, line termination amplifier is shown in Fig. 4c. The external series-shunt feedback applied for the buffered common gate amplifier completely modifies the properties of the input impedance as well as noise behaviour leaving the signal gain unchanged with respect to open loop configuration [12]. Now the transistor M_1 contributes with its g_m as a series noise source (!) which allows for longer shaping and noise minimization, without affecting the input impedance which after applying the feedback, depends mainly on the feedback and load resistors ($r_{in} = 1/g_{m1} + R_L R_1/(R_1 + R_2)$) if g_{m1} is sufficiently high (true since g_{m1} is driven to minimize the noise). One difficulty in the presented approach might be the design of the very high performance unity gain buffer which must drive a relatively low feedback impedance with good linearity if one wants to keep a high dynamic range i.e., a reasonable value of load resistor R_L.

The practical implementation of this architecture is shown in Fig. 6. It shows a low noise, line termination preamplifier designed for the upgrade of ATLAS Liquid Argon Calorimeter designed in the IBM8WL 130 nm SiGe process chosen for its radiation tolerance and low value of base spread resistance of BJTs, keeping the noise contribution from $r_{bb'}$ negligible [16]. The signal formation in the detector is relatively long (400 ns drift time of the generated charges) and the applied shaping time is shorter (50 ns peaking time of the response). Consequently the dynamic range and the input noise have been specified in terms of current amplitude and rms and not charge as for any other presented amplifier. The detector is connected with 5 m long, 25 Ω impedance transmission line and the input impedance of the amplifier matches this value. As an input stage, a super common base configuration has been chosen, allowing for minimization of the power, while driving the equivalent transconductance of the Q_1 and Q_2 pair above 1 S for 6 mA total current. It has been shown [12, 17] that in case of a super common gate/base configuration with series-shunt feedback both transistors contribute with the series noise, however the contribution of Q_2 is attenuated with the gain of the booster built with Q_1 and R_C. In order not to noticeably degenerate gain of the booster and keep good noise performance, R_1 has to be relatively small (1 Ω in the presented design).

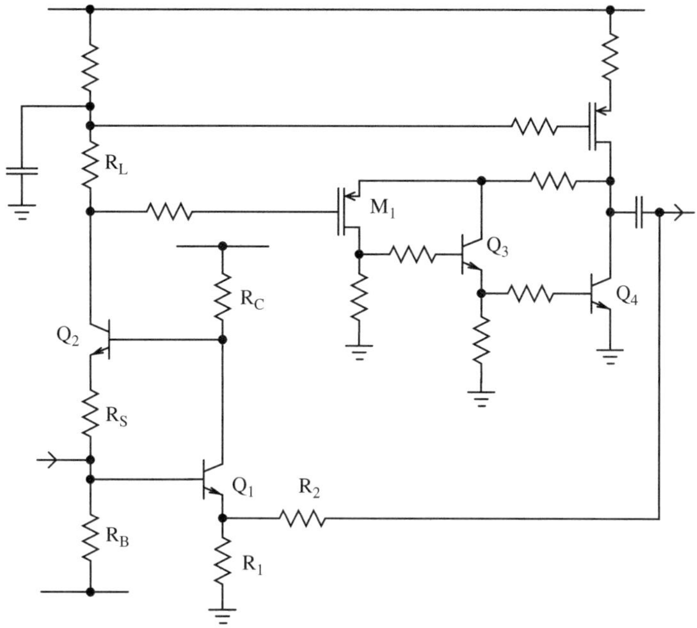

Fig. 6 Low noise, line termination preamplifier for Liquid Argon ATLAS calorimeter

The range of input signals is specified for 10 mA maximum amplitude. This sets the requirements for the load resistor R_L (600 Ω) and R_2 resistor (~30 Ω). As a consequence the unity gain buffer must have very low output impedance and this is provided by employing the White follower configuration (M1, Q3 and Q4). The power consumption is of the order of 40 mW and an equivalent input noise below 100 nA rms for a 1 nF detector capacitance has been achieved. Although the power consumption is relatively high, it is still satisfactory taking into account the relatively low granularity of this system and peripheral position in the ATLAS detector, and one should appreciate the excellent noise performance of this circuit providing covering almost five orders of magnitude dynamic range of the measured detector signals.

4 CSA Built with High Gain Amplifiers and Shunt-Shunt Feedback

While preamplifiers based on a common gate configuration are suitable for readout of relatively high detector pulses (hundreds of femto-Coulombs to pico-Coulomb range), CSA preamplifiers with shunt-shunt feedback are found in applications for readout of extremely low signals from e.g., highly segmented semiconductor detectors (~1 fC) connected closely with the readout using chip-to-chip bonding techniques.

Dealing with very low input signals from detectors associated with a net of parasitic fringing capacitances to the adjacent channels sets two requirements for the CSA amplifier. The first is to drive the closed loop gain of the preamp to minimize noise contribution from the consequent stages, and the second is to lower the input impedance as much as possible in order to improve charge collection efficiency and to lower crosstalk signals to the neighbors. Both requirements are satisfied by pushing up the open loop gain and the Gain Bandwidth Product (GBP) of the amplifier. As already mentioned, semiconductor detectors and their readout electronics installed close to the interaction point of ATLAS and CMS experiments must withstand very high radiation doses. Readout electronics designed for those detectors are implemented in deep submicron CMOS processes, namely 65 and 130 nm nodes, which can tolerate such environment. Designing single stage, high bandwidth and high open loop gain amplifiers in such submicron technologies with low intrinsic gain of the transistors, leads to the use of various configurations of the cascode circuit which automatically cancels the Miller effect.

The argument for increasing the GBP of the amplifier working with shunt-shunt feedback connected to the fast shaper is twofold. First, the position of the high frequency pole responsible for stable operation depends directly on the GBP and the ratio of input and feedback capacitances. In other words for a slow preamplifier we will be obliged to use relatively high value of feedback capacitance which is not optimal for closed loop gain. Second, the bandwidth of the amplifier will constrain the PSRR and input impedance characteristic at high frequency, the latter being responsible for charge collection efficiency and crosstalk problems. Problems of crosstalk signals can be better understood if one analyzes the charge collection process in the time domain.

One can view the detector with its readout immediately after the generation of the signal charge as a capacitive network distributing this charge through the capacitive couplings as the preamplifiers have not yet had time to react. As the preamplifiers settle reestablishing virtual ground at their input the crosstalk due to the capacitive network gradually reduces to zero. Therefore, if the noise shaping filter is sufficiently slow compared to the preamplifier rise time less coupling will be introduced. Having in mind that for ATLAS and CMS electronics the timing response of shaper is constrained, the only way to provide this separation is to speed up the preamp. Some additional benefit can be offered by a transimpedance amplifier which provides lower input impedance in the frequency range of the shaper. Note that this capacitive crosstalk has to be distinguished from crosstalk induced by the motion of signal charge in the detector itself, which is governed by Ramo's theorem: signal charge may introduce during its travel through the detector volume signal transients on electrodes other than the ones on which it finally gets collected. This represents another source of cross-talk if the readout is fast enough to capture these transients.

As already stated, the optimization of the CSA noise by adjusting the time constant of the shaper is restricted for ATLAS and CMS electronics by the required 25 ns timing resolution. In practical cases of the preamplifiers designed for LHC inner tracker detectors optimized for low power, the series noise related to the thermal noise of the input transistor is dominant. Therefore input stages built with NMOS devices are preferable since they provide a more favorable transconductance or lower thermal noise to current ratio for strong and moderate inversion, and allow smaller dimensions i.e., lower gate capacitance and hence lower noise in weak inversion.

4.1 Cascode Amplifiers Used for CSA Input Stages

Figure 7 shows basic configurations of the telescopic cascodes built with NMOS input devices used in the input stages for ATLAS and CMS front ends for silicon trackers. It is worth mentioning here that for the readout of semiconductor segmented detectors which are intrinsically single ended structures, the single ended telescopic cascodes provide lower noise (highest current in the input device, lower currents in the active loads i.e., lower noise contribution) and the same dynamic range as folded cascodes.

We start the review of noise performance of the structures presented in Fig. 7 from the simple telescopic cascode (a). Assuming that the current source in the active load of the cascode is degenerated [18] the dominant contribution to the ENC of the cascode, is the series noise related to the thermal noise of the input transistor M_1. Although the voltage gain seen at the drain of M_1 is practically zero dB, the input-referred noise contribution of the transistor M_2 is attenuated by the intrinsic voltage gain of M_1 being g_{m1}/g_{ds1} (see [12]). Since the transistor M_2 is biased with the same current as M_1, the transconductance is of the same order even if M_2 is biased closer to strong inversion region i.e., its noise contribution can be neglected.

As previously stated, the GBP product of the cascode depends on the transconductance of the first transistor and parasitic capacitance at the cascode output.

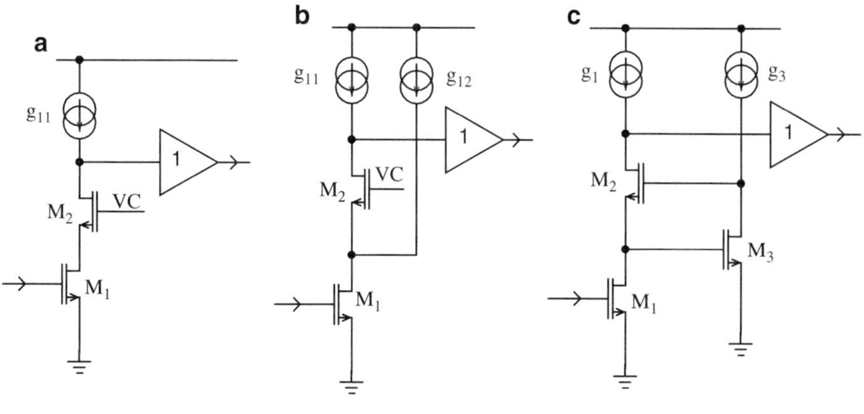

Fig. 7 Basic configurations of the cascode input stages for CSA amplifiers in ATLAS and CMS silicon trackers

This feature is used to drive the GBP for the circuit shown in Fig. 7b. Splitting the active load of the cascode from Fig. 7a into two current sources, one can minimize the dimensions of the cascode active load, i.e., its parasitic capacitance, and effectively increase the GBP. One can also reduce the noise contribution of the current source directly supplying the drain of M_1 by increasing its $V_{gs} - V_T$.

For the preamplifier stages designed for silicon strip detectors with typical capacitances per channel of the order of few to a few tens of pico-farads, the single cascode stage might not provide sufficient open loop gain if it is implemented in deep submicron process with low intrinsic transistor gain. The regulated cascode amplifier resolving this problem is shown in Fig. 7c. The gain of the basic cascode made with transistors M_1 and M_2 is enhanced with the gain of the booster built with transistor M_3. For the regulated cascode amplifier the dominating noise source is still transistor M_1 which contributes with its transconductance as series noise source. The contribution of the cascode transistor M_2 is attenuated by the intrinsic gain of the transistor M_1 multiplied by the gain of the booster amplifier and can therefore be completely neglected (see [12]).

The contribution of the booster transistor M_3 is attenuated by the intrinsic gain of M_1 (g_{m1}/g_{ds1}). All transistors, M_1, M_2 and M_3 contribute to the ENC as a series noise sources. Minimizing overall power consumption, one will have the tendency to run the booster amplifier at a very low current. This must be done carefully to avoid extra noise contribution from the M_3, especially for some deep submicron technologies where the intrinsic gain of M_1 might be relatively low.

4.2 Feedback Circuits

The primary component of the feedback circuit defining the preamplifier signal gain is the feedback capacitor, integrating the charge delivered by the sensor. The baseline restorer circuit must discharge the feedback capacitor preventing saturation of the preamp and in case of a DC coupling to the detector must absorb the thermally generated leakage current which becomes very significant for heavily irradiated sensors. Taking into account the very high signal rate from the sensor (a few percent occupancy is a standard requirement for the inner trackers, meaning that on average each channel is hit once every few tens of 25 ns clock cycles) the discharge of the feedback capacitor should be performed with the time constant comparable to the time constants of filter used in the shapers. In this case the feedback circuit provides high pass filtering function and we say that the preamplifier works in transimpedance mode. Low values of the feedback impedance impose the use of a unity gain buffer at the cascode output. This prevents loading effects and degradation of cascode gain.

Although for the fast electronics for CMS and ATLAS trackers (shaping time around 25 ns) the dominant noise source is the series noise from the input devices, one has to keep the parallel noise contribution from the real part of the feedback impedance at the reasonable level. For the feedback capacitances of the order of

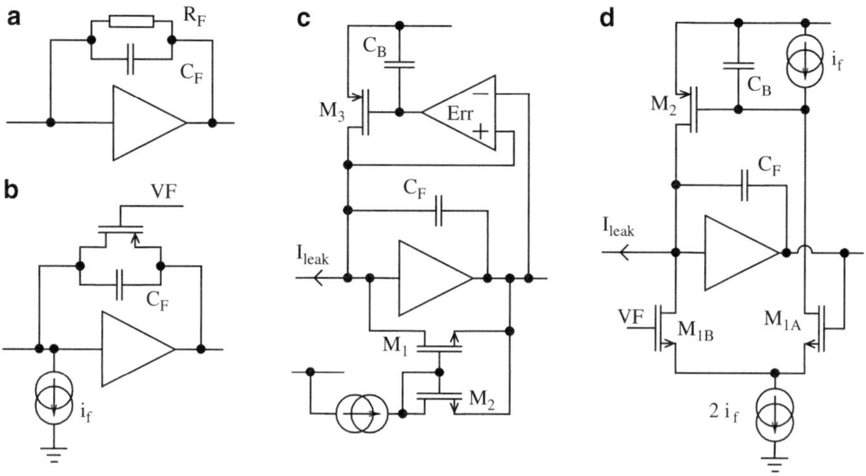

Fig. 8 Selection of feedback configuration for CSA amplifiers for ATLAS and CMS tracker electronics

tens of femto-farads the equivalent feedback resistance should be above 100 kΩ which for a 20–25 ns peaking time usually is not a problem since the series noise contribution is higher. All these statements are true because for the tracker electronics we are not designing for minimum ENC but rather for minimum power, keeping the ENC at the specified level.

The selection of the feedback circuits for CSA amplifiers developed for ATLAS and CMS trackers is shown in Fig. 8. The simplest, purely resistive feedback, is shown in Fig. 8a. For a relatively low value of feedback resistor, the circuit can tolerate some leakage current from the detector but the DC shift at the preamp output reduces dynamic range and creates a problem with DC coupling to the next stage. The feedback resistor contributes to the ENC of preamp as a parallel noise source.

Figure 8b shows an active feedback built with a transistor biased in saturation controlled by the current source. The analytical analysis of this feedback showing different modes of operation can be found in [19]. Again, some low detector leakage can be tolerated but for the LHC experiments, where the sensor radiation damage is severe, the circuit is often used with AC coupled detectors. For the nominal range of the detector signals (Minimum Ionizing Particle MIP) the circuit provides discharge with the equivalent impedance equal to $1/g_m$ (the feedback can be modeled as a common gate amplifier built with feedback transistor with its input connected to the preamp output). For very large signals generated from time to time in the sensors due to the nuclear reactions, the circuit will provide quadratic compression substantially limiting the dead time (true for a PMOS feedback transistor and n-type silicon detectors planned for the upgrade of ATLAS and CMS inner trackers). Potentially the circuit can have better noise performance than simple resistor feedback if the feedback transistor is in weak inversion. In practice for single ended amplifiers with NMOS input transistor, the DC at the preamp input is relatively low and the current source, i_f, is difficult to degenerate, resulting in an extra noise contribution from it and making the overall noise performance of this feedback slightly worse than the simple resistor feedback. The big advantage of this structure is that the gate voltage of the feedback transistors controls the DC potential at the preamp output and facilitates DC connection to the next stage.

Another approach for the active feedback circuit is shown in Fig. 8c. The feedback transistor M_1 biased in the linear range is controlled by the gate voltage of M_2 biased in saturation with the current source. The g_{ds} of the transistor M_1 for steady state i.e., $V_{DS} = 0$ and $I_{DS} = 0$, is equal to the g_m of the transistor M_2. For the MIP signals the feedback transistor typically operates in linear mode providing equivalent feedback resistance equal to $1/g_{ds}$. For very high input charges the feedback transistor might enter temporarily the saturation region and provide the same large signal behavior as circuit from Fig. 8b i.e., quadratic compression. From the noise perspective view it will contribute as a parallel noise source with its g_{ds} conductance i.e., the noise contribution will be equal to the contribution from the equivalent resistor. Since the feedback transistor works in the linear range, it is possible to add the leakage compensation circuit by means of an error amplifier measuring the voltage across the feedback transistor and driving the gate of compensation transistor M_3 (in fact the same architecture of leakage compensation will work for circuit Fig. 7a with a feedback resistor). Since the error amplifier runs at very low bias and in addition the output is filtered with the blocking capacitor C_B, the circuit is sensitive to the slow changes of leakage current only. Because the feedback transistor works in linear mode with zero voltage across it (V_{DS}), the DC potential at the preamp output is fixed by the internal bias i.e., for the single ended cascode with NMOS input transistor by its V_{GS} voltage which creates similar problems with DC coupling to the next stage as in case of circuit Fig. 7a.

In some cases, nonlinear feedback can be used deliberately to vary the output pulse duration as a function of input signal [20] and to obtain information about the signal amplitude through the time the signal is above threshold.

An interesting feedback circuit was proposed by Krummenacher [21] shown in Fig. 8d, which provides a fast feedback for the return to baseline, and a slow feedback to absorb the leakage current while maintaining a well-defined baseline. The fast feedback consists of a differential amplifier built with M_{1A} and M_{1B}. The discharge of the feedback capacitor C_F is provided through transistor M_{1B} with the current controlled by M_{1A} (sum of currents constant) being driven by the preamp output. The time constant of this fast feedback is defined by the transconductance of the transistors in the differential pair. The voltage at the preamp output is fully controlled by the feedback i.e., it is equal to the reference voltage applied to the gate of M_{1B}. One critical point of the design is related to the extra pole and zero created by the parasitic capacitance at the sources of the transistors M_{1A} and M_{1B} and its transconductance. For good stability this pole must be located much above the pole formed by the feedback capacitor C_F and the g_m of M_{1A}. Consequently the parasitic capacitance at the sources of M_{1A} and M_{1B} must be minimized by proper sizing of the transistors and careful layout. The slow changes of the voltage at preamp output caused by detector leakage current are amplified by M_{1A}, filtered with C_B

and applied to the gate of M_2 supplying the compensation current to the preamplifier input. As in the case of circuit from Fig. 8c the leakage compensation circuit works for one polarity of the leakage current (example shown for n-type detectors as planned for ATLAS and CMS upgrades). In order to provide the compensation for the opposite polarity of leakage current one has to change all transistors in the feedback from NMOS to PMOS and vice versa.

4.3 Examples of the CSA Preamplifiers for ATLAS and CMS Tracker Detectors

In this section we present a few examples of the input stages designed for front end chips for the upgraded ATLAS and CMS inner tracker systems. Figure 9a shows the preamplifier developed for the ABC130 front end chip intended to be used in the upgraded inner tracker for the ATLAS experiment. The chip has been implemented in the IBM CMOS 130 nm process. The detailed description of the front end architecture can be found in [22]. The circuit is designed to work with the AC coupled silicon strip detectors of moderate length with a total capacitance up to 5 pF but it can tolerate longer strips with capacitances up to 10 pF. The input stage is built with the regulated cascode amplifier with NMOS input transistor. It is biased with two regulated cascode current sources built with PMOS transistors degenerated for low noise performance with the resistors. With the regulated cascode structure the open loop gain of the amplifier is around 80 dB with the Gain Bandwidth Product of 2 GHz. The input transistor can be biased with currents ranging from 80 to

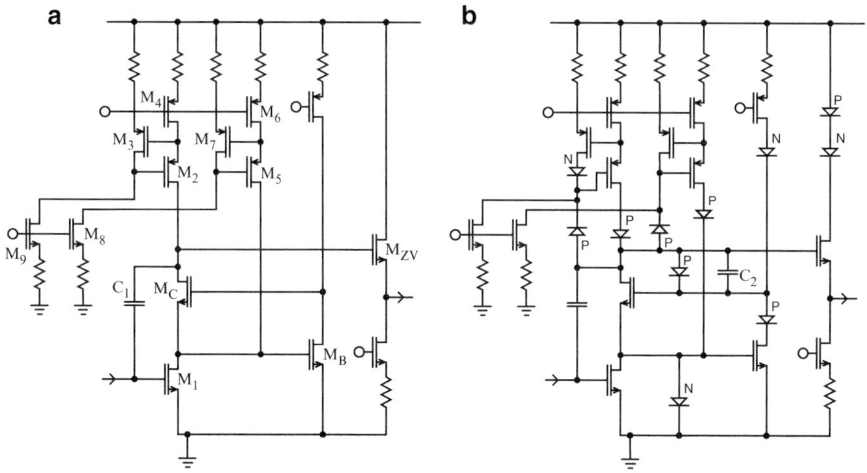

Fig. 9 (**a**) ABC130 preamplifier for ATLAS silicon strip tracker and (**b**) BCM1F preamplifier for CMS beam condition monitoring system

140 μA depending on the detector capacitance. In the ABC130 chip the preamp is followed with the low power (12 μA) shaper and discriminator. The preamplifier is enclosed with the active feedback loop described in Fig. 9b. The overall shaping of the front end is CR-RC^2 with an intrinsic peaking time of 22 ns and a gain around 90 mV/fC. The linear dynamic range of about 10 fC is typical for tracking applications, but the circuit has to be able to recover in a reasonable time (μs) from large input signals (pC) caused by nuclear interactions in the sensor to avoid dead time by compression or clipping in the feedback as explained earlier. An ENC below 800e$^-$ for the detector capacitance of 5 pF and 80 μA input transistor current has been demonstrated, well inside the specifications for tracker electronics for heavily irradiated silicon detectors.

The cascode amplifier is buffered with a simple source follower built with a native NMOS transistor biased with only 4 μA current. Although the buffer output impedance provides sufficient driving capability, it might create instabilities for low input capacitances due to the extra high frequency pole it introduces. A simple Miller compensation is provided by the C_1 capacitor. The overall power consumed by the full chain is between 200 and 280 μW depending on the bias setting of the input transistor.

Figure 9b shows the preamplifier designed for the CMS beam condition monitoring [23]. The architecture of the input stage is a modified structure of ABC130 front end enabled for high voltage supply allowing for higher bandwidth by increasing the drive ($V_{gs} - V_T$) of the transistors. A number of safety diodes built with small NMOS and PMOS devices have been added protecting the chip during the transients caused by overdrive signals from the detector and during power switch-on. The limited number of chips connected to 5 pF diamond detectors and installed close to the interaction point has to distinguish the beam halo, so the requirements for time resolution are very strict. The peaking time and Full Width at Half Maximum (FWHM) should be below 10 ns which was a driving parameter for this design. The input transistor is supplied with 300 μA bias current and the noise for a 5 pF detector capacitance is below 800e$^-$. The bandwidth of the preamp was further improved by applying feedforward compensation using the C_2 capacitor. As for ABC130 design the preamplifier is enclosed with the active feedback loop described in Fig. 8b.

Figure 10a shows a simplified schematic of the preamp for the CBC chip [24] designed for silicon strip detectors for the CMS tracker. The chip has been implemented in IBM 130 nm process.

The specifications for detector capacitance, power and noise performance are nearly the same as for the ABC130 design. The input stage is built with a telescopic cascode with NMOS input transistor and extra current source driving the GBP. The initial intention was to use the chip with DC coupled n-type and p-type detectors. For that reason a simple resistive feedback has been applied (Fig. 8a) which has an impact on dynamic range as well as on the type of the coupling to the next stage (AC).

Figure 10b shows the simplified schematic of preamplifier for the MPA front end chip [25]. The chip is implemented in commercial 65 nm process and it is intended to be used for readout of long silicon pixels (strixels) in CMS tracker. Although

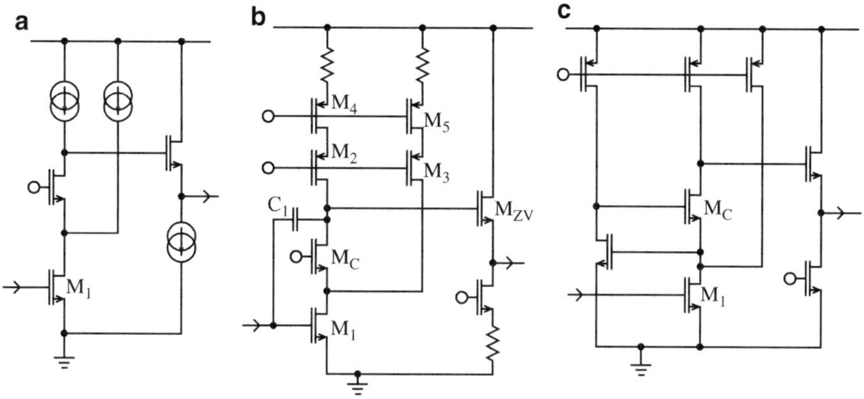

Fig. 10 (**a**) CBC preamp for CMS silicon strip tracker, (**b**) MPA preamp for strixel (long pixels) detector of CMS, and (**c**) FEI4 preamp for the ATLAS pixels

the detector capacitance is specified for 300 fF only, since the chip is intended for low cost bump bonding, this capacitance doubles because of big input pad and ESD structure applied. The input transistor is biased with 16 μA current what for the peaking time of 24 ns and nominal input load provide the noise below 300e$^-$. Again the input stage is based on the telescopic cascode stage with NMOS input device and extra current source optimizing the bandwidth. For the relatively low biases a GBP above 2 GHz has been achieved thanks to high f_T of the process used. A relatively moderate open loop gain being of the order of 55 dB is still sufficient for this application. Circuit is intended to support DC coupled n-type detectors and therefore it employs the Krummenacher feedback (Fig. 8d).

Figure 10c shows the simplified schematic of the preamplifier for ATLAS pixel front end chip FEI4 implemented in IBM 130 nm process. The new pixel layer using this front end chip has been installed recently in ATLAS in the scope of high energy upgrade. The analog part of the single pixel consumes about 10 μA and for nominal 500 fF capacitance the ENC is specified to be below 300e$^-$. As in the other examples the input stage is built with telescopic cascode with NMOS input device. The use of regulated cascode has been justified by the possibility of reduction of one bias line which facilitate layout and had positive impact on the PSRR. Circuit uses active feedback described in Fig. 8c.

5 Future Trends: Active Sensors in Commercial CMOS

Reverse biased, fully depleted silicon p-n diode arrays fabricated on high resistivity (kΩ cm) wafers in dedicated processes have become the standard sensors of the inner detector layers. For a typical thickness of a few hundred microns, they deliver a signal charge of a few fC upon a particle traversal, resulting in a collected charge over input capacitance (Q/C) ratio of $\sim$0.1 mV for silicon strip detectors and up to $\sim$10 mV for pixel detectors leading with the present granularities to power densities of $\sim$10 mW/cm^2 and several hundred mW/cm^2.

Standard CMOS processes are now receiving interest also for the sensor fabrication not only because of the lower cost per unit area, but also because the readout or part of it can be integrated with the sensor. So far only two high energy physics experiments installed monolithic detectors: Depleted P-channel Field Effect Transistor (DEPFET) [26] pixels in the Belle-II experiment at KEK in Japan (not based on a commercial CMOS technology) [27] and CMOS Monolithic Active Pixel Sensors (MAPS) in STAR at Brookhaven National Laboratory in the US [28]. ALICE is the first LHC experiment to adopt MAPS for an important upgrade. It plans to fully replace its Inner Tracking System (ITS) [29, 30] in 2018 with a new tracker covering 10 m^2 constructed out of 4.5 cm^2 chips each containing $\sim$500,000 channels (see Fig. 11 for a die picture) [31, 32].

In many MAPS devices (Fig. 12a) including the ones for ALICE and STAR the epitaxial layer from which signal charge is collected is not fully depleted leading to an important diffusion component in the charge collection and hence only moderate radiation tolerance. For ATLAS and CMS collection by drift from a depleted layer is mandatory to meet radiation tolerance requirements.

Traditional MAPS devices often limit the in-pixel circuit to a few transistors of a single type (Fig. 12b), and combine it with a rolling shutter readout, incompatible with the time resolution requirements for ATLAS and CMS, which require more complex full CMOS in-pixel circuitry.

For ALICE a MAPS technology has been adopted offering this possibility by shielding the Nwells containing PMOS transistors with a deep Pwell implant from the epitaxial layer preventing them from collecting any signal charge [33]. This way the signal charge is collected (by diffusion) from underneath the circuitry, detecting particle traversals everywhere in the pixel, even in the area occupied by the circuit. Such full efficiency over the pixel surface is essential for high energy physics.

Fig. 11 Die picture of the first large scale prototype for the ALICE ITS upgrade. The chip measures 3 cm by 1.53 cm and contains 500,000 pixels [32]. The chip is intended for flip-chip bump-bonding on a carrier, but has been wire-bonded for the first tests

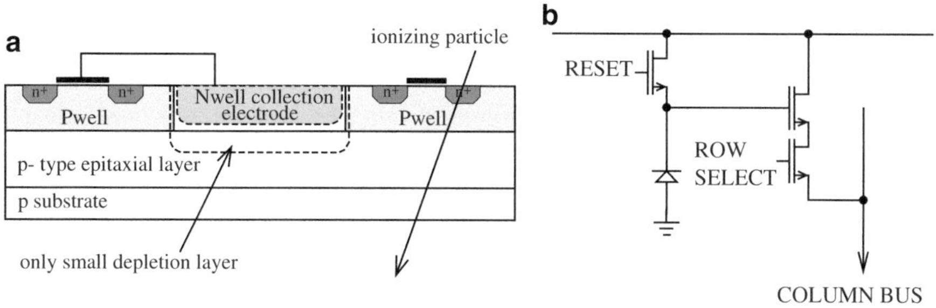

Fig. 12 (**a**) In traditional MAPS devices only a small fraction of the sensitive epitaxial layer is depleted, resulting in a significant diffusion component in the charge collection and hence only moderate radiation tolerance, and (**b**) the in-pixel circuit is often limited to only a few transistors of the same type

For the ALICE ITS upgrade a capacitance reduction at the input node to about 2.5 fF yielded Q/C values above 80 mV spread over only few pixels. This favorable value has allowed the development of an amplifier-comparator only consuming about 40 nW for a few μs peaking time, corresponding to a low analog power density of about 5 mW/cm^2. This was reached in this first version with an open loop amplifier, as often used with MAPS, operating deep into weak inversion. A PMOS input transistor was used to take advantage of the non-linear behavior caused by signals of many tens of mV. Challenges are discrepancies of mismatch in weak versus strong inversion as for instance observed also in [34], modeling issues more in general, random telegraph noise in the very small devices where also new results still become available [35], etc. This is clearly work in progress and we still hope to make further improvements. Nevertheless, the prototype chip satisfies the specifications for the ALICE experiment, and the result obtained illustrates the importance of optimizing Q/C, and hence the need for a small collection electrode.

Preserving a small collection electrode and efficiency over the full pixel surface for more complex readout circuitry and combining it with charge collection from a depleted layer has remained a challenge in commercial CMOS technologies (some functional devices have been demonstrated, but required ‘exotic’ process steps like double sided processing [36]), but several promising developments exist (for a few examples see the CPIX workshop [37]) and it is likely some solution will be found in the near future for monolithic detectors. It may also be that with progress in interconnect technology (through silicon vias, wafer bonding, ...) some form of hybrid technology would also provide an economically viable solution.

Future upgrades should not increase the material in the detector as this would increase the probability that particles interact with this material, scatter and deviate from their trajectory degrading the quality of the measurements. Cables providing power and cooling pipes removing it dominate this material. Therefore, to replace the outer tracker layers of strip detectors, where the economic impact is the largest due to the large area, matching or staying below their power density of a few tens of mW/cm^2 is essential. A further Q/C improvement would help to reduce the analog power, and would practically eliminate it if a few hundred mV can be reached on a single pixel [38]. Such improvement may not be totally out of reach: a fully depleted epitaxial layer would reduce the number of pixels in a typical cluster to close to 1. A further factor 3 or 4 improvement by lowering C in a finer technology or increasing Q by a thicker sensitive layer would bring the 80 mV value obtained for the ALICE ITS upgrade to these few hundred mV. This value and the non-linearity offered by the weak inversion region would then also be sufficient to decrease the reaction time of the circuit to values compatible with 25 ns time resolution as it would allow a particle hit to turn on a transistor to the μA level.

Of course power is not only consumed in the analog part but also in the digital circuitry gathering the data on-chip and transmitting it off-detector. It is clear that distributing a clock to every pixel in a highly granular detector ($10 \times 10\ \mu m^2$) is prohibitive for a target power density of ~10 mW/cm^2, and new hit-driven architectures will be essential to enable such a detector to replace the outer tracker layers. If this materializes, the pressure for low power will only increase as the position resolution offered by such detector can only be fully exploited if the detector material is further decreased.

6 Conclusions

Although the number of the presented designs is limited in view of the variety installed in the CMS and ATLAS experiments at the LHC, we hope that the presented selection is still representative. Implementations might differ especially when the specifications can provide more freedom in terms of power or radiation levels. There is a clear tendency to use in the input stages the telescopic cascodes with NMOS input device for the front end electronics built for inner tracker systems, where the driving key is the power for a specified speed and noise. The examples of the designs for the ATLAS and CMS inner tracker detectors are the successors of the chips presently operating in the experiments implemented in older technologies. Although for the moment they are chosen as a baseline, it is still not clear they will be used in the final systems without any modifications. A number of scenarios are possible, including a complete change of the front end circuit design—and the construction of the full inner tracker detectors—if one takes into account recent developments of radiation hard monolithic CMOS sensors and the further Q/C optimization they may offer.

Acknowledgement The authors would like to thank the conference organization for the invitation to the conference and the colleagues from the ESE group and the ATLAS, CMS and ALICE collaborations with whom they have been working over the years.

References

1. CMS Collaboration (2008) J Instrum 3:S08004, ISBN 978-92-9083-336-9
2. ATLAS Collaboration (2008) J Instrum 3:S08003, ISBN 978-92-9083-336-9
3. Van Vonno N (1996) In: CERN/LHCC/96-39, p 411
4. Liu M (1996) In: CERN/LHCC/96-39, p 407
5. Redolfi J (1995) In: CERN/LHCC/95-56, p 38
6. Dentan M et al (1993) IEEE Trans Nucl Sci 40(6):1555. doi:10.1109/23.273505
7. Saks NS et al (1984) IEEE Trans Nucl Sci 31(6):1249. doi:10.1109/TNS.1984.4333491
8. Saks NS et al (1986) IEEE Trans Nucl Sci 33(6):1185. doi:10.1109/TNS.1986.4334576
9. Mavis, D.G., Alexander, D.R. (1997) Employing radiation hardness by design techniques with commercial integrated circuit processes. In: Digital Avionics Systems Conference, 1997. doi: 10.1109/DASC.1997.635027
10. Anelli G et al (1999) IEEE Trans Nucl Sci 46(6):1690. doi:10.1109/23.819140
11. Bonacini S et al (2012) J Instrum 7:P01015. doi:10.1088/1748-0221/7/01/P01015
12. Kaplon J, Kulis S. PH-EP-Tech-Note-2015-001. http://cds.cern.ch/record/2002443?ln=en
13. Anghinolfi F et al (2004) IEEE Trans Nucl Sci 51(5):1974. doi:10.1109/TNS.2004.836048
14. Blin S et al (2010) In: Nuclear science symposium conference record (NSS/MIC). IEEE, pp 1690, 1693. doi:10.1109/NSSMIC.2010.5874062
15. Lucotte A et al (2004) Nucl Instrum Methods A 521:378. doi:10.1016/j.nima.2003.10.104
16. Dressnandt N et al (2009) In: Proceedings of TWEPP2009, September 21–25, Paris, France. doi:10.5170/CERN-2009-006.132
17. Chase RL, Rescia S (1997) IEEE Trans Nucl Sci 44:1028. doi:10.1109/23.603798
18. Bilotti A, Mariani E (1975) IEEE J Solid State Circuits SC-10:516. doi:10.1109/JSSC.1975.1050652
19. Jarron P et al (2006) Nucl Instrum Methods A 377:435. doi:10.1016/0168-9002(95)01454-3
20. Peric I et al (2006) Nucl Instrum Methods A 565:178. doi:10.1016/j.nima.2006.05.032
21. Krummenacher F (1991) Nucl Instrum Methods A 305:527. doi:10.1016/0168-9002(91)90152-G
22. Kaplon J, Noy M (2012) IEEE Trans Nucl Sci 59:1611. doi:10.1109/TNS.2012.2200503
23. Przyborowski D et al (2014) Design of a front-end ASIC for single crystal diamond sensors application as beam condition monitors in CMS and LHC. In: Presented on ACES 2014—fourth common ATLAS CMS electronics workshop for LHC upgrades
24. Raymond M et al (2012) J Instrum 7, C01033. doi:10.1088/1748-0221/7/01/C01033
25. Ceresa D et al (2014) J Instrum 9, C11012. doi:10.1088/1748-0221/9/11/C11012
26. Kemmer J et al (2007) Nucl Instrum Methods A 253:365. doi:10.1016/0168-9002(87)90518-3
27. Marinas C et al (2011) Nucl Instrum Methods A 650:59. doi:10.1016/j.nima.2010.12.116
28. Dorokhov A et al (2011) Nucl Instrum Methods A 640:174. doi:10.1016/j.nima.2010.12.112
29. Musa L et al (2012) Tech. rep. CERN-LHCC-2012-013. LHCC-P-005, CERN, Geneva. http://cds.cern.ch/record/1431539?ln=en
30. Musa L (2013) In: CERN-LHCC-2013-024; ALICE-TDR-017
31. Keil M (2014) In: Presented at PIXEL 2014, Niagara Falls
32. Yang P et al (2014) In: Presented at PIXEL 2014, Niagara Falls
33. Senyukov S et al (2013) Nucl Instrum Methods A 730:115. doi:10.1016/j.nima.2013.03.017
34. Pineda de Gyvez J et al (2014) IEEE J Solid State Circuits 39:157. doi:10.1109/JSSC.2003.820873
35. Banaszeski da Silva M et al (2014) In: IEDM 2014
36. Snoeys W et al (1994) IEEE Trans Nucl Sci 41(6):903–912. doi:10.1109/16.293300
37. https://indico.cern.ch/event/309449/
38. Snoeys W (2014) Nucl Instrum Methods A 765. doi:10.1016/j.nima.2014.07.017

Part III
Low-Power RF Systems

The third part of this book is dedicated to recent developments in wireless communication systems, with a particular focus on achieving better power efficiency. The emergence of multi-standard mobile devices and internet of things drive the need for power optimization and interference resilience. The six chapters discuss a variety of RF circuits and systems, including clock references, receiver and transmitter components, as well as integrated SoCs for wearable, medical and automotive applications.

The first paper, by Raghavasimhan Thiruarayanan et al., describes the problem of energy overhead in duty-cycled transmitters due to the slow start-up of conventional crystal oscillators. The use of FBAR resonators is proposed, resulting in a frequency synthesizer that wakes up in 3 μs, while also supporting data rates up to 16 Mbps and high frequency agility.

In the second paper, by Alan Wong, the development of an SoC for wireless body area networks is described. The system includes a wireless frontend, sensor interfaces and power management. After an overview of wireless standards, the requirements, architecture and implementation of the SoC are discussed. The SoC in 65 nm CMOS operates down to 1.1 V and consumes 0.5 μW of power on average.

The third paper, by Eric Klumperink et al., gives an overview of advances in N-path filters. This type of filters provides high linearity, high Q, programmability, and is well scalable in CMOS technologies. These properties are advantageous for multi-standard interference-resilient radios. The paper describes the general concept and modeling methods of N-path filters, and also reviews recent developments and state-of-the-art implementations.

The fourth paper, by Patrick Reynaert and Brecht Francois, gives an extensive overview of RF and mm-wave power amplifiers in CMOS, as used in wireless transmitters. Various amplifier classes, power combining strategies and multi-path techniques are described and analyzed, where the main focus is on energy efficiency and linearity. A fully-integrated Doherty RF PA with on-chip power combiner is presented, achieving high output power, linearity and efficiency.

In the fifth paper, Yao-Hong Liu describes the design and implementation of an energy-efficient phase-domain receiver. After presenting an overview of existing receiver architectures, a sliding-IF receiver with phase-to-digital converter is proposed. The phase-domain architecture is favorable for low-voltage operation in scaled technologies. Fabricated in 90 nm CMOS, the RX achieves an efficiency of 1.2 nJ/bit, with 2 Mbps data rate and -92 dBm sensitivity.

The sixth paper, by Jérémie Chabloz et al., describes a low-power versatile CMOS transceiver for automotive applications, which can be used for remote keyless entry and ignition, tire pressure monitoring systems and immobilizer functionality. The system architecture and circuit details are described. The implemented chip in 0.18 μm CMOS provides a versatile, cost-effective and low-power solution for the advent of wireless communication interfaces in cars.

PLL-Free, High Data Rate Capable Frequency Synthesizers

Raghavasimhan Thiruarayanan, David Ruffieux, and Christian Enz

Abstract The PLL based frequency synthesizer has been the main impediment to achieve a low energy dissipation in the radios employed in Wireless Sensor Networks (WSN); in spite of duty cycling. This is due to the crystal oscillator reference in the frequency synthesizer which dissipates significant energy during its long wakeup phase. To address this issue and thereby obtain the full advantage proffered by duty cycling, an overview of FBAR based synthesizers which can wake up in just 5 μs is presented. In addition, these synthesizers can also support high data rates as compared to the PLL based radios, thereby offering the possibility to increase the rate of duty cycling and thus further lowering the average energy dissipation.

1 Introduction

The evolution of radios over the past few years has seen a continuous increase in the maximum data rate that can be achieved as well a reduction in the power dissipation during communication. State-of-the-Art (SoTA) radios in the 2.4 GHz ISM band achieve 2 Mbps peak data rate while consuming 5.4 mW of power [1]. But in reality, such high peak data rates are needed only for continuous communication like video streaming. In the case of Ultra-Low Power (ULP) systems like WSN, much lower data rates in the order of a few kbps would suffice. Such systems are usually battery powered and therefore there is a need to improve the energy autonomy. For example, let us consider a wireless motion tracking system

R. Thiruarayanan (✉)
EPFL, Neuchâtel, Switzerland

CSEM, Neuchâtel, Switzerland
e-mail: raghavasimhan.thirunarayanan@epfl.ch;
Raghavasimhan.THIRUNARAYANAN@csem.ch

D. Ruffieux
CSEM, Neuchâtel, Switzerland

C. Enz
EPFL, Neuchâtel, Switzerland

K.A.A. Makinwa et al. (eds.), *Efficient Sensor Interfaces, Advanced Amplifiers and Low Power RF Systems*, DOI 10.1007/978-3-319-21185-5_13

with 10 degrees of freedom (DoF) (three gyroscope, three magnetometer, three accelerometer and one pressure sensor). Such a sensor networks need to achieve a data rate of 10 kbps while consuming 10 μW to improve the battery life (with a CR2032 button cell battery, it amounts to 2.5 years of lifetime). But unfortunately the radios that can achieve such low data rates consume about 1 mW (100×) more power. Therefore, duty cycling of high data rate capable radios has emerged as the preferred technique to achieve extremely low energy dissipation in Wireless Sensor Networks (WSN). In this method, the radio is asleep most of the time and only wakes up intermittently to communicate data in short bursts at high data rates. Since the radio is powered off most of the time (i.e., during the sleep phase), this leads to a reduction of the energy dissipated in the system. The main deterrent to this process of duty cycling is the non-deductible energy overhead of the system, whose main source is the crystal oscillator (XO) reference present in Phase Locked Loop (PLL) based frequency synthesizers. The XO consumes around 50–100 μW power for continuous operation, which is greater than the power spent to communicate. On the other hand, even if the XO is duty-cycled, it takes about 1 ms to wake up and consumes around 0.5–1 mW of power for this purpose, during which time, the radio remains idle and cannot communicate. Thus, this energy spent in wake up is effectively wasted. The common Energy dissipated per bit communicated (Energy/bit) figure of Merit (FoM) that is used to quantify the energy efficiency of the SoTA radios fails to take into account this energy wasted during wakeup and thus does not reflect on the true energy efficiency of the system. As the maximum data rate that the radio can handle increases, the duration of the packet communicated as well as the energy spent for communication decrease and subsequently the latter becomes comparable to the energy wasted as XO overhead. Figure 1 depicts this

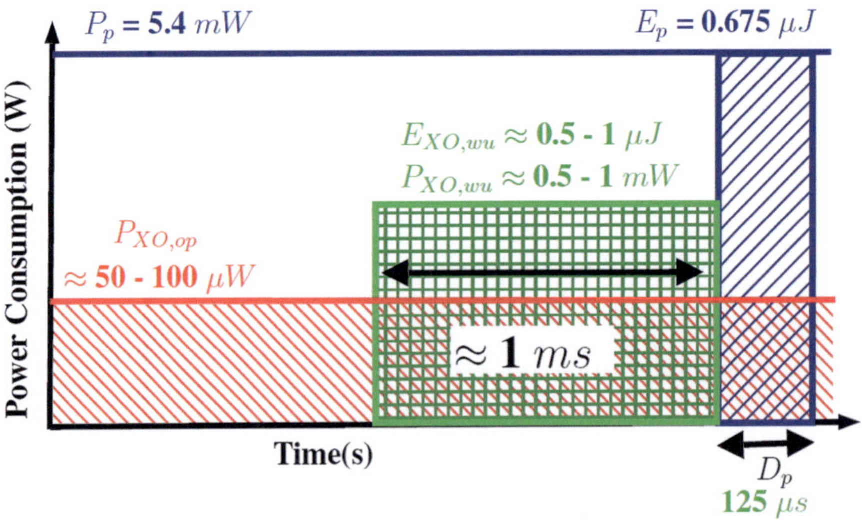

Fig. 1 PLL based transmitter duty cycling showing crystal oscillator wake up overhead

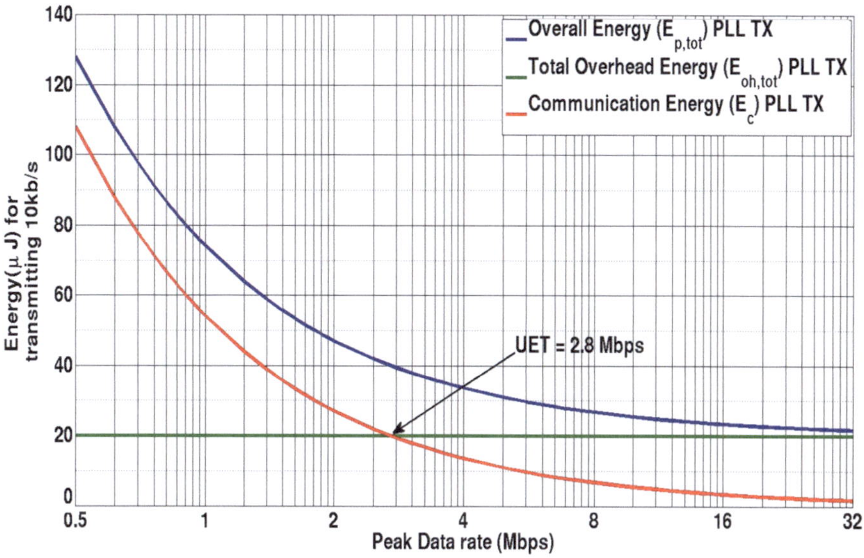

Fig. 2 Energy dissipation of a 10 kbps ULP system with the PLL based TX of [1]

scenario where a radio consuming 5.4 mW transmits a data packet of duration 125 μs (corresponding to a 32 byte packet: 2 byte per DoF plus CRC and packet header for the motion tracking system mentioned earlier). Here, the energy spent for communicating one packet is 0.675 μJ, while the energy wasted in the XO wake up is of the same order (0.5 μJ). Therefore, there exists a crossover point of the energy spent for communication and the energy wasted called the *Useful Energy Threshold* (UET) of the system below which any increase in data rate is projected to have limited impact on the actual energy efficiency. This is depicted in Fig. 2 which shows the energy plots for aforementioned example (i.e. 10 kbps average data rate) with the UET being identified as 2.8 Mbps.

Therefore, the only way to keep the migration towards higher peak data rates meaningful is to reduce the wake up time of the XO which is used as the frequency reference. The wake-up time of the XO (subsequently the radio) is directly proportional to the Q-factor of the crystal. Typical crystals have a Q in the order of 10^5 and therefore need a long time to wake up. Thus there is a need for an alternative to the crystal which can wake up in a short time span. To aid this fast wake up, any alternative should have lower Q as compared to the crystal but at the same time Q should not be too low as to affect the phase noise performance of the synthesizer. Thin-Film Bulk Acoustic Resonators (FBAR) [2, 3] which belong to the category of Micromachined resonators, offer a solution to this problem since they have Q factor in the order of 500–2000 and thus possess wake up times in the order of a few μs. In addition, the frequency of the FBAR is in GHz (1–7 GHz) range and hence are conducive to loop-free architectures at these frequencies; which facilitates higher data rates as compared to the PLL based synthesizers. Further details about the FBAR and the evolution of the FBAR based synthesizer architecture are given in the following sections.

2 FBAR

FBARs, which are a class of Bulk Acoustic Wave (BAW) resonators have found widespread use in duplexers due to their small size, high rejection and low insertion loss [4]. They consist of a piezoelectric material (typically AlN) sandwiched between two electrodes as shown in Fig. 3. When an electric field is applied between these electrodes, it causes mechanical deformation of the piezoelectric material. This causes an acoustic wave that travels in the direction of the thickness of the piezoelectric film for a particular orientation of electric field. The acoustic wave is reflected back at the film interface with air due to impedance mismatch. When the thickness of the film equals an integer multiple of half wavelength, a standing wave is created by the forward travelling wave and the reflected wave. This acoustic wave in turn modifies the electric field distribution inside the piezoelectric film which changes the electrical impedance of the device. Thus, the electrical impedance of the resonator changes with frequency [5]. The electrical equivalent of this circuit is given by the Butterworth Van Dyke model (Fig. 4) and is similar to that of a quartz crystal [6]. It consists of a series RLC network along with a parallel capacitance. This gives rise to two resonance frequencies one corresponding to the series RLC branch (series resonance) and the other corresponding to the total resonator itself (parallel or anti-resonance). The relation between these frequencies is given by

$$\omega_s = \frac{1}{\sqrt{L_m C_m}}$$

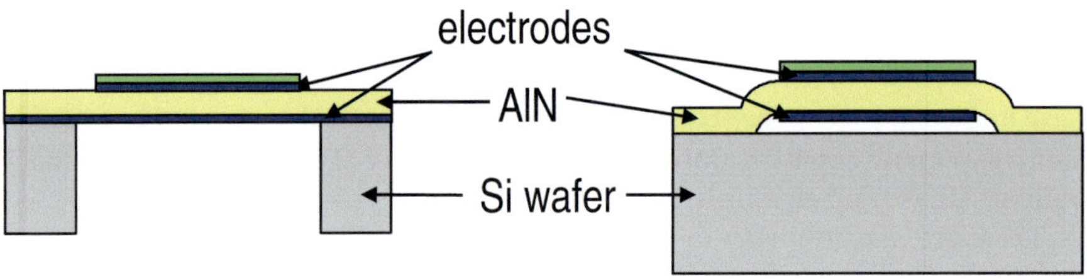

Fig. 3 Cross section of an FBAR

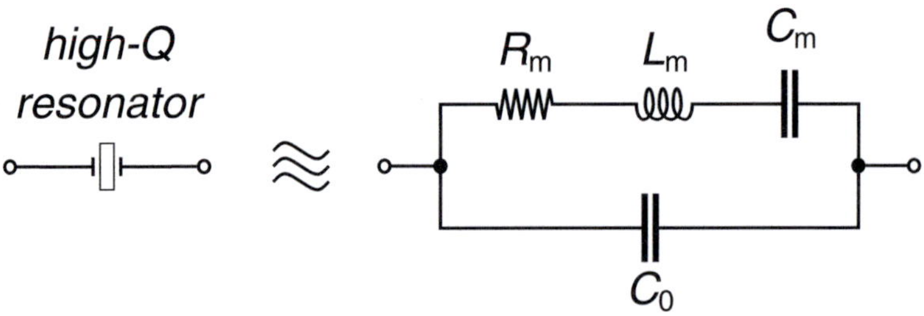

Fig. 4 Equivalent circuit of an FBAR

$$\omega_p = \omega_s \sqrt{1 + \frac{C_m}{C_p}} \tag{1}$$

The main performance parameters of the FBAR are the effective electromechanical coupling coefficient k_{eff} and the Q factor which specifies the energy loss in the resonator material. The coupling coefficient relates to ratio of the stored mechanical energy or in other words is the ratio of the current in the motional branch (series branch) to the parallel branch. The relationship between this coupling coefficient and the resonance frequencies is given by

$$k_{eff}^2 = \frac{\pi^2}{4}\left(\frac{\omega_p - \omega_s}{\omega_p}\right) \tag{2}$$

Thus, the coupling coefficient determines the interval of the resonance frequencies and therefore the tenability of the FBAR. Typical values of the coupling coefficient for an FBAR is about a few % and typical values of Q are in the range of 500–2000.

3 FBAR-Based Synthesizer Architecture

Figure 5 depicts the block diagram of the FBAR based synthesizer architecture. It begins by a temperature compensated FBAR DCO providing the LO signal which is then divided by a fixed value to produce the IF signal i.e., $f_{IF} = f_{LO}/N$; following which the IF is upconverted by the LO in a mixer to produce the wanted frequency ($f_{RF} = f_{LO} + f_{IF}$). The channel selection is done by tuning the FBAR frequency as

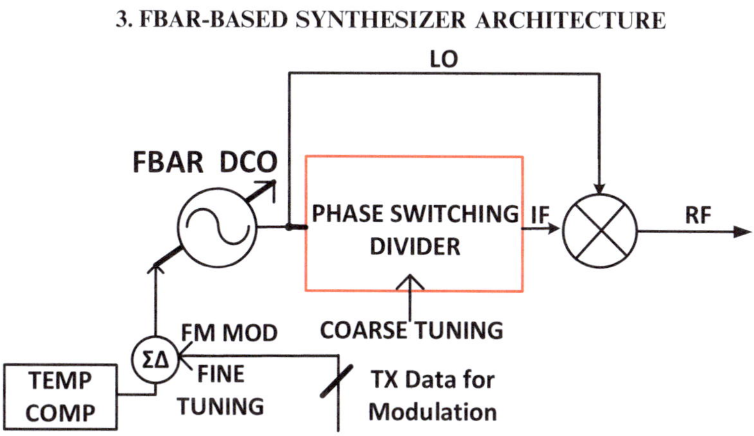

Fig. 5 Block diagram of the FBAR based synthesizer architecture

well as adjusting the value of N. The centre frequency of the FBAR is chosen to make sure that the IF harmonic spurs fall outside the band of interest. For instance, with the band of interest in this case being $f_{RF} = 2.36$–2.5 GHz and the channel to be addressed is at 2.36 GHz, if f_{LO} is chosen to be greater than or equal to 2.2 GHz, the second harmonic IF spur will be located at frequencies $\leq$2.5 GHz which is within the ISM band. Therefore the constraint on the LO is given by $f_{LO} < 2.2$ GHz. If the synthesizer is used in a Transmitter (TX), then modulation can be performed by varying the DCO frequency.

The divider for producing the IF is implemented as a Phase Switching Divider (PSD) with a division step of 0.2. This is due to the limited FBAR tuning available to address all channels in a given band. Any DCO utilizing this synthesizer architecture must satisfy the condition given in (3) to make sure that the centre frequency corresponding to all channels within a given frequency band can be synthesized

$$\Delta f_t \geq \frac{k \cdot f_{LO}}{N_L^2 + (k+1)\,N_L + k} \tag{3}$$

where Δf_t is the tuning range required on the FBAR, N_L is the lowest division ratio required for IF generation (which corresponds to the highest IF) and k is the division ratio step. For instance, considering the aforementioned example with f_{LO} being 2.2 GHz, if the aim is to address the 2.36–2.5 GHz band, the lowest division ratio N_L is 7.33 (for $f_{RF} = 2.5$ GHz and $f_{IF} = 300$ MHz). If a divider with a step size $k = 1$ is used, the tuning range required to address all the channels according to (1) is $\Delta f_t = 31.6$ MHz. This is impossible to achieve for an FBAR thus making this architecture unsuitable to cover a wide frequency range. This indeed was the issue with the synthesizer presented in [3]. In addition to this, some of the tuning ($\simeq$1 ‰) needs to be allocated to account for the PV variations of the FBAR [7]. Whatever tuning range is left after the allocation for channel coverage and PV variations is used for modulation (in the case of a TX). Thus, the only option available to relax this tuning bottleneck and thereby increase the maximum possible data rate is to reduce the value of k in (1). This has been accomplished by using a PSD with a division ratio step of 0.2 which reduces the tuning range required to cover the entire 2.36–2.5 GHz band to $\Delta f_t = 7$ MHz, which is in the order of the nominal tuning range of an FBAR [8].

Therefore, for a typical FSK modulation with a fixed modulation index, the tuning range of the FBAR dictates the maximum achievable data rate of the TX while being able to address all the channels within a given frequency band of interest. This maximum data rate is given by

$$Max.\ Data\ Rate\ \ (all\ \ channel\ \ coverage) = \frac{(TR - PV - \Delta f_t)}{MI} \tag{4}$$

where *TR* is the maximum available FBAR tuning range, *PV* is the tuning range required to compensate for the Process and Voltage variations and *MI* is the FSK modulation index. The maximum data rate can go above this value given in (4), but in that case only a few frequency channels can be addressed.

With the system level considerations of the synthesizer having been dealt with, the description of the different synthesizer blocks is given in the following subsections.

3.1 FBAR DCO

FBAR DCOs have the advantage of providing LO signals with very low phase noise along with fast wake up [9]. The transistor level schematic of an FBAR DCO is shown in Fig. 6. The DCO is implemented using a complementary structure in order to halve the power dissipation [10]. It consists of two cross-coupled pairs providing negative conductance to compensate for the energy loss in the FBAR. To avoid the latch-up of the circuit at start-up, the cross coupled pairs are ac coupled at

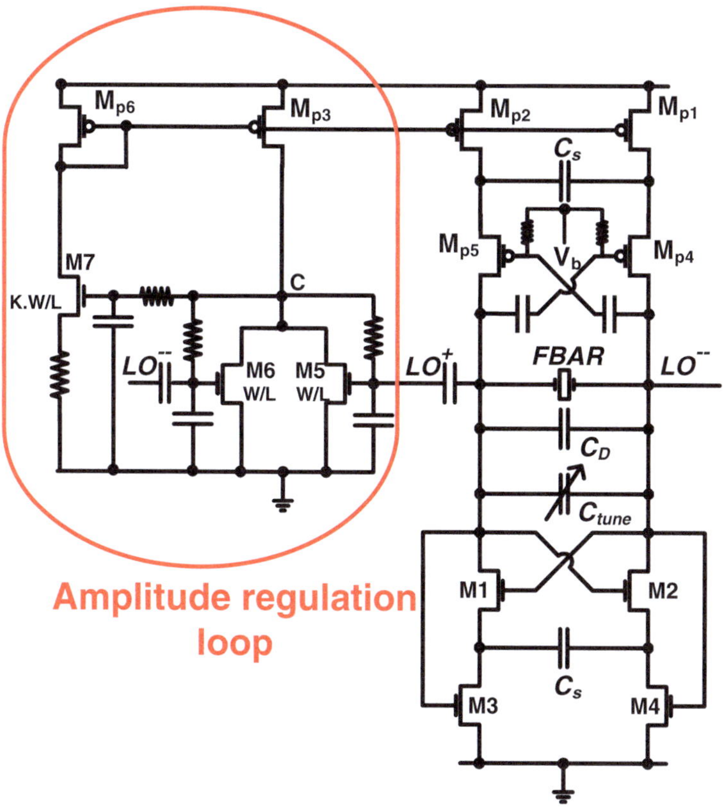

Fig. 6 FBAR DCO schematic

their sources with common mode feedback provided by two NMOS transistors (M3 and M4) are implemented below the NMOS cross coupled pair. The circuit also includes an amplitude regulation loop (M5–M7) based on the concept of a PTAT current reference [11]. A bank of 31 pairs of depletion/inversion MOS capacitors coarsely tune the DCO. Coarse tuning is also accomplished by changing the division ratio of the PSD. The DCO also has three other MOS capacitances driven by the output of a 7-bit second order $\Sigma\Delta$-modulator enabling fine tuning. These two tuning mechanisms are cumulatively shown as C_{tune} in Fig. 6. The frequency resolution of this synthesizer is around 0.9 ppm and the maximum tuning range is set by the resonator itself.

3.2 Phase Switching Divider, Mixer and PA

The PSD block diagram is shown in Fig. 7. It is made up of a five stage ring oscillator (ILRO) that is injection locked to the FBAR DCO.

This ring oscillator produces five phases at the LO frequency that are spaced $0.2/f_{LO}$ apart in time. The PSD also consists of a finite state machine (FSM) that

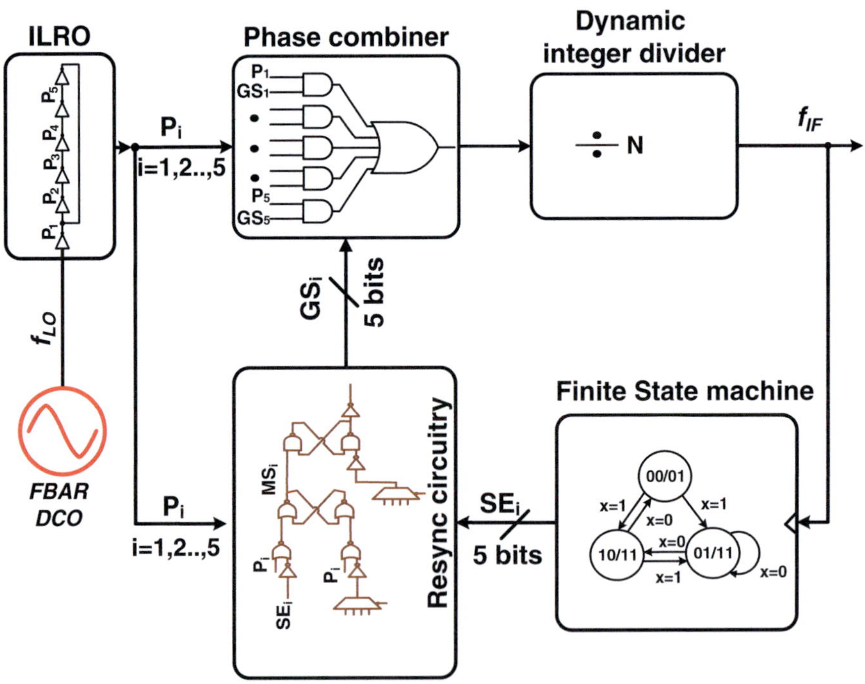

Fig. 7 Schematic of the PSD

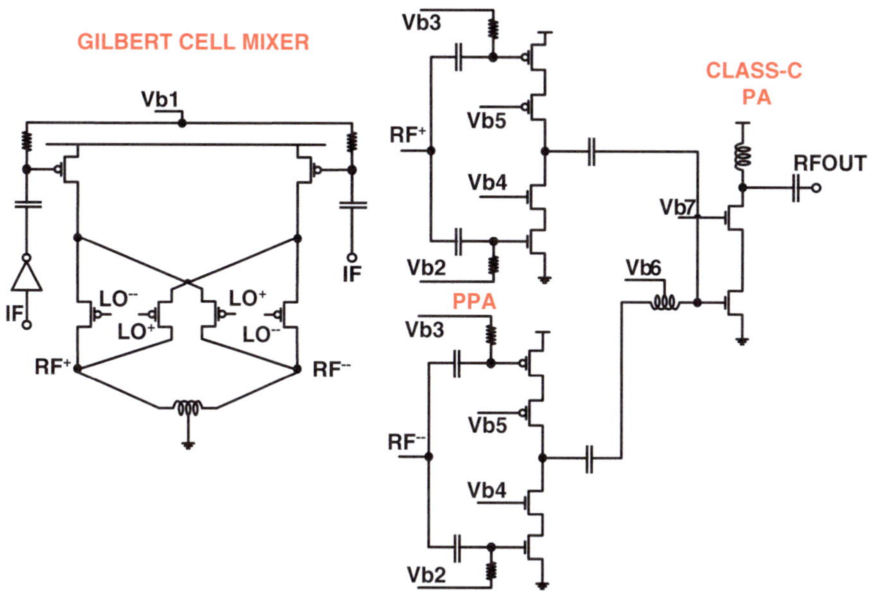

Fig. 8 Gilbert Cell Mixer and Class-C PA

produces the select signals with each of them corresponding to a phase signal from the ILRO. The select signals are resynchronized using the edges of the phase signals following which the multiplication of the select signal and its phase occurs in a phase combiner. The resulting signals are then summed together using OR gates. This summed signal is then fed to an integer divider that is set to divide by the nearest integer of the division ratio required. The output of this integer divider is the wanted IF signal which also clocks the FSM. The mixer is implemented as a single-balanced Gilbert cell with resonant load at 2.44 GHz to reduce the synthesizer loading. It is followed by a class-C Power Amplifier section as shown in Fig. 8.

4 Measurements

This synthesizer was implemented in a high data rate TX system. The FBAR used was measured to have a maximum tuning range of 4.9 ‰. After accounting for PV variations and tuning range to cover the entire frequency band (2.4–2.48 GHz ISM band), the maximum achievable data rate was found from (4) to be 12 Mbps (FSK). If only a few select channels need to be addressed, then this figure rises to 16 Mbps. The eye diagrams of the high data rates are given in Fig. 9.

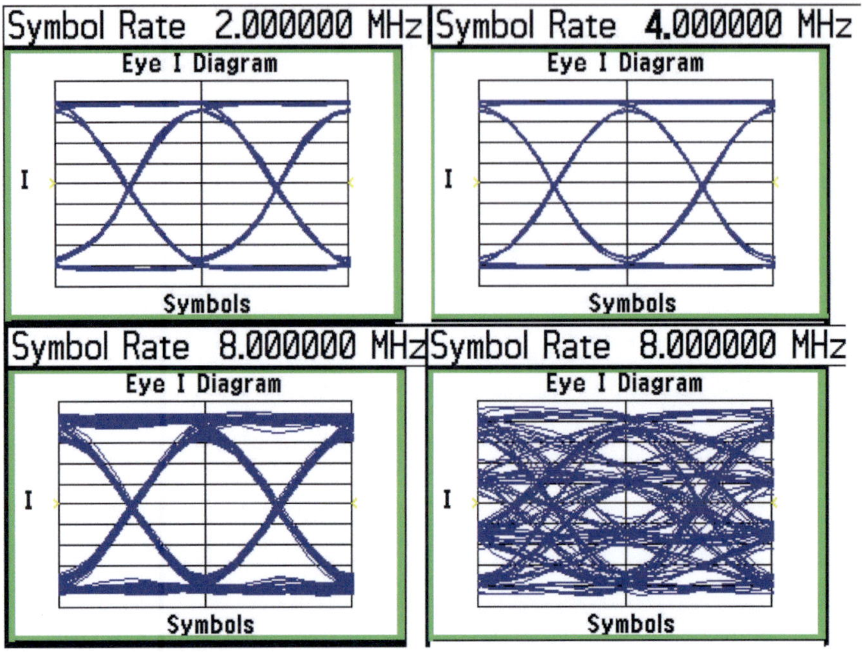

Fig. 9 GFSK eye diagrams at 2 Mbps (*top left*), 4 Mbps (*top right*), 8 Mbps (*bottom left*) and 16 Mbps (8 Mbps – 4 FSK) (*bottom right*)

The frequency agility of the TX is shown in Fig. 10 which depicts the time domain TX output versus frequency with 1 Mbps FSK with a modulation index of 0.5. From the current profile it is clear that the TX takes only 5 μs to start transmission after wake-up; with the synthesizer itself (FBAR DCO) starting up in just 2 μs. The rest of the circuit takes 3 μs after which the sample data pattern "0110 1010" can be seen. The TX can be turned off in 3 μs while channel switching also can be performed in just 3 μs. This frequency agility is one of the main advantages of this TX since it allows us to perform frequency hopping to any channel within the band in a span of 3 μs, while still being a narrow-band system.

The TX consumes 9.2 mA from a 1.2 V supply. Based on this data, the energy consumption of this TX system was calculated akin to the introduction and plotted in Fig. 11, with the synthesizer overhead energy being reduced to 36 nJ (down from 1 μJ) which is a 34× improvement over the XO based PLL system. This comparison is plotted in Fig. 11 which also shows the improvement (7×) in the overall energy dissipation figures.

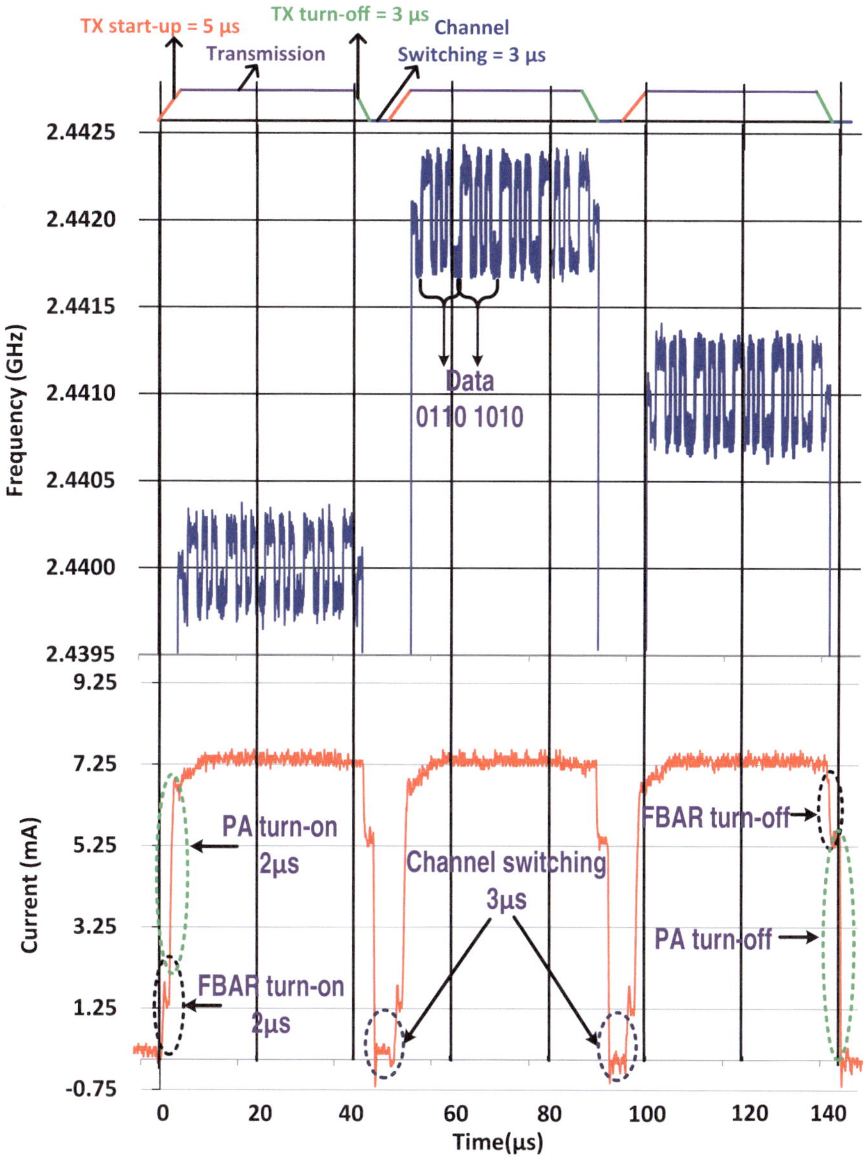

Fig. 10 Frequency agility of the FBAR based synthesizer (TX)

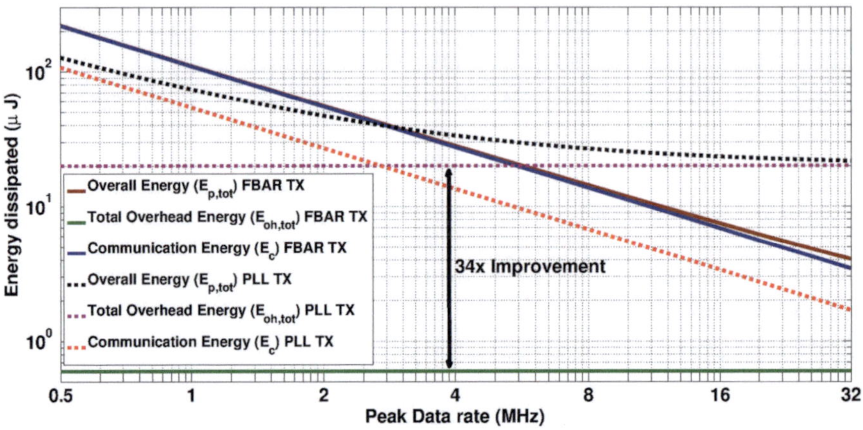

Fig. 11 Energy dissipation variation in the PLL based TX and the FBAR based TX with increasing peak data rate

5 Conclusions

An overview of the advantages and design of an FBAR based fast wake-up frequency synthesizer has been presented herein, along with an application in a transmitter. Taking advantage of the fast wake-up of the FBAR in 3 μs, this synthesizer achieves a substantial (34×) reduction in energy overhead, while also supporting very high data rates up to 16 Mbps. Combined with fast channel switching ability, this synthesizer serves as an ideal candidate for minimizing power consumption in Ultra Low Powered systems such as WSN and BAN.

References

1. Liu Y (2013) A 1.9 nJ/b 2.4 GHz multi standard transceiver for personal/body-area networks. In: ISSCC digest of technical papers, pp 446–447
2. Heragu A (2013) A 2.4 GHz MEMS-based PLL-free multichannel receiver with channel filtering at RF. IEEE J Solid State Circuits 48:1689–1700
3. Larson JD et al (2000) Power handling and temperature coefficient studies in FBAR duplexers for the 1900 MHz PCS band. In: Ultrasonics symposium, 2000 IEEE, vol 1, pp 869–874
4. Zhang Y et al (2013) Multilayer integrated film bulk acoustic resonators. Springer, Berlin
5. Vittoz E (2010) Low-power crystal and MEMS oscillators—the experience of watch developments. Springer, Netherlands
6. Wang K (2014) A 1.8 mW PLL-free channelized 2.4 GHz ZigBee receiver utilizing fixed-LO temperature-compensated FBAR resonator. In: ISSCC
7. Gilbert S (2013) Sub-10 fs jitter S-band oscillators and VCOs in a 1 × 1 × 0.23 mm^3 chip scale package. In: Proceedings IEEE frequency control symposium
8. Thirunarayanan R (2014) A 700 pJ/bit, 2.4 GHz, narrowband, PLL-free burst mode TX based on an FBAR with 5 μs startup time for highly duty-cycled systems. In: RFIC symposium

9. Shi J, Otis B, et al (2011) A sub-100 μW 2 GHz differential Colpitts CMOS/FBAR VCO. In: Proceedings IEEE custom integrated circuits conference
10. Thirunarayanan R (2012) Complementary BAW oscillator for ultra-low power consumption and low phase noise. J Analog Integr Circuits Signal Process 73(3):769–777
11. Vittoz E et al (1977) CMOS analog integrated circuits based on weak inversion operations. IEEE J Solid State Circuits 12(3):224–231

Ultra Low Power Wireless SoC Design for Wearable BAN

A.C.W. Wong

Abstract This paper discusses the key features and technical considerations in the design of ultra-low power wireless system-on-chips (SoC) for wireless body area networks (WBAN). The requirements of these sensor nodes, primarily for wearable professional medical monitoring applications, together with the available protocols governing the over-the-air communications are introduced. Furthermore, analysis of the major facets of the SoC system architecture and circuit building blocks will be covered, and presented in conjunction of a practical case-study of a complete multi-standard WBAN SoC fabricated in 65 nm CMOS technology operating in the 2.36–2.5 GHz frequency band.

1 Introduction

In recent years there has been significant interest and growth in low-power wireless technologies into medical applications and WBAN [1]. Medical WBANs consist of human body worn or implanted sensor nodes used to monitor vital signs (such as temperature, heart rate and electrocardiogram) or to stimulate normal bodily functions (such as pacemakers or cochlear implants).

To date, application specific wireless propriety solutions and protocols for specific biomedical devices or products have been the norm, limiting the extent of any one BAN. Hence the WBAN community worked together to develop a wireless communication standard IEEE802.15.6 to promote interoperability between all devices in and around the body. Furthermore the IEEE802.15.6 standards are optimised for ultra low power device implementation, while offering the levels of quality of service (QoS) and security required for personal medical data. In parallel, the consumer electronics industry has migrated existing standardised wireless protocols for personal area networks such as Bluetooth to meet the demanding

A.C.W. Wong (✉)
Frontier MicroSystems, Toumaz Ltd, 115 Olympic Avenue, Building 3, Milton Park, Abingdon, Oxfordshire OX14 4SA, UK
e-mail: Alan.Wong@toumaz.com

K.A.A. Makinwa et al. (eds.), *Efficient Sensor Interfaces, Advanced Amplifiers and Low Power RF Systems*, DOI 10.1007/978-3-319-21185-5_14

low energy yet robust needs of WBAN. Having interoperability between ultra-low power sensor nodes allows efficient medical sensor data processing and data fusion, ultimately leading to less intrusive "smarter" products and so improved quality of life and clinical outcomes for patients.

2 Wireless Protocols and Standards

It is important to understand the attributes of a body area network, in order evaluate the suitability of various WBAN standards for professional medical applications. The network needs to be secure and able to respond immediately to changes i.e. alarm conditions. Nodes are miniature and battery powered, and as such body sensor nodes are severely resource and power constrained. The nodes communicate with a central hub which is less resource and power constrained. The hub may exist in a mobile phone or a wireless router which allows connectivity to a local or wide area network. Although nodes predominately transmit, the links must be bidirectional to support QoS features such as acknowledgements, as well as the transmission of feedback data (e.g. stimulation signals and control signals) to the body sensor node, allowing all nodes in the network to be centrally managed. Since the sensor data is used by physicians in a medical setting, it needs to be accurate and reliable.

2.1 IEEE802.15.6

The IEE802.15.6 working group was formed to develop an international standard for short (i.e., human body range), low power and highly reliable wireless communication for use in close proximity to, or inside, a human body. The WBAN was designed to support a combination of reliability (QoS), low power, data rate and noninterference requirements to broadly address the breadth of body area network applications.

The resulting WBAN standard IEEE 802.15.6 was ratified in February 2012 [2], and consists of a single Medium-Access-Controller (MAC) which supports a star or two hop star network topology as shown in Fig. 1.

The MAC supports QoS, Medical Implant Communication Service (MICS) band communication for body implanted devices, and emergency communications. Built-in to the MAC is strong security, macroscopic and microscopic power management, and coexistence and interference mitigation. Access methods include scheduled, random and improvised, allowing for a flexible WBAN.

In addition to the MAC, there are three possible physical (PHY) layers:

1. Narrow Band (NB) PHY: Optimized WBAN for ultra low power biomedical applications;

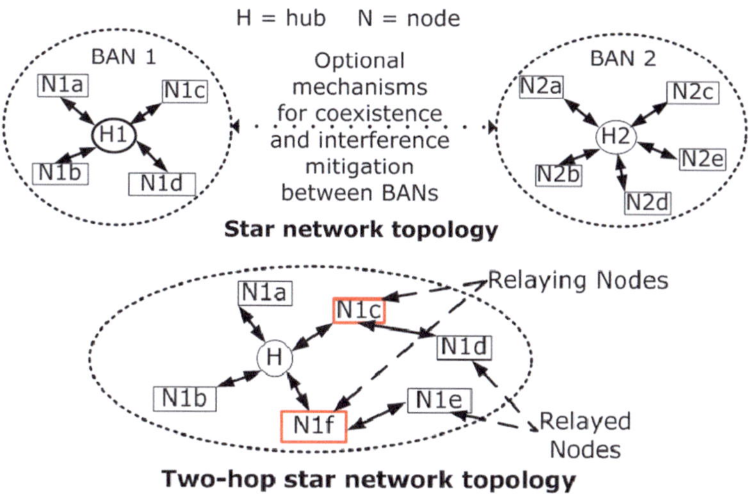

Fig. 1 IEEE802.15.6 MAC network topologies

2. Ultra Wide Band (UWB) PHY: WBAN for higher data rate entertainment applications;
3. Human Body Communications (HBC) PHY: utilizing the human body as the channel.

Only the NB PHY is architected specifically for biotelemetry applications.

2.2 Bluetooth

The Bluetooth Special Interest Group identified that although both classic Bluetooth (Version 1) and Enhanced Data Rate Bluetooth (EDR: Version 2) could address some WBAN applications, both exhibit peak power and average power consumption values which are too high for the majority of small form factor medical applications. As such the Low-Energy (LE) standard (Version 4.0) [3] was adopted for low data rate (small packets) and infrequent communication from a node, where the critical low power requirement is needed, to a hub in a mobile phone or web service. It was not designed for streaming applications requiring high reliability and has inefficient support for multiple nodes/networks. Recently rebranded as Bluetooth SMART, some of these technical limitations of v4.0 have been addressed somewhat in v4.2 extension [4] which adds optional features such as:

1. LE Secure Connections
2. LE Data Length Extension; for improved data throughput

Feature	Bluetooth LE	IEEE802.15.6 NB PHY	Comment
Frequency	2.4-2.4835GHz	2.4-2.483GHz 2.36-2.4GHz (US) *(400/868/915/950MHz)*	
Symbol rate	1Msps	600ksps *(250ksps &187.5ksps)*	
Data rate 2.4GHz	1Mbps	121.4kbps to 971.4kbps	Lower data rates provide a more robust link
Channels 2.4GHz	40 (37 data)	39 in MBANs, 79 in ISM	39 quiet channels for 802.15.6
FEC	None	BCH 51,63	
App Data thru put	260kbps	~750kbps	
Max payload	0-216 bits	0-2040bits	Maximum LE payload is short
Range: Specification	10m LOS	40m to 110m LOS	Based on specification. TX=-10dBm
Range: TX=-10dBm	75m LOS	100m to 260m LOS	Assuming the "reference receiver": NF=10dB, Implementation loss=6dB
Link Margin	+8.2dB	+10.2 to +18.7dB	-10dBm TX power @3m with 20dB fade margin
Modulation	GMSK	Rotated DBPSK/DQPSK	DPSK allows a range of data rates

Fig. 2 Bluetooth LE 4.0 & IEEE802.15.6 NB PHY comparisons

2.3 Standards Comparison

The key advantage of standardization is the enabling of interoperability of various body sensors, as such any standard has an advantage over a proprietary [5], or "closed" system from that aspect.

Figure 2 outlines the key comparisons between IEEE 802.15.6 NB and Bluetooth LE 4.0 PHY.

The IEE802.15.6 NB PHY has specific advantages over LE in WBANs in that it can utilize multiple frequency bands, in particular the quiet MBANs spectrum allocated to medical devices only in the U.S., with extension in the EU ongoing as shown in Fig. 3. Furthermore it has more available RF channels and higher data throughput with the ability to scale data rates for improved range and robustness. In terms of link budget IEEE802.15.6 NB also wins assuming the same output power and data rate. This is true both for the performance specified in the standards and if one was to apply a reference 10 dB noise figure with 6 dB implementation loss "reference receiver" to both modulations.

Bluetooth LE has advantages in terms of power consumption for episodic data transmissions, partly due to a simpler-to-implement amplitude modulation (AM) free GFSK modulation, compared to the IEEE 802.15.6 requirement for rotated DPSK. Figure 4 shows the modulation spectrum comparison, where some power is sacrificed in IEE802.15.6 NB PHY to implement DPSK resulting in better channel efficiency (and so less inference into adjacent bands), and improved SNR performance. The major advantage for Bluetooth is that it has significant market penetration due the fact that most mobile platforms are expected to have at least Bluetooth Version 4.0 functionality and hence basic LE functionality.

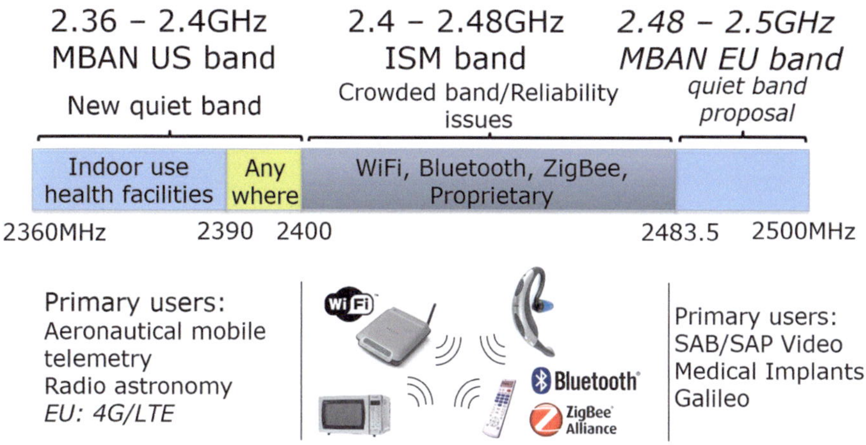

Fig. 3 2.4 GHz license-free frequency spectrum

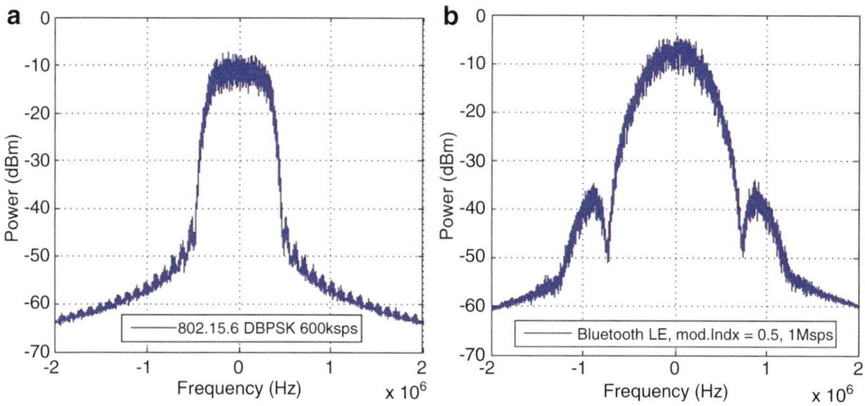

Fig. 4 Modulation comparison for (**a**) IEEE 802.15.6 NB and (**b**) Bluetooth LE

3 SoC Architecture

The key features of a fully integrated WBAN SoC [6] are shown in Fig. 5, where all components are optimized for low power. Modern CMOS process technologies allow the integration of all these features on a single die for very low cost and small form factor.

The peak power consumption of a WBAN SoC is the wireless PHY layer, so methods to reduce radio duty cycle are important. In a medical WBAN, the signals being monitored are typically physiological in nature and of low frequency, hence suitable for local processing on the sensor node to reduce data transmission over the air. Examples would be locally processing of ECG signals to detect heart rate or specific QRS waveforms and only transmit over-the-air the instantaneous heart rate or QRS template numbers to be matched on the hub node.

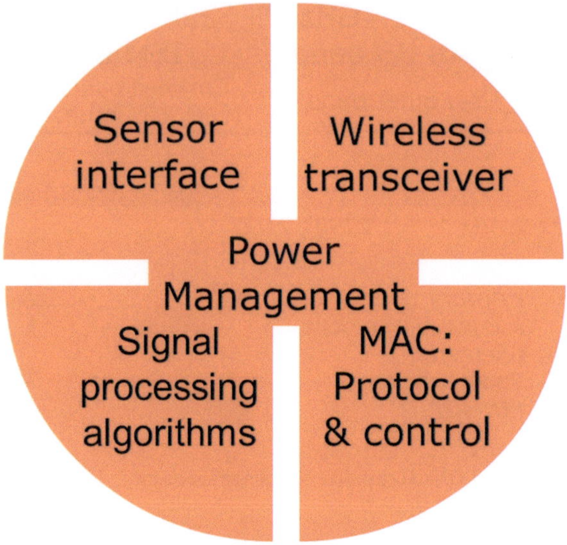

Fig. 5 Key features for a WBAN SoC

Average power consumption, if the radio operates on a low duty cycle, is dominated by the sensor interface circuitry, which may be active continuously or pseudo continuously with a single sensor interface ADC sampling multiple sensor inputs in a TDMA approach. The power management block needs to be operating with a minimum of power over-head whilst controlling the SoC optimally, both in terms of maximum and minimum power modes, but also to minimize wake-up and mode transition times.

4 Wireless Architecture

Since the radio will dominate SoC peak power, it is critical to minimize the power consumption. This is particularly true for body worn sensors where small battery technologies typically have low voltage output and high internal resistance. Current draws beyond several mA can cause battery voltage collapse and so limit how much instantaneous power is available to the SoC.

4.1 Wireless Receiver Architecture

For more than a decade, CMOS wireless receivers for digital modulations have typically adopted a zero or low IF architecture (Fig. 6a). The requirement for a quadrature LO signal at or near the RF frequency (f_{RF}) to drive the complex down conversion mixer is a challenge for power consumption, requiring a VCO which

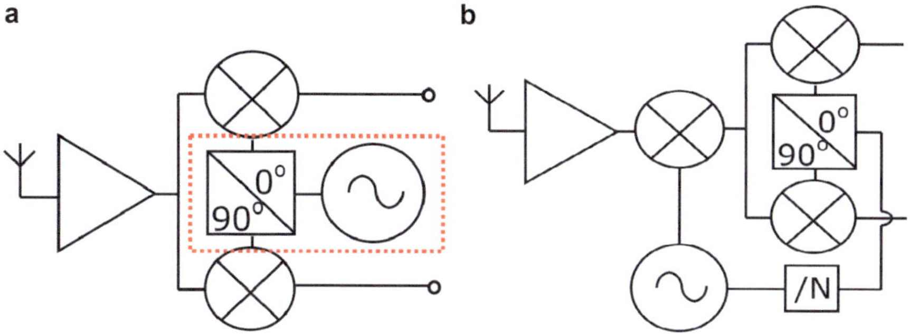

Fig. 6 Radio RF front-end architecture using (**a**) zero/low-IF and (**b**) sliding IF

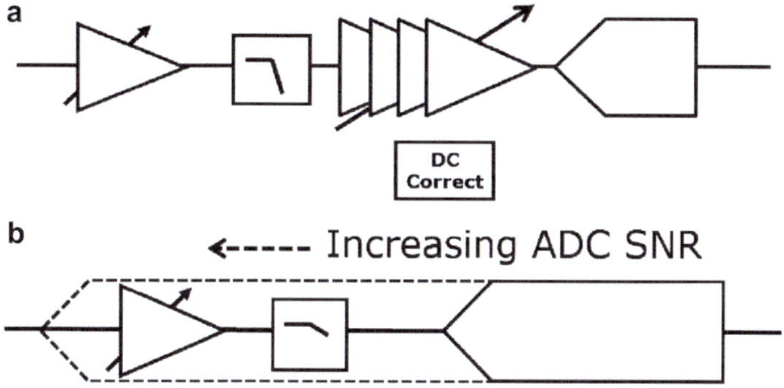

Fig. 7 Radio back-end architecture using (**a**) low SNR and (**b**) high SNR ADC

either has a quadrature output directly or operates at $N * f_{RF}$ followed by a frequency divide-by-N. Alternatively using a polyphase filters in either LO or RF path is problematic as these passive networks typically need low impedance drive at f_{RF} which is power hungry.

In the sliding-IF [7] architecture, a dual conversion is used where first and second LO are harmonically related. Quadrature LO is generated at a much lower $f_{RF}/(N+1)$ frequency and the first mixer is also driven at a frequency lower than f_{RF}. This allows for significantly lower power in the LO generation. Impacts traditionally problematic for zero/low-IF receivers such as DC offsets and flicker noise are reduced due to the sliding-IF dual conversion approach. However the non-complex first mixer does create an image frequency at $f_{RF} - (2 * f_{SLIDINGIF})$ which needs filtering. In a narrow-band system such as 802.15.6 or Bluetooth such filtering can be achieved in tuned circuits in the LNA at a minimal increase in silicon area or BOM cost.

The receiver back-end interface into the digital demodulation/subsystem typically consists of an ADC as shown in Fig. 7. A low dynamic range ADC needs signal amplification and filtering and DC correction, whilst a high dynamic range ADC can move up the receiver chain and take signal directly from the down-conversion mixer without the need for additional analog signal processing. Whilst a high dynamic range ADC can save silicon area and costs, it usually requires oversampling to achieve the dynamic range. Given signal bandwidths in the MHz range, this leads to 10's MHz clock frequency requirement just for the ADC. This clock could be derived from the PLL needed to generate LO signals, but a channelized radio means the ADC clock rate and sample rate would be channel dependent which is not optimal for the digital subsystem. Deriving this clock directly from the crystal reference is possible, but this limits the oversampling rate that can be achieved. Low SNR ADCs can be forgone for direct demodulation [8] and specific modulations lend themselves to specific ADCs [9], all of which can achieve very low power.

4.2 Wireless Transmitter Architecture

The radio transmitter architecture again has several system approaches which can yield energy efficiency as shown in Fig. 8. The traditional zero/low-IF IQ modulator suffers the same power constraints as the zero/low-IF receiver, requiring quadrature LO drive near f_{RF}. The sliding-IF's final "real" mixer produces a double sideband output, so half its energy is wasted in the unwanted sideband, which needs filtering. The loop modulator architecture [10] can save silicon area as only a transmitter buffer/PA is needed in addition to direct data modulation inside the PLL. However to avoid VCO/PA interaction which can cause transmitter performance degradation, typically a divider [11] is inserted between VCO and PA. Since the output power is limited to relatively low levels by WBAN standards and the available power from the body worn battery, it is possible to achieve the required loop modulator performance with the VCO operating at f_{RF} [12].

Another interesting transmitter architecture suitable for low power WBAN applications are direct phase selection transmitters [13]. However generation of multiphase LO is again an issue, but with novel techniques like injection locking [14–16] ultra-low power and acceptable performance is possible.

4.3 Synthesizer Architecture

Whilst ultra-low power radios without a crystal locked synthesizers exist, it is difficult to use such architectures to achieve the required frequency accuracy and performance to meet 802.15.6 and Bluetooth requirements. When architecting a WBAN SoC careful selection of the RF PLL is critical as it is major contributor to active radio power consumption. Figure 9 shows a typical PLL using a traditional analog/digital PFD-Charge-pump versus an ADPLL (all digital PLL) [17].

The key limitation for the ADPLL is the TDC (time-to-digital) conversion block, which for comparable resolution to a tradition PFDCP typically requires

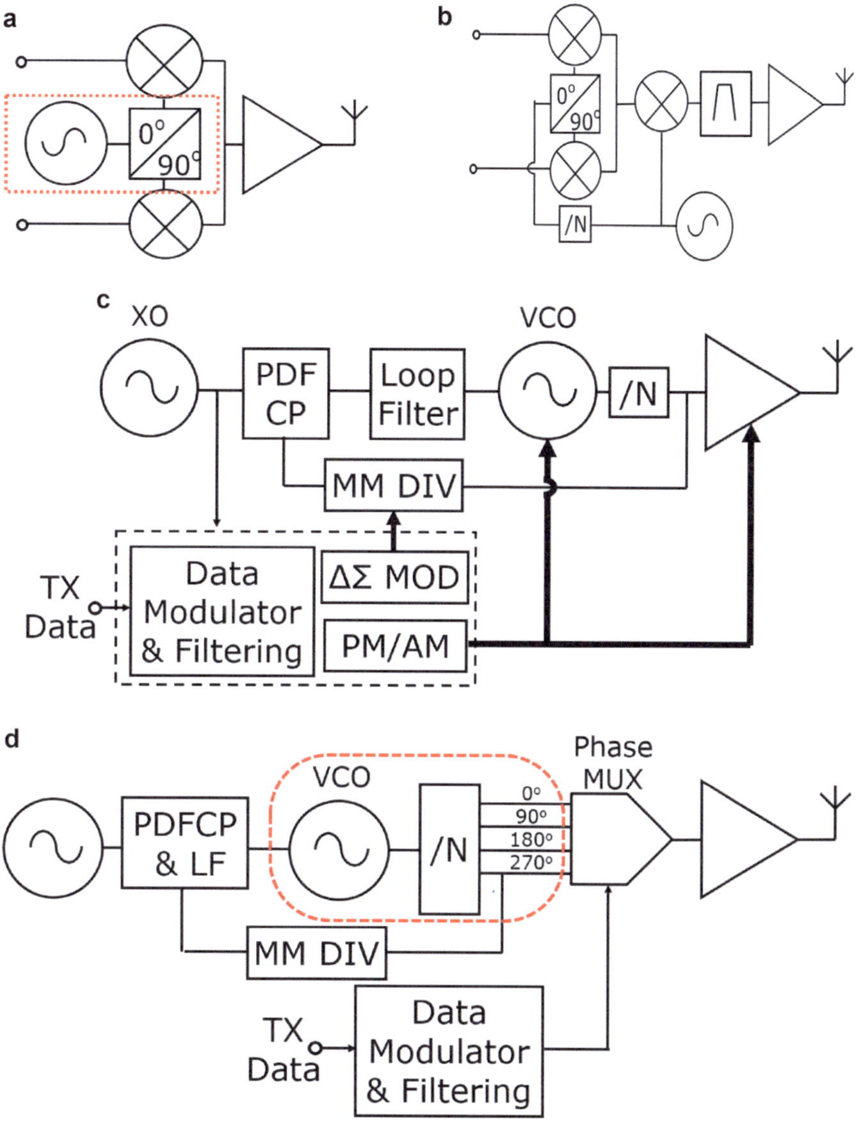

Fig. 8 Radio transmitter architecture using (**a**) zero/low-IF, (**b**) sliding-IF, (**c**) loop modulator and (**d**) direct phase selection

significant power. Research in this area [18, 19] to replace the TDC with something of comparable performance and power consumption to a PFDCP is ongoing, and the use of ADPLL in ultra-low power WBAN transceivers is gaining traction if the CMOS technology node is sufficiently small scale.

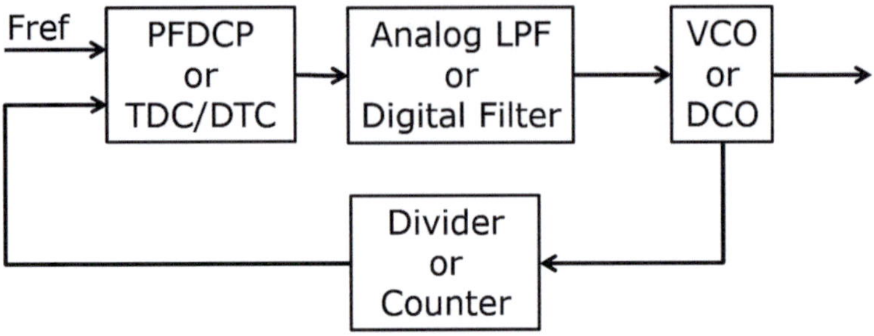

Fig. 9 Synthesizer architecture: traditional vs. ADPLL

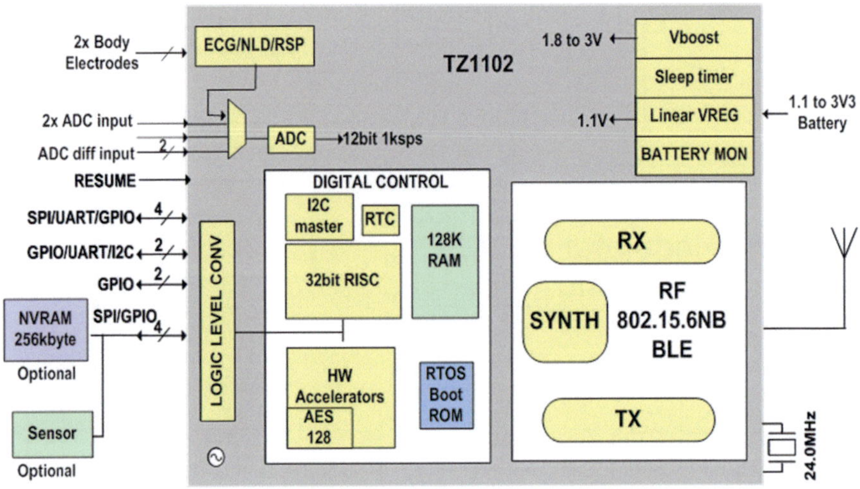

Fig. 10 WBAN SoC block diagram

5 WBAN SoC Case Study in 65 nm CMOS

As a case study, a full SoC fabricated in 65 nm CMOS technology is presented which aims to be interoperable between both IEEE 802.15.6 and Bluetooth LE standards. The SoC block diagram is shown in Fig. 10 and comprises all the key parts as previously shown in Fig. 5.

5.1 Wireless Transceiver

The wireless transceiver [20] architecture is shown in Fig. 11 and comprises a sliding-IF receiver and loop modulator transmitter with two point frequency and phase modulation points in the PLL/VCO and an AM modulation port on the PA to minimize SoC peak power consumption.

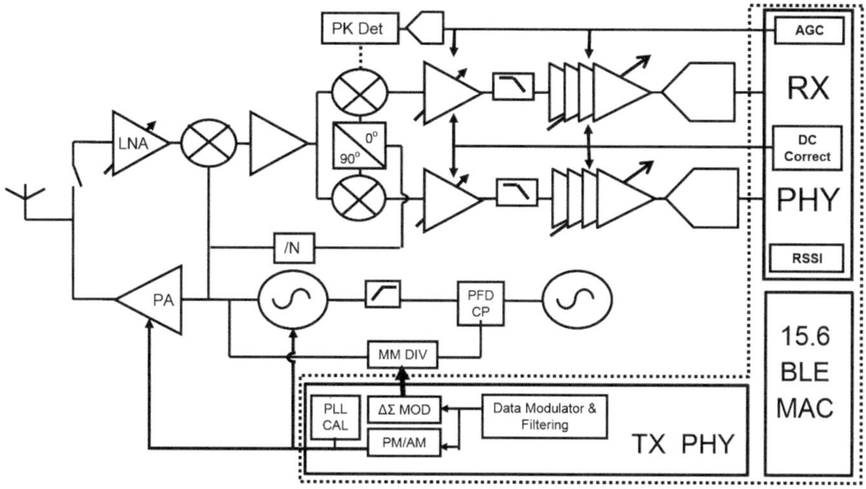

Fig. 11 Wireless transceiver architecture

5.2 *Digital Subsystem and Protocol*

The IEEE802.15.6 MAC is implemented in hardware to save power and includes RX/TX packet handlers, timing synchronization and direct memory access. For Bluetooth LE, the link layer is split between hardware and software.

The digital core is based upon a 32-bit RISC processor with peripherals such as 2 x UART, 2 x SPI, I2C and JTAG interfaces. The whole digital subsystem runs at a maximum speed of 24 MHz directly off the crystal oscillator used for the RF. Since little overhead is placed upon the uP to maintain the communication link due to dedicated hardware, it is freed up to perform locally processing tasks such as data compression or analysis.

The SoC digital fabric is carefully partitioned and multiple power islands and extensive clock gating are used to optimize the power dissipation.

Security is implemented with a 256-bit strength elliptic Diffie Hellman secure key exchange hardware and AES-128 CCM hardware for message security. An accompanying random number generator circuit for seed generation is shown in Fig. 12, which consists of a white noise generator (a combination of substrate pnp shot and resistor thermal noise) followed by voltage amplifier and limiter. To maintain a flat band and minimize sensitivity to common-mode noise, a differential topology is employed and a DC correction loop rejects DC and low frequency noise.

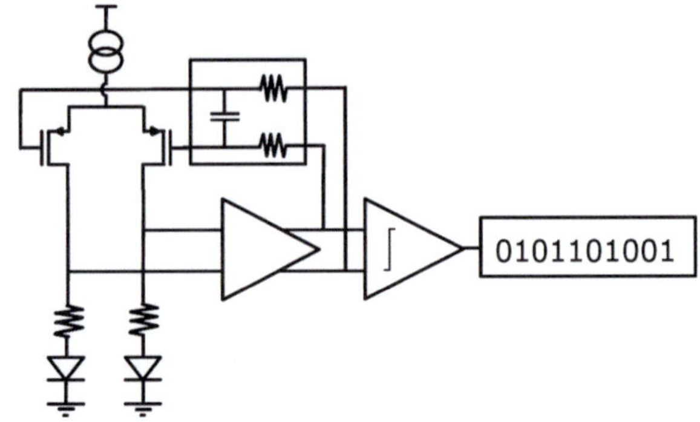

Fig. 12 Random number generation for secure key generation

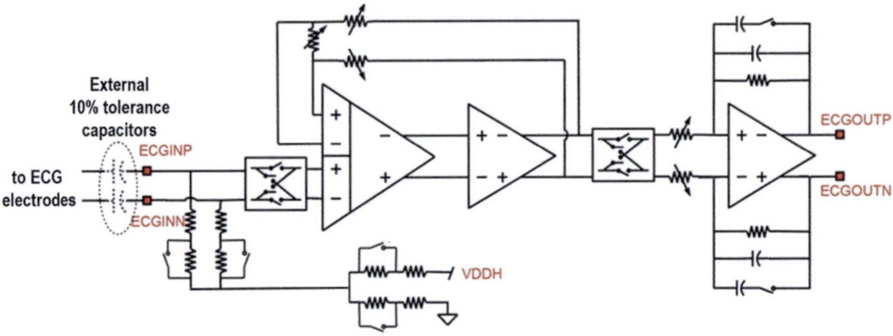

Fig. 13 ECG circuit

5.3 *Sensor Interface*

The sensor interface section compromises an ultra low power 12-bit ADC sampling up to 1 ksps. Also integrated are various sensor interfaces including an ECG front-end circuit [21], which uses a 3 V interface to the body. The elevated voltage significantly reduces issues such as motion artifacts on the ECG signal quality. The ECG circuit comprises a DDA (differential-difference-amplifier) [22] for superior CMRR as shown in Fig. 13 and the whole ECG sub-system meets the ECG Medical Device Standards [23].

5.4 *Power Management*

The implementation of multiple voltages levels and power domains on the SoC significantly complicates the power management unit (PMU) which is directly connected to the battery. Inside the PMU resides the PSU (power supply unit) which comprises multiple voltage regulators and supplies, since the SoC operates from a single 1.1–3.3 V battery source.

All main circuits can be separately power gated, including the crystal oscillator, so the PMU is clocked alternatively from a nano-power sleep timer when the crystal oscillator is off. The nano-power sleep timer is used by both PMU and sensor interface ADC and is continually on in "deep sleep" mode between radio transactions to keep time. It is periodically calibrated to the crystal oscillator and is able to keep 100 ppm accuracy whilst only consuming less than 400 nW of power.

During radio power-off and since the digital subsystem is also completely power gated without non-volatile memory, the radio protocol state parameters are stored in the always-on PMU power domain i.e. directly connected to the battery. On radio wake-up, whilst the crystal oscillator is starting-up a third instant-on-clock is involved to reload radio state parameters into the digital subsystem at a much faster rate than would be possible with the sleep timer. As such system latency is minimised which in turn keeps the power optimised.

5.5 *Implementation and Performance*

The SoC is fabricated in 65 nm CMOS and occupies 7.84 mm^2 (Fig. 14) and is packaged in a $5 \times 5 \times 0.9$ mm^3 VQFN/MLF package.

Figure 15 shows example radio performance and Fig. 16 summarizes the power consumption in various SoC modes, where in IEEE802.15.6 mode the uP need not be clocked faster than 4 MHz.

Fig. 14 SoC die photograph

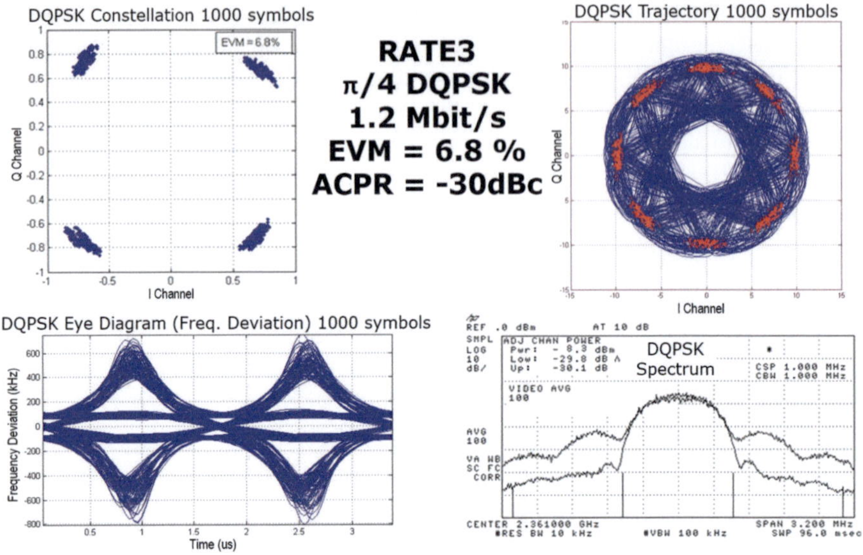

Fig. 15 IEEE802.15.6 NB PHY performance at 2.4 GHz in terms of constellation, trajectory, frequency deviation and ACPR (AM off and on)

Mode	Radio	CPU	XO	RAM	Sleep Timer	Sensor Interface	Idc (Typ)	Units
Radio TX	ON	24MHz	ON	ON	ON	ON	6.7	mA
Radio RX	ON	24MHz	ON	ON	ON	ON	7.0	mA
Radio TX	ON	4MHz	ON	ON	ON	OFF	5.0	mA
Radio RX	ON	4MHz	ON	ON	ON	OFF	5.4	mA
Processor	OFF	24MHz	ON	ON	ON	OFF	2.0	mA
Processor	OFF	1MHz	ON	ON	ON	OFF	0.2	mA
ECG/SIADC	OFF	OFF	OFF	OFF	ON	ON	20	uA
Deep Sleep	OFF	OFF	OFF	OFF	ON	OFF	0.4	uA
Hibernate	OFF	OFF	OFF	OFF	OFF	OFF	0.1	uA

Fig. 16 Power consumption performance from 1.1 V in various modes. Radio on mode in this case is IEEE802.15.6 NB PHY RATE3

6 Conclusions

More interoperability between WBAN nodes can yield more data for physicians allowing opportunities for data fusion between sensors which in turn leads to better clinical outcomes for patients. Fully integrated SoCs in CMOS technology can achieve these goals at very low cost. Key to the durability and patient experience is small form factor, and since the body worn battery source is very limited, ultra-low-power operation is a must.

By careful design choices in architecture and implementation, power can be optimized whilst performance maintained to adhere to standardized radio protocols (such as IEEE802.15.6 NB PHY and Bluetooth LE), and bio-telemetry requirements.

A case study was presented of a multi-standard wireless SoC solution implemented in 65 nm CMOS technology, which achieves average active power consumption tending to 0.5 μW (at low radio duty cycle) with a dedicated fast response application specific PMU. Peak SoC power is dominated by the 2.4 GHz wireless transceiver section when active, and this peak complete SoC power can be kept below 6 mW whilst maintaining maximum link range and 1.2 Mbit through-the-air data rate in IEEE802.15.6 mode. Low active radio power is achieved by the VCO & RFPLL operating at or below the RF frequency, a sliding-IF receiver and polar loop transmit modulator. The complete solution can operate from a single voltage down to 1.1 V.

References

1. Yoo H-J et al (2011) Body area network: technology, solutions, and standardisation. In: ISSCC digest of technical papers, p 531
2. IEEE 802 LAN/MAN Standards Committee (2012) IEEE standard for local and metropolitan area networks—part 15.6: wireless body area networks
3. Bluetooth SIG (2014a) Specification of the bluetooth system v4.0. www.bluetooth.org
4. Bluetooth SIG (2014b) Specification of the bluetooth system v4.2. www.bluetooth.org
5. Omeni O, Wong A, Burdett AJ, Toumazou C (2008) IEEE Trans Biomed Circuits Syst 2(4):251–259. doi:10.1109/TBCAS.2008.2003431
6. Wong ACW et al (2008) A 1 V wireless transceiver for an ultra-low-power SoC for biotelemetry applications. IEEE J Solid State Circuits 43(7):1511–1521
7. Chen M et al (2003) A CMOS bluetooth radio transceiver using a sliding-IF architecture. In: IEEE custom integrated circuits conference proceedings, pp 455–458.
8. Shakeri K, Hashemi H, Parsa A, Fotowat A, Rofougaran R (1999) A 1 V CMOS 2/4-level FSK demodulator for pager applications. In: 42nd IEEE midwest symposium on circuits and systems, vol 1, pp 219–222
9. Masuch J et al (2013) A 1.1 mW-RX-81.4 dBm sensitivity CMOS transceiver for bluetooth low energy. IEEE Trans Microwave Theory Tech 61:1660–1673
10. Oshmia T, Kokubo M (2005) Simple polar-loop transmitter for dualmode bluetooth. In: IEEE international symposium on circuits and systems (ISCAS), vol 4, pp 3966–3969
11. Weber D et al (2008) A single-chip CMOS radio SoC for v2.1 bluetooth applications. In: IEEE ISSCC digest of technical papers, pp 364–365
12. Wong ACW et al (2012) A 1 V 5 mA multimode IEEE802.15.6/Bluetooth low-energy WBAN transceiver for biotelemetry applications. In: ISSCC digest of technical papers, pp 300–301
13. Liu Y-H, Li C-L, Lin T-H (2009) A 200-pJ/b MUX-based RF transmitter for implantable multichannel neural recording. IEEE Trans Microwave Theory Tech 57:2533–2541
14. Diao S et al (2012) A 50-Mb/s CMOS QPSK/O-QPSK transmitter employing injection locking for direct modulation. IEEE Trans Microwave Theory Tech 60(1):120–130
15. Cheng S et al (2013) A 110pJ/b Multichannel FSK/GMSK/QPSK/π/4-DQPSK transmitter with phase-interpolated dual-injection DLL-based synthesizer employing hybrid FIR. In: ISSCC digest of technical papers, pp 450–451

16. Rahman M, Elbaray M, Harjani R (2014) A 2.5 nJ/bit multiband (MBAN & ISM) transmitter for IEE 802.15.6 based on a hybrid polyphase-Mux/ILO based modulator. In: RFIC symposium, digest of papers, pp 17–20
17. Staszewski RB, Balsara PT (2006) All-digital frequency synthesizer in deep-submicron CMOS. Wiley, New York. ISBN 0471772550
18. Chillara VK, Liu Y-H, Wang, B et al (2014) An 860 μW 2.1-to-2.7 GHz all-digital PLL-based frequency modulator with a DTC-assisted snapshot TDC for WPAN (Bluetooth Smart and ZigBee) applications. In: ISSCC digest of technical papers, pp 172–173
19. Tasca D, Zanuso M, Marzin G et al (2011) A 2.9-4.0-GHz fractional-N digital PLL with bang-bang phase detector and 560-fsrms integrated jitter at 4.5-mW power. IEEE J Solid State Circuits 46:2745–2758
20. Devita G, Wong A, Dawkins M, Glaros K, Kiani U, Lauria F, Madaka V, Omeni O, Schiff J, Vasudevan A, Whitaker L, Yu S, Burdett A (2014) A 5 mW multi-standard Bluetooth LE/IEEE 802.15.6 SoC for WBAN applications. In: ESSCIRC, pp 283–286
21. Ng KA, Chan PK (2005) A CMOS analog front-end IC for portable EEG/ECG monitoring applications. IEEE Trans Circuits Syst I Reg Papers 52(11):2335–2347
22. Sackinger E, Guggenbuhl W (1987) A versatile building block: the CMOS differential difference amplifier. IEEE J Solid State Circuits SC-22(4):287–294
23. IEC Medical Device Standards 60601-1, 2-25, 2-27, 2-47, 2-49. 2-51. www.iec.ch

Towards Low Power N-Path Filters for Flexible RF-Channel Selection

Eric A.M. Klumperink, Michiel C.M. Soer, Remko E. Struiksma, Frank E. van Vliet Nauta, and Bram Nauta

Abstract N-path filters can offer high-linearity high-Q channel selection filtering at a flexibly programmable RF center frequency, which is highly wanted for Software Defined Radio. Relying on capacitors and MOSFET switches, driven by digital non-overlapping clocks, N-path filters fit well to CMOS and benefit from Moore's law. The basis of this filtering is the linear periodically time variant (LPTV) behaviour of a switch-R-C series circuit, which realizes frequency translated filtering, where a baseband filter characteristic is shifted around the switching frequency. This paper reviews the basic concept of N-path filters and recent developments, with special attention to possibilities to reduce power consumption by increasing the impedance level. The basic operation of the switch-R-C kernel that is at the core of N-path filtering is reviewed in terms of transfer function and noise performance.

1 Introduction: N-Path Filter Concept

In order to realize a software defined radio receiver, flexibly tunable RF channel-selection filters with high Q, high linearity and strong blocker handling capability are wanted. Recently, passive switched capacitor circuits have been proposed as a solution direction to this challenging problem, and high-Q band-pass and notch filters at low GHz frequencies have been demonstrated [1–16]. The filter concept is known under different names, as "N-path filter", "commutated network", "frequency translated filter", "impedance frequency translation" or "Linear Periodically Time Variant" (LPTV) network. It can be traced back to the 40s of the previous century [17–21].

Figure 1 shows the basic concept of an N-path band-pass filter consisting of N identical signal paths [21]. Each path has a switch for frequency down-conversion, followed by a low-pass filter for baseband filtering and a second switch for frequency up-conversion. Overall, this results in band-pass filtering, where the pass-band bandwidth is defined by the low-pass filter and clock duty-cycle, while the center frequency is equal to the clock or switching frequency f_s. The resulting

E.A.M. Klumperink (✉) • M.C.M. Soer • R.E. Struiksma • F.E. van Vliet Nauta • B. Nauta
University of Twente, Enschede, The Netherlands
e-mail: e.a.m.klumperink@utwente.nl

K.A.A. Makinwa et al. (eds.), *Efficient Sensor Interfaces, Advanced Amplifiers and Low Power RF Systems*, DOI 10.1007/978-3-319-21185-5_15

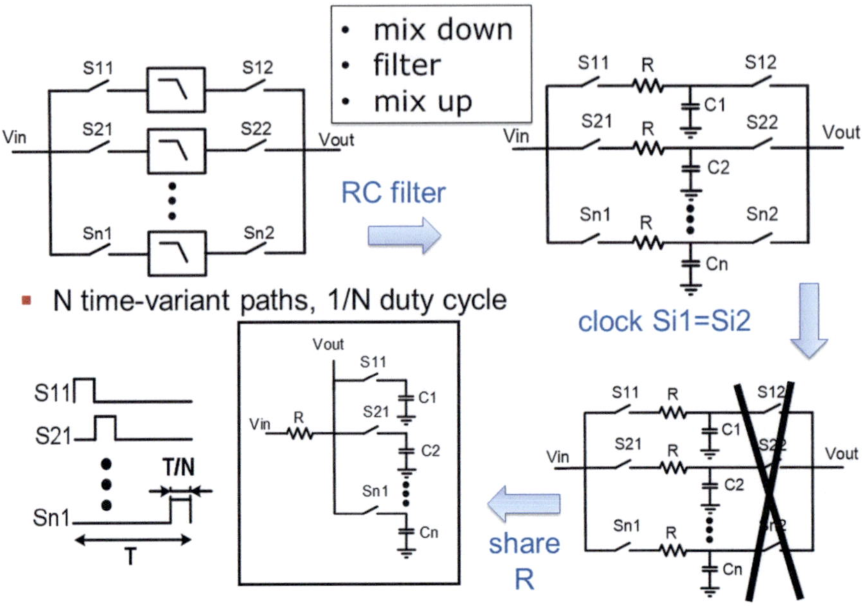

Fig. 1 N-path filter architecture evolution to a simple switch-R-C circuit

band-pass filter $Q = \frac{f_s}{BW_{-3dB}}$ can be high. E.g. $f_s = 1$ GHz and $BW_{-3dB} = 1$ MHz renders a Q = 1000, much higher than feasible with on-chip LC filters, in which the inductor quality factor is typically limited to 10 . . . 15. Note also that the filter center frequency tracks the digital clock frequency, which can be programmed flexibly.

The two sets of N switches S_{11}–S_{1n} and S_{21}–S_{2n} are driven by multi-phase non-overlapping digital clocks in a polyphase fashion: each clock has 1/N duty-cycle and a phase that increase from 0 to 2π, with increments $2\pi/N$. The low-pass filter can simply be realized as a passive R-C filter. Further simplification is possible, realizing that one set of switches can be removed if the clock of the two sets of switches is identical (see Fig. 1). Also, one resistor can be shared by all the paths, resulting in a very simple circuit with one resistor and N switched capacitors, where V_{out} is the output.

The operation of such circuits is explained in the time-domain in Fig. 2. When the signal and switching frequency are the same (Fig. 2a), each of the capacitors, which is connected to the input for $1/N$th of the period, always 'sees' the same part of the input sine. Assuming that the RC time-constant is much larger than the switching period, it is charged to the average of the source voltage during that interval. For frequencies other than the switching frequency (Fig. 2b), the baseband capacitors 'see' a different part of the sinusoid each period, the average of which is zero. Hence a signal close to the switching frequency f_s is passed to the output, while a signal at $1.5f_s$ is rejected.

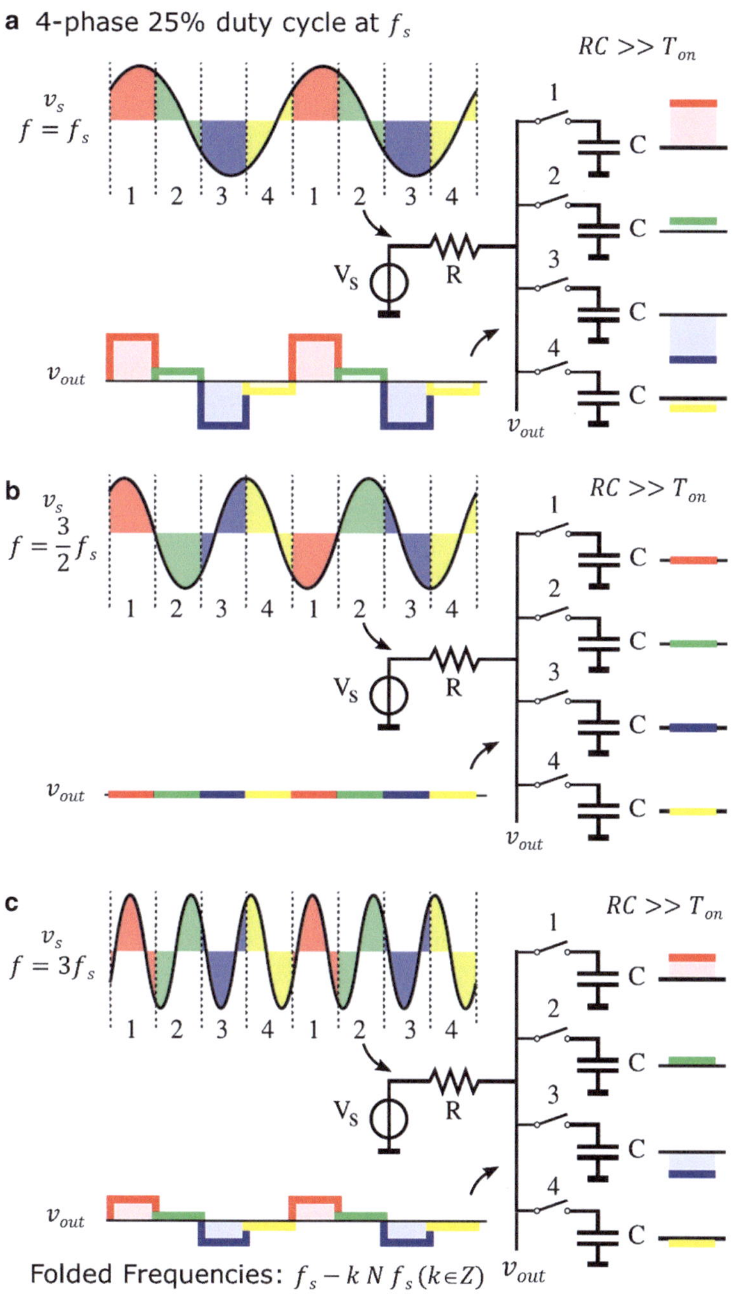

Fig. 2 Time domain waveforms illustrating filtering in a 4-path filter for a sinewave input: (**a**) at $f = f_s$ producing a response at the output V_{out}; (**b**) at $1.5f_s$ resulting in rejection; and (**c**) at $3f_s$, rendering an undesired folding response

Note that the signal in the pass-band is a kind of stair-case approximation of the sine wave input signal. Hence, not only the fundamental appears at the output, but also higher order harmonics. Also, responses at harmonics of the switching frequency occur ("comb filter"). Moreover, signal down-folding occur as illustrated in Fig. 2c. If a higher value of N and hence more phases are used, the filter output will more closely resemble the actual shape of the input signal in the pass-band. This simple time domain view illustrates all of these filter properties, but it is less suitable to predict the exact frequency domain transfer function and filter shape. In Sect. 2 we will review analysis methods to analyze the transfer function and noise figure. Also, we will review N-path filter developments over the past few years and discuss developments to improve their transfer function in Sect. 3. In Sect. 4, recent developments towards low power N-path filters are discussed, while Sect. 5 draws conclusions. The appendix summarizes the key analysis results obtained from Linear Periodically Time Variant analysis of a switched-RC loop, that is the basis for N-path filters.

2 Analysis Methods of N-Path Filter Properties

Although the N-path filter circuit is simple, the exact mathematical treatment of periodically time variant networks can be quite involved, and approximations are very much desired to make analysis tractable. We will briefly discuss three main ways that have been proposed to analyze N-path filters. In this paper we will restrict ourselves to N-path filters with a time continuous input and output. These N-path filters actually predate N-path filters with a time discrete input and output, as the ones discussed in [22], which can be analyzed much easier in the z-domain. In contrast, traditional N-path filters continuously integrate the input signal on capacitors, albeit different ones, and have a continuously defined output signal, assuming non-overlapping clock with a duty cycle 1/N are used.

A common simplification is the one visualized in Fig. 2. Already in [18, 19] the transfer function of a "comb filter using synchronously commutated capacitors" was derived, assuming that $RC \gg T_{on}$, so that the voltage during the on-time is approximately constant. The derived −3 dB RF-bandwidth is:

$$BW_{-3dB} = \frac{1}{\pi NRC} \tag{1}$$

The same assumption has been used in [1, 23], where the energy dissipated during one period was used to derive an approximation for the input impedance of the N-path filter close to the switching frequency $\left(f_{RF} \approx f_s\right)$. A resistive impedance results, which models power that is actually "reflected" by the N-path filter at harmonics of the switching frequency and dissipated in the resistor of Fig. 2 (v_s is a sinewave, while v_{out} is a staircase waveform with harmonics). This is sometimes referred to as the "transparency property of a switch": in contrast to a unilateral active mixer, a passive mixer "works" in both directions: it both down-converts the RF signal to baseband, but also up-converts the (almost) DC-voltages on the capacitors to RF (see Fig. 2).

Another analysis procedure in [24] assumes four baseband low-pass impedances that are driven by four switches with 25 % duty-cycle. The RF-source is modelled as a current source with parallel impedance Z_s (e.g., $Z_s = R$ in Fig. 2). Input frequencies are assumed to be confined to $\frac{f_s}{2}$ and $\frac{3f_s}{2}$, and the current in the 4-path filter is approximated by a Fourier series. In this way an approximation of the RF-current in the form of an infinite sum is found, which holds if $Z_{BB}(k \cdot f_s) \ll |Z_S(f_s)|$ (k is a nonzero integer). Also, several imperfections like clock mismatches, phase noise are analyzed. However, due to the input frequency limitation $\frac{f_s}{2} < f < \frac{3f_s}{2}$, harmonic responses and signal folding from higher frequencies cannot be analyzed, while the baseband impedance assumption may also introduce inaccuracies.

In contrast, the analysis method in [4, 12, 25] does not need upfront assumptions on input frequency. It assumes a series connection of a voltage source, resistor, switch and capacitor ("switched-RC kernel" or switched-RC loop as in Fig. 9 to be discussed later) and *holds for arbitrary values of N, R, C, f_s and duty-cycle.* As no upfront assumptions on the input frequency are made, not only the transfer function and noise can be analyzed, but also harmonic responses and harmonic folding. Based on this analysis method, a simplified equivalent R-L-C filter valid close to the switching frequency can be derived [4, 26] for arbitrary N. Figure 3 shows the model equations and compares simulation and model calculation results for a differential 4-path band-pass filter [2, 4]. Clearly, the fit is very good, except for harmonic responses that are neglected in this simplified R-L-C model. With slight adaptations, this model can also be used for N-path notch-filters, see the 8-path notch filter in [6, 26].

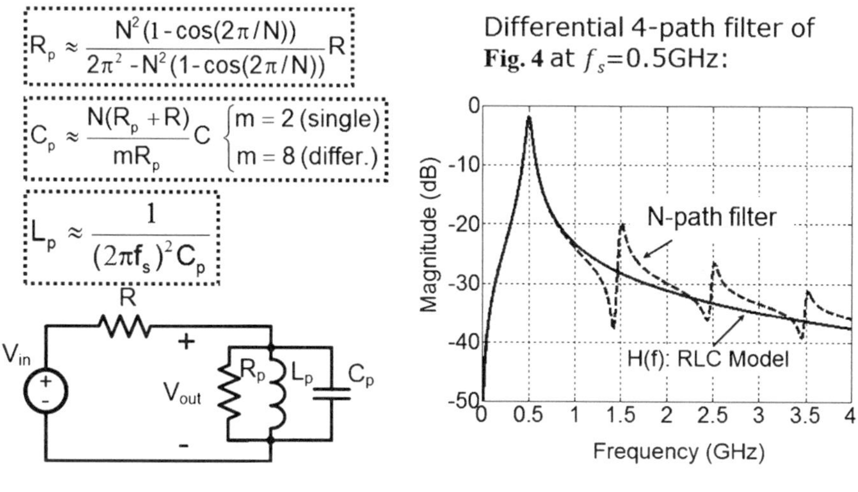

Fig. 3 RLC model for an N-path filter valid around $f = f_s$ [4]

3 Filter Shape Limitations and Improvements

The switch resistance was neglected so far. However, it will play a role especially when an N-path filter works in a low-ohmic environment, e.g., to directly filter the antenna signal. Ideally, the capacitors shunt all signal to ground outside the pass-band, but switch resistance will limit the achievable impedance to ground to R_{SW} (see Fig. 4). The achievable filter attenuation can be estimated with a simple resistive divider. For a switch resistance of 5 Ω and R = 50 Ω, this will limit the achievable rejection of the filter in Fig. 4 to about 20 dB. Another worry may be clock overlap. As shown in Fig. 4, clock overlap destroys the pass-band transfer to below −10 dB for 30 % duty-cycle instead of 25 %, due to charge sharing between the signal paths, so it should be avoided. Instead it is better to slightly lower the duty-cycle, as this slightly improves pass-band loss, although it will degrade the maximum rejection (see the 20 % duty-cycle curve in Fig. 4).

When the switching frequency is changed, the center frequency shifts, but the shape and bandwidth remains the same, essentially equal to a real-pole response shifted around the switching frequency. If the baseband impedance is changed, e.g., via transconductors coupling signals from one capacitor to another, the filter

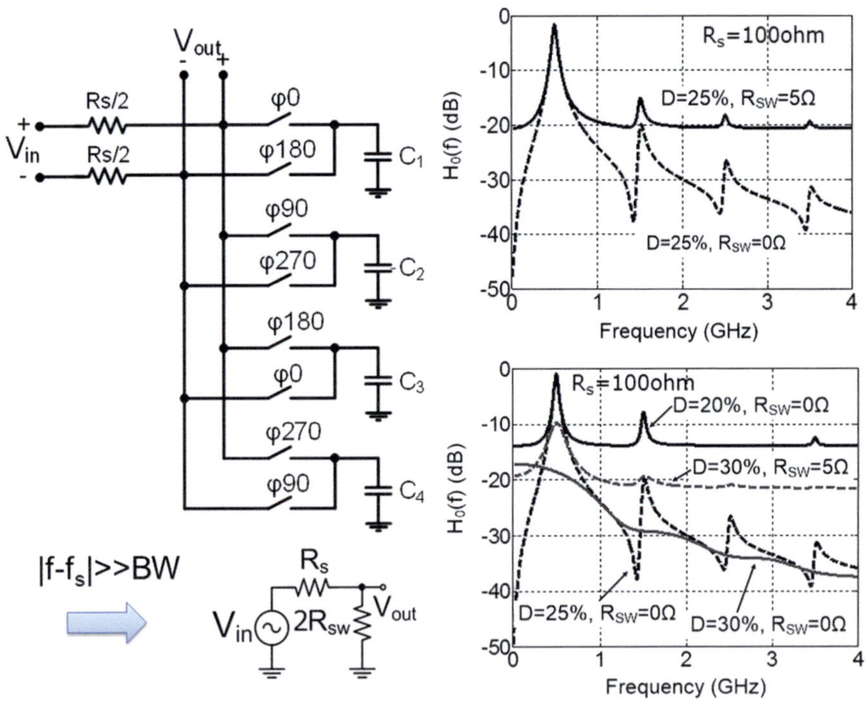

Fig. 4 Effect of switch resistance R_{SW} and duty cycle D on the transfer of a differential 4-path filter driven from a 100 differential source [4]

peak can be shifted in frequency, but essentially still has the "real-pole" shape [9]. However, if two N-path filters with shifted center frequencies are used, the order of the passband filter is doubled and a "flat-top shape" can be approximated [7]. A disadvantage of this technique is the 1/f noise of the transconductors that operate at baseband. To avoid this, it would be better to use transconductors operating at the RF-frequency. Note however, that the model in Fig. 3 only holds for a pure switched-RC network. If we add capacitance to the shared switch-node in Fig. 2, charge sharing between capacitors will take place, which results in extra dissipation, extra noise and destruction of the high-Q N-path filter shape. An approach to realize higher order N-path filters is to use gyrators to couple N-path band-pass filter sections, as has been proposed by Heinlein for 3-phase filters [27]. Figure 5 shows the block schematic of an 8-path 6-order band-pass filter with gyrator coupling between the two last switched capacitor band-pass sections [10]. Together with a first section, a flat-top band-pass filter shape is approximated.

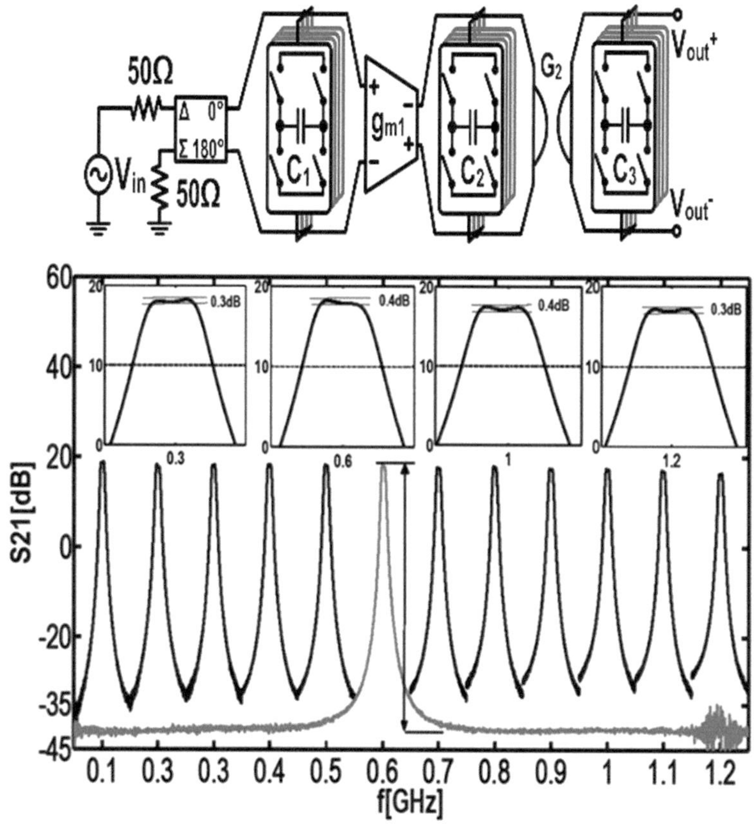

Fig. 5 8-Path sixth-order band-pass filter with flat-top filter shape and more than a decade tuning range [10]

Moreover, as the voltage over the capacitor is sensed by transconductors (g_{m1} and two transconductors constituting gyrator G_2), switch resistance no longer limits the achievable filter rejection [10]. Note that this filter is tunable over more than a decade, achieving 19 dB gain with about 3 dB Noise Figure. The filter has 8 MHz bandwidth and shows an out-of-band IIP3 of +26 dBm and compression point of +7 dBm at 50 MHz frequency offset. This high out-of-band linearity and compression point is an important asset of such filters, together with their high Q and flexibly controllable center frequency.

Although the results shown are promising, there are several challenges that may impede broad adoption of N-path filters in radio receivers. One important limitation is the reciprocal mixing of phase noise with strong blockers. In [24] the phase noise effects are analyzed, which are challenging but feasible for the requirements in [28]. Another worry can be the LO-radiation to the antennas that can however be reduced by calibration techniques [29]. Another problem is the signal folding that occurs in N-path filters. This can be analyzed in more detail with the analysis method of [4], where it is shown that folding occurs from frequencies $k\,(N \pm 1)\,f_s$ and equations quantify the strength. Hence, for higher N, the first folding frequency is further away and the conversion gain is lower, at the cost of more clock phases and paths. To suppress the folding, a linear time invariant pre-filter can be used, e.g., a low-pass filter. It has recently been shown that a series inductor between the resistor R and switched capacitors in Fig. 2 renders a low-pass pre-filter which improves N-path filter transfer and noise figure [30]. More work is needed in this domain.

4 Towards Lower Power N-Path Filters

As the bandwidth BW_{-3dB} according to (5) is inversely proportional to $N\cdot R\cdot C$, there is a clear trade-off between the total required capacitance, N · C, and bandwidth. When directly operating in a 50 environment, quite big capacitance would be required to realize a narrow bandwidth, e.g., 32 nF total capacitance would be needed for 200 kHz bandwidth. Moreover, to minimize noise and mitigate switch resistance related limitations, switch resistance should be much lower than source resistance R. As lowering the switch resistance requires wider switch transistors with larger gate capacitance, this increases the dynamic power consumption. Using a newer technology significantly decreases power, almost by a factor 2 for 28 nm CMOS compared to 65 nm CMOS (see Table I in [31]). However, there are also other ways to reduce power, most notably by increasing the impedance level of an N-path filter. This can be realized by combining an N-path filter with an LNA. Two recent proposals that do so are described in the next two paragraphs and in Fig. 6.

It has been proposed to use an LNA with voltage gain before the N-path filter [32]. Basically the output resistance of the LNA now acts as R of the N-path filter following the LNA as shown (Fig. 6a). If down-conversion is wanted anyway, the baseband voltages can directly be used as output, as in the mixer-first receiver in [33]. Still the LNA output voltage is filtered by the N-path filter, improving

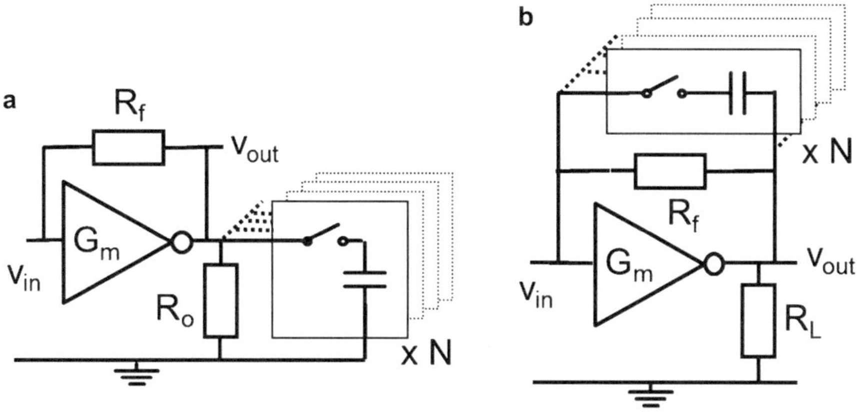

Fig. 6 Increasing the N-path impedance level by: (**a**) transconductor with load resistance $R_o > R$; (**b**) exploiting the Miller effect ("Gain-boosted N-path filter" [11, 13]

interference robustness. To achieve voltage gain, the typical output resistance level of an LNA is higher than 50, e.g., 500. As a result, the N-path filter can have 10× lower capacitance for the same bandwidth, 10× higher switch-resistance, 10× smaller switch width and hence 10× lower dynamic power consumption. However, as blockers may now be amplified in the LNA despite of the N-path filtering, a higher supply voltage is desired to allow for enough voltage swing without clipping [32]. To maximise the output resistance and still cope with strong out-of-band blockers, an LNTA can also be used as has been proposed in [15, 34]. Now the *parasitic* output resistance of the LNTA acts as resistor R for the N-path filter. As an N-path filter is high-ohmic in-band, there will be significant voltage gain in-band [15, 34]. Hence the LNTA effectively works as LNA with voltage gain in-band. However, out-of-band, the N-path filter will be low-ohmic, and the LNTA operates as V-I converter with a low output voltage swing. Hence high linearity and compression are still possible at sufficient frequency offset from the N-path filter centre frequency. Overall, significant power consumption savings are possible, as the N-path filter switches can be much smaller than in a 50 environment, see for instance the power comparison in [35].

An alternative approach is shown in Fig. 6b [11, 13]. Here an N-path notch-filter [4] is put in the feedback path of a voltage amplifying LNA, with transconductance G_m and load resistance R_L. Such notch-filtering in the feedback path of an amplifier overall results in a band-pass filter (close to the notch frequency, the feedback is weak and peaking in the forward gain occur). More importantly, the input signal V_i is filtered now by an N-path filter, but in a "gain-boosted" way. Comparing the gain boosted filter in Fig. 6b to a traditional N-path filter in Fig. 2 driven by the same resistor R_s, the gain boosting approach has two key advantages:

1. Due to the voltage gain of the LNA, the feedback capacitors now see $(A_{v0} + 1)$ times more voltage than the input voltage v_{in}, which produces $(A_{v0} + 1)$ times

more input current (miller effect). In other words, the N-path filter impedance seen from the input is lowered by a factor $(A_{v0} + 1)$, requiring $(A_{v0} + 1)$ times less capacitance in the N-path filter.

2. For the same N-path filter performance, the on-resistance of the switches can be increased by a factor $(A_{v0} + 1)$, i.e., smaller switch width can be used, saving considerable dynamic power consumption in the buffer circuits driving the switch transistors.

An interesting recent idea combines this gain boosted N-path filtering in the LNA with frequency down-conversion, re-using the same capacitor voltages [36].

A more detailed analysis of the properties of gain-boosted N-path filters and low power implementation aspects can be found in [11, 13]. When comparing the experimental results of the fully passive N-path filters with the N-path filters that are combined with an LNA, the main trade-off is between linearity on the one hand and power consumption and chip area on the other. The best reported out-of-band IIP3 for a gain-boosted N-path filter is in the order of 15–17 dBm [13, 37], which is about 10 dB lower than IIP3 results reported for passive N-path filters. This is because the non-linear properties of amplifiers degrade the linearity performance compared to what is possible with passive switch-R-C circuits.

5 Conclusions

In this paper recent developments in CMOS N-path filters for high-Q RF-channel selection at low GHz frequencies were reviewed. The behaviour of a switched-RC loop, that is the basic building block for N-path filters was analyzed in different ways, both in the time and frequency domain. It was shown that a "slow loop" is wanted with $RC \gg T_{on}$ both for high-Q filtering and for minimizing Noise Figure. Also, it was shown that a lower duty-cycle renders less conversion loss. Proposal to improve the transfer function of N-path filters have been briefly reviewed. Also, recent proposals to reduce power consumption and reduce total required capacitance by increasing the impedance level were described.

Appendix: Summary of the Analysis of a Series Switched-RC Loop

This appendix will analyse the signal transfer properties and noise behaviour of a single switched-RC loop, which is the building block of an N-path filter. It summarizes the unified LPTV analysis of switched-RC passive mixers and samplers in [12]. Interestingly, we will see that this circuit can both behave like a mixer but also as a sampler, depending on the on-time of the switch T_{on} compared to 2RC.

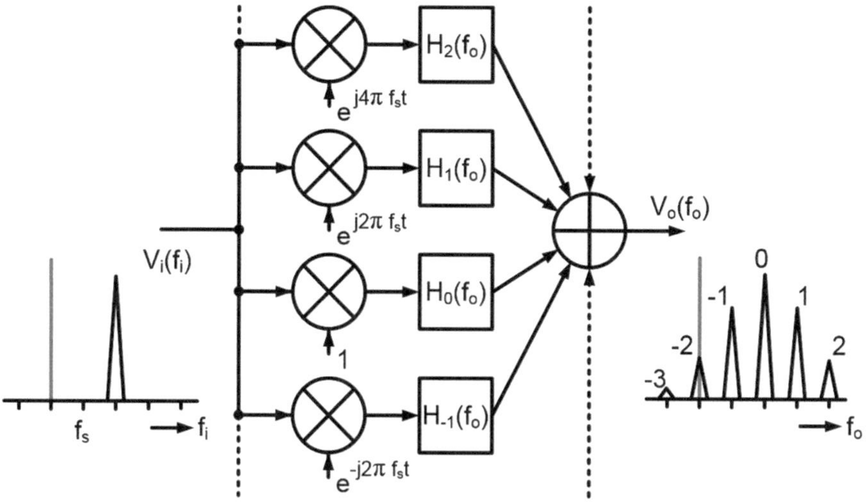

Fig. 7 Generic frequency domain model of an LPTV system with harmonic transfer function H_n

LPTV Systems Modelling

In a Linear Time Invariant (LTI) network, a sinusoidal input signal with frequency f_i results in a sinusoidal tone at its output with the same frequency, and only the amplitude and phase change as described by one transfer function.

In contrast, a sinusoidal input frequency will generate a multitude of sinusoidal output signals in a Linear Periodically Time Variant (LPTV) network. Figure 7 shows a generic model for such a system. There are several paths from input to output, each consisting of a frequency shift and LTI filter H_n. An input spectrum can thus be translated to multiple locations in the output spectrum, with frequency shifts $n \cdot f_s$, where n is an integer and f_s is the switching frequency, i.e., the inverse of the time period T_s. This statement can be captured into the equation [38]:

$$V_o(f_o) = \sum_{n=-\infty}^{\infty} H_n(f_o)\, V_i(f_o - nf_s)\,. \tag{2}$$

$H_n(f_o)$ constitute the harmonic transfer functions (HTFs) that define the response of the LPTV system. Due to the frequency shifts, the frequency variable f_i in the input spectrum V_i has to be separately defined from the frequency variable f_o in the output spectrum V_o. For each frequency shift, with harmonic index n, the relation between input and output frequency is defined as:

$$f_o = f_i + nf_s \tag{3}$$

For random noise signals, the relation between the input power spectral density (PSD), N_i, and output PSD, N_o, in LPTV systems is [39]:

$$N_o(f_o) = \sum_{n=-\infty}^{\infty} |H_n(f_o)|^2 N_i(f_o - nf_s). \tag{4}$$

Hence output noise around f_o can originate from input noise N_i at many frequencies $(f_o - nf_s)$: all contributions are shifted in frequency and summed together, but now according to the absolute HTF squared, as the equation is expressed in terms of power.

Mixing and Sampling in LPTV Systems

The frequency shifting property of LPTV systems allows for implementing mixers. One example is given in Fig. 8a, where a multiplication is performed of the input signal with a rectangular clock signal.

This multiplication in the time domain is equivalent to a convolution in the frequency domain with the Fourier transform of the clock wave. For a periodic square wave clock with 50 % duty cycle, this transform consists of a train of impulses weighted with the sinc function[1] [40], i.e.,:

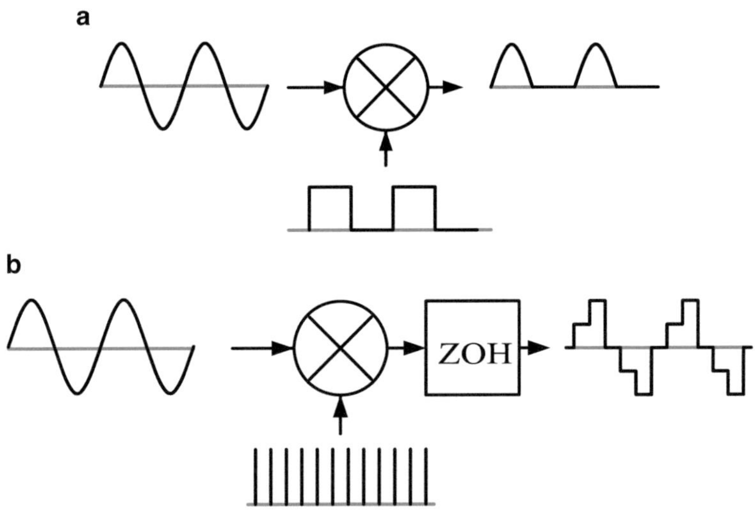

Fig. 8 Time domain example of (**a**) switching mixer and (**b**) sample and hold

[1] In particular, the normalized sinc function $sinc(x) = \frac{\sin(\pi x)}{\pi x}$.

$$H_{n,mixer} = \frac{1 - e^{j\pi n}}{j2\pi n} = sinc\left(\frac{n}{2}\right) e^{\frac{j\pi n}{2}} \tag{5}$$

where $H_{n,mixer}$ is a HTF as in (2). Note that this equation does not depend on frequency but only on frequency-shift (n). In a mixer design, one of the frequency shifts, with index $n = w$, is the desired conversion. For frequency downconversion, this is usually $n = -1$ and for upconversion n = 1. Other shifts should then be suppressed, either by filtering or multi-path polyphase techniques. If downconversion and upconversion occur simultaneously, the frequency shift can be 0, which is the intended functionality of an N-path filter.

Switching circuits can also be used as sampler in discrete-time systems. Figure 8b gives an example of a sample-and-hold system. The sample process is represented by multiplying the input signal with an impulse train in the time domain, after which a zero-order-hold (ZOH) holds the sample till the next is available.

As with the mixer, the multiplication with an impulse train gives rise to the frequency shifting property of LPTV systems. For the sampler however, the ZOH performs a convolution of the sampled input signal with a rectangular pulse in the *time* domain. This is equivalent to a multiplication in the frequency domain with the Fourier transform of a rectangular pulse [40], and the HTFs can be written as:

$$H_{n,sampler} = \frac{1 - e^{-j\,2\pi\frac{f_o}{f_s}}}{j\,2\pi\frac{f_o}{f_s}} = sinc\left(\frac{f_o}{f_s}\right) e^{\frac{-j\pi f_o}{f_s}} \tag{6}$$

Comparing* (5) *and* (6)*, it is seen that mixing with a rectangular clock results in HTFs weighted by the sinc as function of harmonic index n (frequency-shift), while performing a sample-and-hold results in HTFs weighted by the sinc as function of the (absolute) output frequency f_o.

Note that the mixing properties of a system are defined in the *frequency domain* and linked to the frequency shifting properties of LPTV systems. On the other hand, sample-and-holding refers to a *time domain* quality, which can, *but need not*, give frequency shifting. We will show later that this depends on how the circuit is designed, but also on how it is used. As discussed in [41], the circuit in Fig. 9 can both be used as a mixer with frequency shifting, but also as a sampler without frequency shifting. To understand this, the classification in [41] may be instrumental (see Table 1). It is based on the type of signal (continous-time versus time-discrete) and the "observation rate" of the input and output signal (time-continuous observation $\Longleftrightarrow$ infinite observation rate). Now, we can either choose to observe the output signal with the same observation rate as the input, but also choose to evaluate the output at a lower observation rate. The latter use case corresponds to sampling (or re-sampling if the signal is already time-discrete). The first case (no change in observation rate) corresponds to the mixer use case.

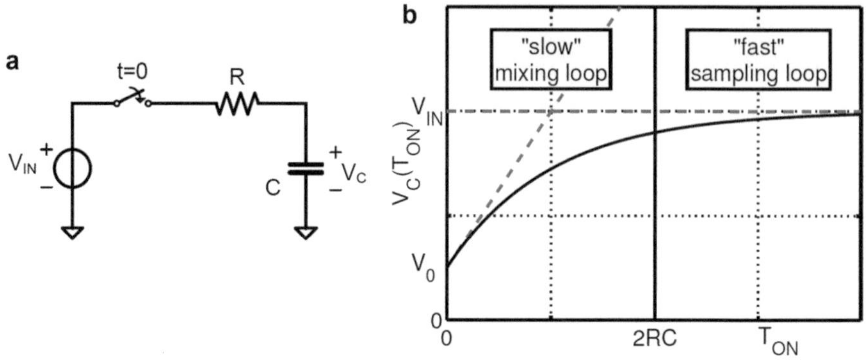

Fig. 9 A single switched-RC loop (**a**) with its time response (**b**)

Table 1 Classification of the operation modes of the circuit in Fig. 9 [41] based on the type of input signal and the observation rate (see text)

Signal operation	Type of input signal	
	Analog-CT (continuous time)	Analog-DT (discrete time)
Mixing (same "observation rate)	CT mixing	DT mixing
Sampling (reduce observation rate)	CT-to-DT sampling	DT re-sampling

Time Domain Behavior of a Switched-RC Loop

Apart from the way we use a circuit, the design of the circuit in terms of RC time strongly affects circuit behaviour. To understand this, we will analyze the time domain behaviour of a switched-RC loop (or "Single Ended switched-RC kernel" [12]). Consider the step response of the RC network in Fig. 9 during a single clock interval. Assume that the capacitor is initially charged to voltage V_0 and a voltage V_{IN} is applied during time T_{on}.

If the switch now opens much later than twice [12] the *RC* time constant, the capacitor voltage has time to settle to the input voltage. The kernel operates in the sampling region and is defined as being a *fast* loop. The charge q_C that is transferred from the source to the capacitor is equal to the capacitance times the voltage difference:

$$q_C\,(t \gg 2RC) \approx (V_{IN} - V_O) \cdot C \tag{7}$$

Hence the amount of transferred charge does not depend on the resistance or switch-on time, as long as the switch is closed long enough to ensure settling.

On the other hand, if the switch opens much earlier than twice the *RC* time constant, the capacitor voltage has not had enough time to settle to the input voltage. The kernel operates in the mixing region and is defined as being a *slow* loop. It can

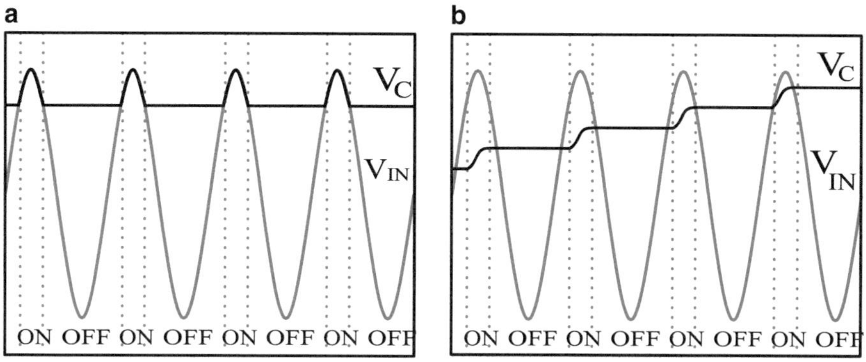

Fig. 10 Time domain behavior of the switched-RC loop for (**a**) a fast loop with and (**b**) a slow loop with Γ « 2

be seen from the time response that the capacitor voltage rises approximately linear, with the asymptote proportional to the resistance in the loop. In this case, the charge transferred to the capacitor is:

$$q_C\,(t \ll 2RC) \approx (V_{IN} - V_O) \cdot \frac{t}{R} \tag{8}$$

In this case, the transferred charge is proportional to the time and resistance, while being (almost) independent of the capacitance.

Expanding the circuit to a periodically operating switch with time period T_s and duty cycle D, in each cycle the switch closes for an interval equal to DT_s and remains open for a time interval equal to $(1-D)T_s$. From the basic LTI behavior of the RC circuit within one clock interval, it can be expected that the circuit responds differently for different ratio's between the switch-on time T_{on} and the RC time constant, which we will call Γ:

$$\Gamma \equiv \frac{D \cdot T_s}{R \cdot C} = 2\pi D \frac{f_{RC}}{f_s} \tag{9}$$

where $f_{RC} = \frac{1}{2RC}$ and $f_s = \frac{1}{T_s}$. Figure 10 shows for a sinewave with $f = f_s$ the difference in behavior between, a *fast* loop, and, a *slow* loop. A fast loop will settle to the input voltage within the clock interval, therefore performing a sample-and-hold operation. In contrast, a slow loop will make a small increment/decrement to the capacitor voltage in each cycle. The source can now be modeled as a current source and the clock waveform effectively multiplies this current with the clock waveform. Hence, a slow loop behaves like a mixer.

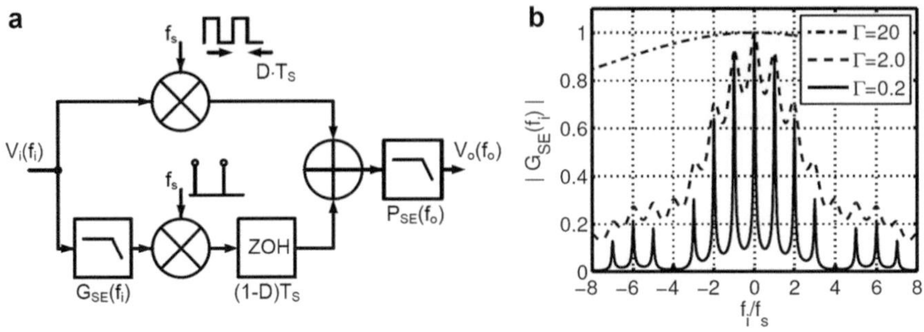

Fig. 11 (**a**) Block diagram equivalent to (10) and (**b**) shape of the filter function of a single ended (SE) switched-RC kernel $|G_{SE}|$

Frequency Domain Behavior of a Switched-RC Loop

In [12], by frequency domain analysis the exact HTF is derived *for arbitrary duty cycles D and RC bandwidths* f_{RC} for a single ended switched-RC loop:

$$H_{n,SE}(f_o) = P_{SE}(f_o)\left(\frac{1-e^{j2\pi Dn}}{j2\pi n} + \frac{e^{\frac{j2\pi(1-D)\cdot f_o}{f_s}}-1}{\frac{j2\pi\cdot f_o}{f_s}} G_{SE}(f_o - nf_s)\right) \tag{10}$$

where P_{SE} is a filter as function of the *output* frequency, and G_{SE} can be rewritten as a filter function of the *input* frequency and Γ as:

$$P_{SE}(f_o) = \frac{1}{1+j\frac{f_o}{f_{RC}}} \tag{11}$$

$$G_{SE}(f_i) = \frac{e^{\frac{j2\pi Df_i}{f_s}} - e^{-\Gamma}}{e^{\frac{j2\pi f_i}{f_s}} - e^{-\Gamma}} \cdot \frac{1}{1+j\cdot\frac{f_i}{f_{RC}}} \tag{12}$$

The behaviour of this switched-RC loop can be represented by the equivalent block diagram in Fig. 11. Note that this diagram *both* contains a mixer branch as in Fig. 8a, as well as the sampler of Fig. 8b, illustrating that both mixing and sample and hold operation is possible. In front of the sample and hold function, there is a filter function $G_{SE}(f_i)$, for which the amplitude transfer is plotted in Fig. 11 for various values of Γ. After summing of the mixer and sample and hold contribution, a low-pass filter function $P_{SE}(f_o)$ filters the combined output signal.

Although this block diagram only describes the downconversion from RF to the baseband capacitor, and not the upconversion that also takes place in an N-path filter, this analysis still highlights some of the key properties of an N-path filter. Clearly, slow loops with $\Gamma \ll 2$ render very narrowband filtering, as is desired for high-Q N-path filters. Also, plots of the conversion gain and noise figure for downconversion

Fig. 12 Contours for (**a**) equal conversion loss in dB and (**b**) equal NF in dB for downconversion with frequency shift—fs

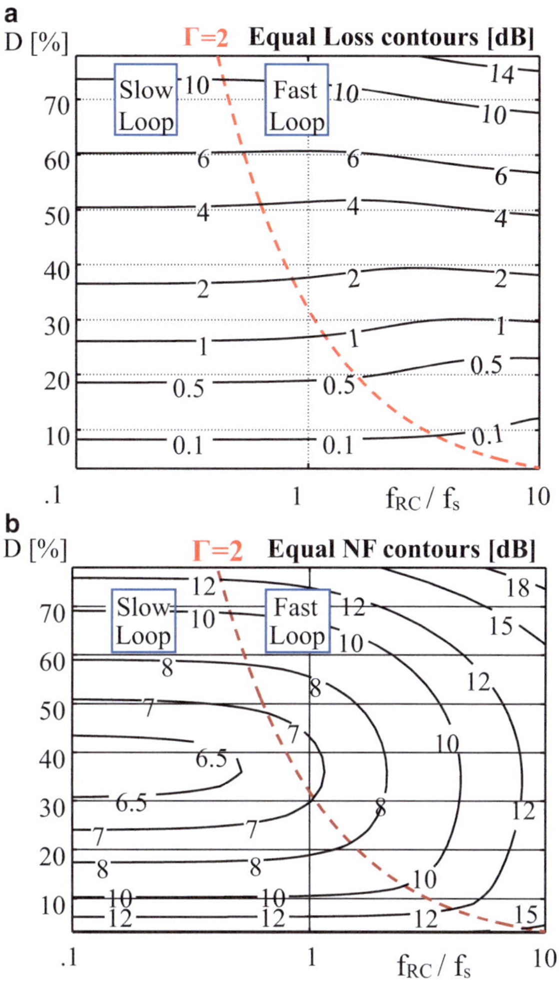

(Fig. 12) reveal that a lower duty cycle results in lower loss, and that fast loops have much higher noise figure, due to more noise folding, than slow loops.

In the limit, approximations to the exact HTF can be made for the two regions [12]. Here we focus on the slow loops ($\Gamma \ll 2$), which gives:

$$H_{n,SE} \approx \frac{sinc(Dn)}{1 + \frac{jf_0}{Df_{RC}}} e^{-j\pi Dn} \tag{13}$$

From this equation we can conclude that the voltage transfer to the capacitor has a lowpass filter with bandwidth D times the RC filter pole, with the amplitude of the harmonic n equal to $sinc(Dn)$ and a phase shift.

N-Path Filter Function

The N-path filter can be constructed from multiple Single Ended (SE) kernels, as indicated in Fig. 1, with the transfer function of a single kernel equal to (13). For ideal switches, the input node is alternately connected to each of the output nodes for a 1/N fraction of the clock period. In the time domain therefore, the voltage on the V_{out} node is the sum of the voltages on the capacitor nodes times the respective clock waveform.

The different timing of the clocks for the N paths results in an incremental phase shift in the frequency domain for each of the paths. Together with the realisation that a multiplication with a rectangular clock wave in the time domain is a convolution with a sinc function in the frequency domain, the transfer function $H_{in}(f)$ from the input (source) voltage node $V_{in}(f_i)$ to the output $V_{out}(f_i)$ in Fig. 1 becomes:

$$H_{in}(f) = \sum_{i=0}^{N-1}\left[e^{-\frac{j2\pi\cdot i}{N}} \cdot \sum_{k}\left(sinc(Dk)\cdot H_{n,SE}\left(f_i - kf_s\right)\cdot e^{-\frac{j2\pi\cdot i\cdot n}{N}}\right)\right] \tag{14}$$

This equation describes the effects of the downconversion from the input to the capacitors, followed by the upconversion from the capacitors to the output node. All harmonics for both up- and downconversion are included.

In order to derive the N-path bandpass filter shape, the harmonic responses except for a single pair are discarded in order to arrive at the RLC model in Fig. 3. The strongest harmonic response is the term associated with $n = -1$ and $k = 1$, which corresponds to a single downconversion with f_s, followed by a single upconversion with f_s.

Applying $n = -1$ and $k = 1$ gives the transfer function to the output, which gives a good approximation of the filter shape around the first clock harmonic f_s as:

$$H_{in}(f) = \frac{V_{out}\left(f_i\right)}{V_{in}\left(f_i\right)} \approx \frac{sinc^2(D)}{1 + \frac{j(f_i - f_s)}{Df_{RC}}} \tag{15}$$

As expected, the low pass filter shape is converted to a bandpass filter around f_s. The double application of $sinc(D)$ indicates the effect of the two frequency translations from input to capacitors to output. Mapping on an RLC resonator network results in the equations in Fig. 3. It is also seen that increasing the number of paths N and thus decreasing the duty cycle D reduces the loss of the filter. The resulting noise figure of the filter is also reduced.

References

1. Cook BW, Berny A, Molnar A, Lanzisera S, Pister KSJ (2006) Low-power 2.4-GHz transceiver with passive RX front-end and 400-mV supply. IEEE J Solid State Circuits 41:2757–2766
2. Ghaffari A, Klumperink EAM, Nauta B (2010) A differential 4-path highly linear widely tunable on-chip band-pass filter. In: 2010 IEEE symposium on radio frequency integrated circuits (RFIC), pp 299–302
3. Mirzaie A, Yazdi A, Zhou Z, Chang E, Suri P, Darabi H (2010) A 65 nm CMOS quad-band SAW-less receiver for GSM/GPRS/EDGE. In: 2010 IEEE symposium on VLSI circuits (VLSIC), pp 179–180
4. Ghaffari A, Klumperink EAM, Soer MCM, Nauta B (2011) Tunable high-Q N-path band-pass filters: modeling and verification. IEEE J Solid State Circuits 46:998–1010
5. Mirzaei A, Darabi H, Murphy D (2011) A low-power process-scalable super-heterodyne receiver with integrated high-Q filters. IEEE J Solid State Circuits 46:2920–2932
6. Ghaffari A, Klumperink E, Nauta B (2012) 8-Path tunable RF notch filters for blocker suppression. In: 2012 IEEE international solid-state circuits conference digest of technical papers (ISSCC), pp 76–78
7. Darvishi M, van der Zee R, Klumperink EAM, Nauta B (2012) Widely tunable 4th order switched Gm-C band-pass filter based on N-path filters. IEEE J Solid State Circuits 47: 3105–3119
8. Darvishi M, van der Zee R, Klumperink E, Nauta B (2012) A 0.3-to-1.2 GHz tunable 4th-order switched gm-C bandpass filter with 55 dB ultimate rejection and out-of-band IIP3 of +29 dBm. In: 2012 IEEE international solid-state circuits conference digest of technical papers (ISSCC), pp 358–360
9. Mirzaei A, Darabi H, Murphy D (2012) Architectural evolution of integrated M-phase high-Q bandpass filters. IEEE Trans Circuits Syst I Regul Pap 59:52–65
10. Darvishi M, van der Zee R, Nauta B (2013) Design of active N-path filters. IEEE J Solid State Circuits 48:2962–2976
11. Lin Z, Mak PI, Martins RP (2014) Analysis and modeling of a gain-boosted N-path switched-capacitor bandpass filter. IEEE Trans Circuits Syst I Regul Pap 61:1549–8328
12. Soer MCM, Klumperink EAM, de Boer PT, van Vliet FE, Nauta B (2010) Unified frequency-domain analysis of switched-series-RC passive mixers and samplers. IEEE Trans Circuits Syst I Regul Pap 57:2618–2631
13. Park JW, Razavi B (2014) Channel selection at RF using miller bandpass filters. IEEE J Solid State Circuits 49:3063–3078
14. Khatri H, Gudem PS, Larson LE (2010) An active transmitter leakage suppression technique for CMOS SAW-Less CDMA receivers. IEEE J Solid State Circuits 45:1590–1601
15. Qazi F, Duong QT, Dabrowski J (2014) Two-stage highly selective receiver front-end based on impedance transformation filtering. IEEE Trans Circuits Syst II Exp Briefs 62:421–425
16. Zhicheng L, Pui-In M, Martins RP (2015) 2.4 A 0.028 mm^2 11 mW single-mixing blocker-tolerant receiver with double-RF N-path filtering, S_{11} centering, +13 dBm OB-IIP3 and 1.5-to-2.9 dB NF. In: 2015 IEEE international solid-state circuits conference (ISSCC), pp 1–3
17. Barber NF (1947) Narrow band-pass filter using modulation. Wireless Eng 24:132–134
18. LePage WR, Cahn CR, Brown JS (1953) Analysis of a comb filter using synchronously commutated capacitors. Trans Am Inst Electr Eng I Commun Electron 72:63–68
19. Smith BD (1953) Analysis of commutated networks. Trans IRE Prof Group Aeronaut Navig Electron PGAE-10:21–26
20. Franks L, Witt F (1960) Solid-state sampled-data bandpass filters. In: 1960 IEEE international solid-state circuits conference digest of technical papers (ISSCC), pp 70–71
21. Franks LE, Sandberg IW (1960) An alternative approach to the realization of network transfer functions: the N-path filters. Bell Syst Tech J 39:1321–1350

22. von Grunigen DC, Sigg RP, Schmid J, Moschytz GS, Melchior H (1983) An integrated CMOS switched-capacitor bandpass filter based on N-path and frequency-sampling principles. IEEE J Solid State Circuits 18:753–761
23. Andrews C, Molnar AC (2010) Implications of passive mixer transparency for impedance matching and noise figure in passive mixer-first receivers. IEEE Trans Circuits Syst I Regul Pap 57:3092–3103
24. Mirzaei A, Darabi H (2011) Analysis of imperfections on performance of 4-phase passive-mixer-based high-Q bandpass filters in SAW-less receivers. IEEE Trans Circuits Syst I Regul Pap 58:879–892
25. Ghaffari A, Klumperink EAM, van Vliet FE, Nauta B (2014) A 4-element phased-array system with simultaneous spatial- and frequency-domain filtering at the antenna inputs. IEEE J Solid State Circuits 49(6):1303–1316
26. Ghaffari A, Klumperink EAM, Nauta B (2013) Tunable N-path notch filters for blocker suppression: modeling and verification. IEEE J Solid State Circuits 48:1370–1382
27. Heinlein W, Moehrmann K, Holmes W (1971) Double-tuned N-path bandpass filters using a single gyrator. IEEE Trans Circuit Theory 18:728–729
28. Mirzaei A, Darabi H, Yazdi A, Zhimin Z, Chang E, Suri P (2011) A 65 nm CMOS quad-band SAW-less receiver SoC for GSM/GPRS/EDGE. IEEE J Solid State Circuits 46:950–964
29. Jayasuriya S, Dong Y, Molnar A (2014) A baseband technique for automated LO leakage suppression achieving <-80 dBm in wideband passive mixer-first receivers. In: 2014 IEEE proceedings of the custom integrated circuits conference (CICC), pp 1–4
30. Duipmans L, Struiksma RE, Klumperink EAM, Nauta B, van Vliet FE (2014) Analysis of the signal transfer and folding in N-path filters with a series inductance. IEEE Trans Circuits Syst I Regul Pap 62:263–272
31. Ghaffari A, Klumperink EAM, van Vliet F, Nauta B (2014) A 4-element phased-array system with simultaneous spatial- and frequency-domain filtering at the antenna inputs. IEEE J Solid State Circuits 49:1303–1316
32. Borremans J, Mandal G, Giannini V, Debaillie B, Ingels M, Sano T et al (2011) A 40 nm CMOS 0.4-6 GHz receiver resilient to out-of-band blockers. IEEE J Solid State Circuits 46:1659–1671
33. Soer M, Klumperink EAM, Ru Z, van Vliet FE, Nauta B (2009) A 0.2-to-2.0GHz 65nm CMOS receiver without LNA achieving >11 dBm IIP3 and <6.5 dB NF. In: IEEE international on solid-state circuits conference digest of technical papers, ISSCC 2009, pp 222–223, 223a
34. Soer MCM, Klumperink EAM, Nauta B, van Vliet FE (2011) Spatial interferer rejection in a four-element beamforming receiver front-end with a switched-capacitor vector modulator. IEEE J Solid State Circuits 46:2933–2942
35. Soer MCM, Klumperink EAM, Nauta B, van Vliet FE (2014) A 1.0-to-2.5 GHz beamforming receiver with constant-Gm vector modulator consuming <9 mW per antenna element in 65 nm CMOS. In: 2014 IEEE international solid-state circuits conference digest of technical papers (ISSCC), pp 66–67
36. Fujian L, Pui-In M, Martins R. A 0.028 mm^2 11 mW single-mixing blocker-tolerant receiver with double-RF N-path filtering, S_{11} centering, $+13$ dBm OB-IIP3 and 1.5-to-2.9 dB NF. In: 2015 IEEE international solid-state circuits conference digest of technical papers (ISSCC), pp 36–37
37. Zhicheng L, Mak PI, Martins RP (2014) A sub-GHz multi-ISM-band ZigBee receiver using function-reuse and gain-boosted N-path techniques for IoT applications. IEEE J Solid State Circuits 49:2990–3004
38. Vanassche P, Gielen G, Sansen W (2002) Symbolic modeling of periodically time-varying systems using harmonic transfer matrices. IEEE Trans Comput Aided Des Integr Circuits Syst 21:1011–1024
39. Liou ML, Yen-Long K (1979) Exact analysis of switched capacitor circuits with arbitrary inputs. IEEE Trans Circuits Syst 26:213–223
40. Phillips CL, Parr JM (1995) Signals, systems and transforms. Prentice-Hall, Englewood Cliffs
41. Ru Z, Klumperink EAM, Nauta B (2010) Discrete-time mixing receiver architecture for RF-sampling software-defined radio. IEEE J Solid State Circuits 45:1732–1745

Efficiency Enhancement Techniques for RF and MM-Wave Power Amplifiers

Patrick Reynaert and Brecht Francois

Abstract Some important aspects of CMOS RF PA design are discussed. First the reader is confronted with the ever prominent efficiency-linearity trade-off in the design of conventional amplifier classes, such as A and B, as well as the challenge to achieve high output power in low-voltage CMOS. To cope with this challenge, several power combining structures are introduced. Next, some RF PA architectures are presented to improve the efficiency at power back-off. Finally, some recently introduced architectures are discussed that use the advanced signal processing capabilities of CMOS to deal with this efficiency-linearity trade-off in RF PA design.

1 Introduction

The growing demand for wireless communication and the exploding production of ubiquitous mobile consumer applications increased the need for highly integrated low-cost highly efficient RF transceivers. Since CMOS has been proven to be a cost-effective platform to integrate the digital base-band circuitry, it is the most preferred technology to implement a transceiver for wireless consumer applications [1]. Thanks to some recent power combining architectures based on transformers [2, 3], it became possible to integrate Watt-level efficient RF-PAs in CMOS.

Modern and future wireless communication standards like LTE, LTE-advanced, WLAN IEEE802.11ac , and many others, created a growing focus on the efficiency enhancement for integrated PAs to increase the battery life time of the mobile devices. These new communication standards exhibit signals with high PAPR, hence demanding not only a high efficiency at peak output power but also at power back-off. However, these future communication standards support high signal bandwidths to provide high data rates for user applications. Consequently, this brings the RF PA

P. Reynaert (✉) • B. Francois
KU Leuven ESAT - MICAS, Kasteelpark Arenberg 10, 3001 Leuven, Belgium
e-mail: patrick.reynaert@kuleuven.be

K.A.A. Makinwa et al. (eds.), *Efficient Sensor Interfaces, Advanced Amplifiers and Low Power RF Systems*, DOI 10.1007/978-3-319-21185-5_16

designer back to the ancient torment about the efficiency-linearity trade-off in power amplifier design, which is discussed for several RF PA architectures and efficiency enhancement techniques.

2 Class A, B, AB and C

The conventional amplifiers classes, such as class A, B, AB and C, rely on trading linearity for efficiency. This trade-off mainly depends on the transistors' bias point, as summarized in Fig. 1a, illustrating the amount of quiescent current depending on the bias point for each class of operation. The schematic of these amplifier classes is presented in Fig. 1a.

It is well-known that the transistor in a class A amplifier always conducts a DC current, known as the quiescent current. Hence, this drains the battery of our mobile devices. Not conducting current all the time would result in a higher efficiency, e.g. in class B bias, where its quiescent current is zero. The class AB realizes a compromise between the two opposites: the transistor is turned off for less than half a period of the applied input signal. The class AB power amplifier defines all conduction angles between half a period, corresponding to class B, and full period of the input signal, corresponding to class A.

By turning off the transistor and consequently reducing the conduction angle, the shape of the voltage and current waveform are not solely depending on the transistors' operation but also rely on the passive circuitry. On one hand, this changes the voltage-current overlap and is directly related to the efficiency. On the other hand, turning off the transistor is introducing higher harmonics at the drain current and voltage waveforms, causing linearity degradation. Although the distortion of the class AB is greater than the class A amplifier, the class AB power amplifier is often used because its distortion is less than in a class B amplifier [4, 5].

By further reducing the conduction angle of the transistor compared to class B, such that the output current is zero for more than half of the period, the efficiency can be further increased while introducing different harmonic content. This operation is called class C. However, the class C power amplifier cannot be used for linear amplification since turning off the transistor results in increased harmonic distortion, as mentioned before.

Figure 2a,b summarize this efficiency-linearity trade-off for a linear amplifier class for different conduction angles: firstly, Fig. 2b shows that the class B allows us to generate the same amount of fundamental output power while decreasing the DC power consumption by a factor of $\pi/2$. In other words, the efficiency is increased from 50 % in class A to $\pi/4$ in a class B operation while reducing indeed the conduction angle α, as illustrated in Fig. 2a, and maintaining identical fundamental output power level. However it is important to note that to deliver the same amount of RF power as a class A amplifier, a class B amplifier requires twice the voltage swing at the gate of the transistor. This means that the voltage gain of the class B amplifier is half of the voltage gain of a class A amplifier. This gain reduction becomes more prominent while decreasing the conduction angle into the class C operating range.

Secondly, the efficiency could be further increased to 100 % by biasing the transistor in deep class C, as illustrated in Fig. 2a. However, not only the DC power consumption of this class C operation but also the fundamental output power theoretically approaches zero.

Finally, regarding the linearity, in the class AB operating range, the largest component next to the fundamental is the second harmonic. While reducing the conduction angle even further in the class C operating range, all harmonics are present and the harmonic power is certainly not negligible. This indicates that the class C operation can not be employed for linear amplification. Figure 2b illustrates that all odd harmonics become zero exactly at the class B operation, such that only even harmonics remain.

3 Impedance Transformation and Power Combining

CMOS typically uses a low supply voltage. Therefore, it becomes a challenge to deliver sufficient power to a 50 Ω load. It is thus necessary to place an impedance transformation network between the PA and the load. This allows to transform the 50 Ω load into a lower impedance such that a higher power can be produced by the RF PA. Two famous on-chip impedance transformation networks exists: the resonant impedance transformation and the magnetically coupled impedance transformation networks.

The typical LC resonant matching network is maybe an obvious way of impedance matching and only needs two components as depicted on Fig. 3a. The efficiency of this network is written as

$$\eta = \frac{Pout}{Pin} = \frac{I_{IN}^2 R_L}{I_{IN}^2 (R_L + R_{sL})} = \frac{1}{1 + \frac{R_{sL}}{R_L}} = \frac{1}{1 + \frac{Q_L}{Q_{ind}}}. \tag{1}$$

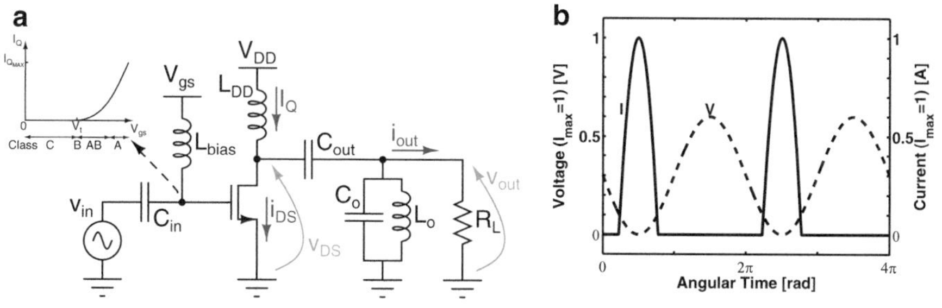

Fig. 1 (a) Simplified schematic of a single-ended linear amplifier and (b) time domain of voltage and current waveforms at the drain of the transistor in a reduced conduction angle implementation

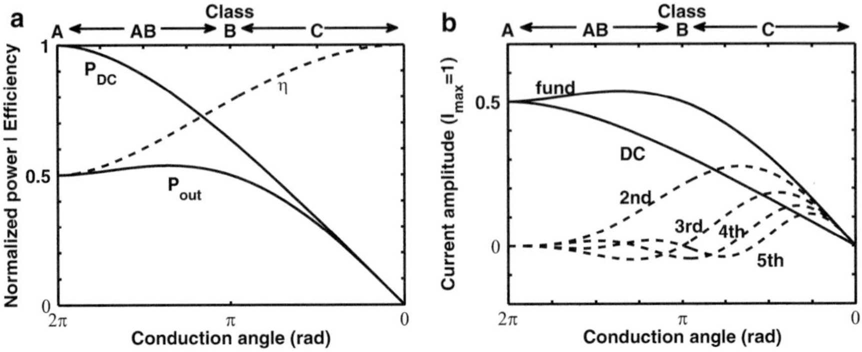

Fig. 2 Efficiency-linearity trade-off in a linear amplifier. (**a**) The output power, P_{out}, the DC power consumption, P_{DC} and the efficiency, η, versus the conduction angle. (**b**) The amplitude of the DC current, the fundamental current and its first four harmonic currents versus the conduction angle

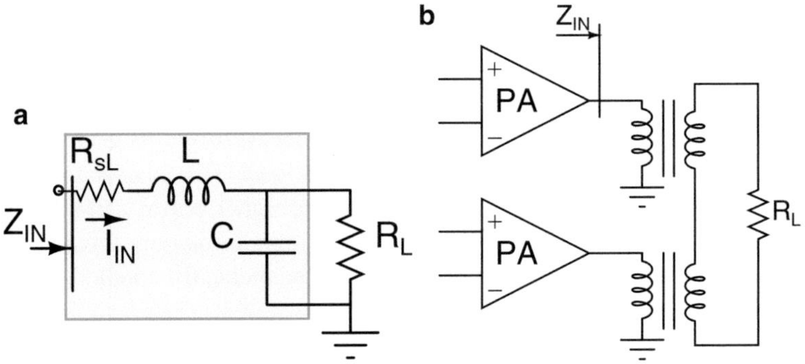

Fig. 3 (**a**) Schematic of an LC matching network and (**b**) principle of power combining

where I_{IN} represents the current delivered into the network as shown in Fig. 3a, R_{sL} is the series parasitic resistance of the inductor and R_L represents the load resistance. In [2], it has been shown that the loss of an LC network is proportional to the impedance transformation ratio, r, defined as

$$r = \frac{R_{load}}{R_{in}} \tag{2}$$

where R_{in} is the transformed impedance, seen at the input of the LC network. Clearly, since the loss of the impedance transformation network is proportional to r, this puts an upper limit on the maximum achievable impedance transformation factor. As such, for low a supply voltage and/or high output power the impedance transformation ratio r needs to be very high and this network will have a very low efficiency.

Another approach for a matching network is to use on-chip transformers. Whereas on-chip LC-match networks suffer significantly from low efficiency due to the low quality factor of the on-chip inductors, the efficiency of an on-chip transformer is independent of the power enhancement ratio (PER) and hence does not reduce for larger PER. However, to achieve Watt-level output power and more with the ever dropping supply voltages in nanometer scale technologies, the PER ratio needs to be much larger than 50 [6]. Increasing the PER even further would result in a transformed impedance with a very low real part experienced by the power amplifier. This makes it very challenging for both the active as the matching network to result in an overall high efficiency.

Therefore, *power combining*, is commonly used at RF and mm-wave frequencies, and encompasses all techniques that allow to deliver more output power than is achievable with only a single active power amplifier device, as shown in Fig. 3b. In general, power combiners can be categorized into two main groups: series power combiners (SCTs) and parallel power combiner (PCTs) [7]. The SCT combiner inherently has a higher power-driving capability and is therefore more favourable for on-chip transformer-based power combiner design. Series power combining transformers (SCT) are divided into two subgroups: SCTs sharing the same primary winding (Fig. 4a) and SCTs employing separate primary winding for each transformer (Fig. 4b). The former was first introduced by Aoki et al. [2] in 2002 as a highly efficient power combining architecture. The latter typically uses a figure-8 shaped transformer architecture to combine the output power of multiple RF power amplifiers, which was introduced several years later. Unlike a figure-8 shaped transformer combiner, in a SCT sharing the same primary winding, high Q slab inductors are used to construct 1:1 transformers resulting in a high passive power transfer efficiency. This structure is called a Distributed Active Transformer or DAT. Each side of the primary (slab) inductor in a DAT-architecture is shared by two different power amplifiers as shown in the 4-way DAT in Fig. 4a [2].

However, the next step in power aware design, is to switch on or off more PAs to change the level of the output power [7, 8]. Since all PAs are coupled to each other

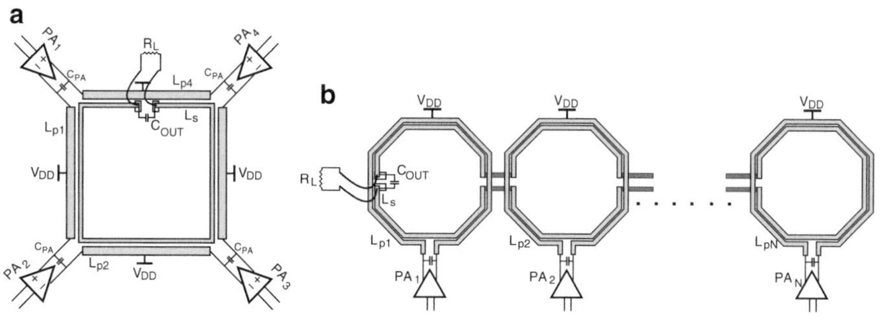

Fig. 4 (**a**) Layout of N-way distributed active transformer (DAT) architecture (i.e. N = 4) and (**b**) layout of a figure-8 shaped transformer-based series power combiner

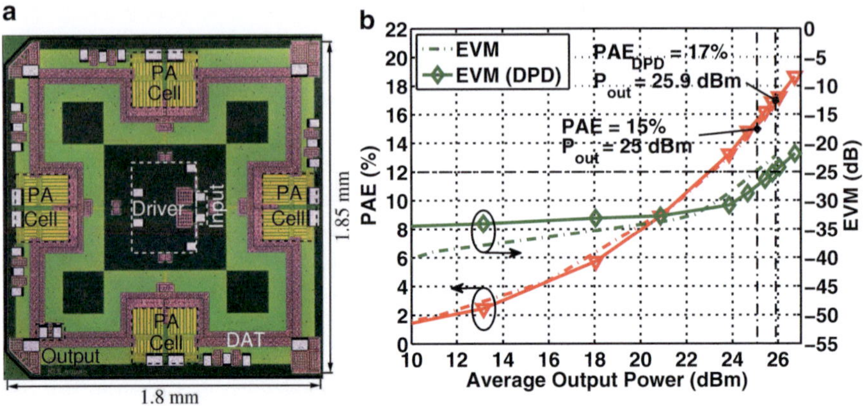

Fig. 5 (**a**) Die photograph of linear class B RF PA with on-chip transformer-based series power combining and (**b**) measured performance with 10 MHz uplink LTE signal (PAPR = 6.92 dB)

through the transformer, this will result in a change of the load impedance as seen by the PAs which is indeed related to (active) load-pull. This approach thus allows to increase the load resistance at power back-off, to further push the efficiency curve upwards [7, 8]. To overcome the limited flexibility of a DAT-series power combiner, SCTs employing separate primary winding for each transformer can come to the designers' rescue at the expense of a slightly lower passive power transfer efficiency, since each RF PA driving its own separate winding is switched off more easily [7]. This is known as flexibility(complexity)-efficiency trade-off in RF PA design exhibiting multiple power amplifiers.

An RF PA with an efficient realization of a transformer-based series power combining (DAT) to deliver the Watt-level output power while not jeopardizing the required linearity and efficiency has been implemented in a standard 90 nm CMOS technology for the extended GSM-band and LTE-band VIII [3, 6]. In this RF PA, the output power of four differential linear class B amplifiers is combined. In addition, the proposed series power combiner allows the designer to integrate both the input and output matching networks as the power combining on the same silicon die. Measurements show that the PA delivers up to 1 W of RF power with 28.4 % drain efficiency and 25.8 % PAE with only a 2 V supply while achieving a high gain of 28 dB. The choice of optimal biasing ensures a very flat gain and small AM-PM distortion up to high output power. While applying an uplink LTE signal with a PAPR of 6.92 dB, the PA produces 25 dBm of average output power with 15 % PAE while obeying the stringent linearity-specifications of the LTE communication standard [6] (Fig. 5).

4 PA Architectures with Multiple RF Paths

So far, only power amplifier topologies employing one single RF path have been discussed. However, some power amplifier topologies decompose a variable envelope signal into several variable envelope signals or into several constant envelope signals or a combination of both. These power amplifier types belong to the class with multiple RF paths and will be described in detail in this section.

The ultimate goal of these techniques is to create a more linear operation starting from a non-linear but highly efficient amplifier, e.g. switching amplifiers class D, E, or vice versa. Therefore, sometimes, these amplifiers are seen as efficiency enhancement techniques or linearization techniques.

4.1 *Doherty Amplifier*

The Doherty amplifier was first proposed by William Doherty in 1936 as an efficiency enhancement or power conservation technique [9]. The basic concept of Doherty is to allow one or more amplifiers to operate at their peak operation and thus their peak efficiency, in order to enhance the overall efficiency, while another amplifier is taking care of the linear amplification.

Originally, the Doherty amplifier consists of two amplifiers and the simplified schematic is presented in Fig. 6a. In a two-stage Doherty, the operation of the two PAs can be described as follows: at low output powers, the auxiliary power amplifier, PA_A, is not active, by removing its input signal or by turning down its bias. Then the linear amplification is performed by the main power amplifier, PA_M. While the input power is increasing, the RF PA approaches its compression, which is around 6 dB below the peak power level, due to the impedance transformation by the quarter wave transmission line [10].

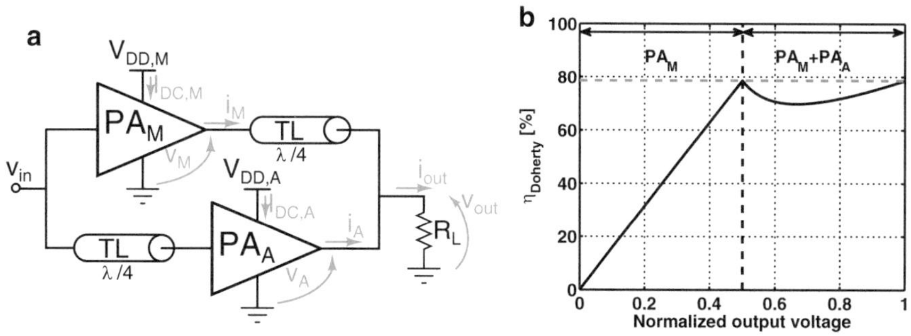

Fig. 6 (**a**) Schematic of a Doherty power amplifier and (**b**) efficiency of a two-amplifier Doherty PA

The basic principle behind the Doherty amplifier is also known as active loadpull [10]. Active load pull means that the impedance seen by one PA can be changed by applying an in-phase current, delivered to the load by a second PA. Consider Fig. 6a, the impedance seen by the auxiliary amplifier would be equal to R_L if the main amplifier was not supplying current. Using Kirchoff, both the current from the auxiliary amplifier, I_A, and the current from the main amplifier passing through the transmission line, I_{M_T}, flow into the load R_L so that the total voltage at the load becomes

$$V_{out} = R_L(I_A + I_{M_T}) \tag{3}$$

In other words, this resembles as if another load was connected to the main amplifier [10], valued as

$$Z_{M_T} = R_L\left(\frac{I_A + I_{M_T}}{I_{M_T}}\right) = R_L\left(1 + \frac{I_A}{I_{M_T}}\right) \tag{4}$$

Simultaneously, the auxiliary amplifier sees a load having a value of

$$Z_A = R_L\left(\frac{I_A + I_{M_T}}{I_A}\right) = R_L\left(1 + \frac{I_{M_T}}{I_A}\right) \tag{5}$$

In other words, from (4), the impedance Z_{M_T} can be transformed into a higher resistive value if I_A is in phase with I_{M_T}. When antiphasing both currents, the transformed impedance reduces. Thus, the impedance by the main amplifier is affected by the operation condition of the auxiliary amplifier and vice versa. This also explains that in the maximum power condition where I_{M_T} and I_A are maximum, the load impedance seen by each amplifier turns into R_L.

The output voltage from (3) can also be obtained in terms of the current in the main amplifier, I_M, as

$$V_{out} I_{M_T} = V_M I_M \tag{6}$$

$$\left(\frac{V_{out}}{I_{M_T}}\right)\left(\frac{V_M}{I_M}\right) = Z_T^2 \tag{7}$$

$$I_{M_T} = \frac{V_{out}}{Z_T} \tag{8}$$

where Z_T represents the characteristic impedance of the quarter wave transmission line.

From Fig. 6b, it can be seen that the efficiency achieves its peak when the output voltage reaches it maximum but also at 6 dB back-off from the peak output power when the condition of $V_{out} = V_{MAX}/2$ is satisfied. If only the main amplifier is operating, the efficiency reaches its peak efficiency of 78.5 % at the transition point. It is illustrated in Fig. 6b that at this second peak the obtained efficiency in a Doherty configuration is 78.5 %, assuming class B amplifier operation. The reduction of the efficiency while both amplifiers are operational is due to the lower efficiency of the auxiliary amplifier. In the upper power region of the Doherty operation, the auxiliary amplifier is not operating full swing yet and reaches it peak efficiency at the peak output power point while the main amplifier constantly is fully saturated and thus operates at its peak efficiency.

Unfortunately, efficiency enhancement always comes at a cost. Indeed, while considering linearity, when ideal linear devices are used to implement both the main and auxiliary PA, the Doherty configuration is behaving linearly. However, in practice, to achieve the high efficiency once the transition point has been past, in the upper 6 dB, the main PA operates in saturation where typically a class B amplifier would be used as a rule of thumb till more or less the P_{-1dB} output power point. In the upper power region, the main PA operates close to its clipping region. This will reduce the slope of the intermodulation distortion curves while backing off compared to the slope of the intermodulation distortion in a conventional class B using identical device [4]. In addition, while operating in the low power region, only half of the matching network and power capability is being used, in comparison with a conventional configuration, e.g. class B. This means that to deliver the same amount of output power as a conventional class, the third order intermodulation distortion would be 6 dB higher in a Doherty [4], in first order approximation.

Anyhow, it is clear that the efficiency characteristic of a Doherty amplifier is its strongest asset. To fully exploit this characteristic in practice, the auxiliary amplifier is often biased in class C operation, as suggested by Doherty, to approach an ideal shut down behaviour of the auxiliary amplifier before the last 6 dB of output power and only starts to operate at that point. Alternatively, adaptive bias networks or power control devices could help to improve the hold-off behaviour of the auxiliary amplifier before the transition point [11]. However, this introduces an envelope signal path in which the circuitry for this control processing exhibits certain delay or speed. On the other hand, unlike some of the rival efficient power amplifier architectures (ET, EER, Polar), there is no need for a power supply modulator to achieve this efficiency enhancement and is therefore also not band limited by the supply modulation bandwidth.

The linearity performance of a Doherty is relatively poor [5] and the use of a quarter wave transmission line makes the Doherty architecture not an integration friendly solution for efficiency improvement. And since there are two or more RF signal paths, the phase matching of the signals generally impedes the design. In addition, for the practical implementation, the designer needs to take into account that the power amplifiers should be able to operate efficiently in a certain load impedance variation region, without reducing the life time of the power amplifier.

Although the integration of a Doherty Power Amplifier introduces quite some challenges as discussed before, a fully-integrated Doherty amplifier is realized in standard 40 nm CMOS for WLAN applications as shown in Fig. 7a. A novel asymmetrical series combining transformer is used to achieve optimal efficiency enhancement offered by Doherty operation. The transformer-based uneven Doherty architecture is analyzed to further improve the back-off efficiency without linearity

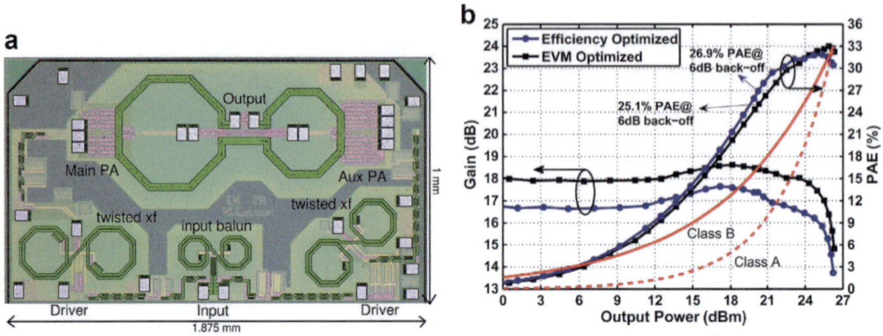

Fig. 7 (**a**) Die photograph of a fully-integrated CMOS two-stage Doherty power amplifier and (**b**) measured large signal performance of the Doherty RF PA at 2.4 GHz

degradation and presented in [12]. The fabricated two-stage uneven Doherty PA achieves a maximum output power of 26.3 dBm at 2.4 GHz with a peak power added efficiency (PAE) of 33 % at 2 V supply voltage. The PAE at 6 dB back-off is still as high as 25.1 %. The Doherty PA is characterized for two different bias points: efficiency optimized bias and EVM optimized bias for WLAN signals. In the efficiency optimized bias point, the PA consumes only 55.5 mA bias current and achieves high back-off efficiency. Figure 7b shows the measured efficiency of the PA at 2.4 GHz and compares with the efficiency of class A and class B power amplifier having the same peak drain efficiency. Figure 7b clearly indicates the back-off efficiency enhancement of the proposed uneven Doherty amplifier. However, slightly increasing the gate bias voltage improves the linearity of the amplifier and allows better performance with the WLAN signal. With a 54 Mbps WLAN 802.11g signal, the Doherty RF PA meets the stringent EVM and spectral mask requirements at 19.3 dBm average output power with a PAE of 22.9 % with no need of predistortion. An open loop digital predistortion is applied to further improve the linearity. The PA satisfies WLAN requirements at 20.2 dBm average output power with a PAE of 24.7 % with predistortion.

In short, some successful attempts of fully-integrated RF Doherty PA in nanometer scale CMOS have been reported and even at mm-wave frequencies [11–14].

4.2 Outphasing/LINC

A highly efficient amplifier, such as class E or class F, in its simplest configuration cannot be used for linear amplification since it delivers a constant envelope RF signal. Now, one could wonder if it is possible to create a variable envelope signal by decomposing it into several constant envelope signals? In 1930 already, Chireix posed himself this identical question. The answer has led to the introduction of the Outphasing technique.

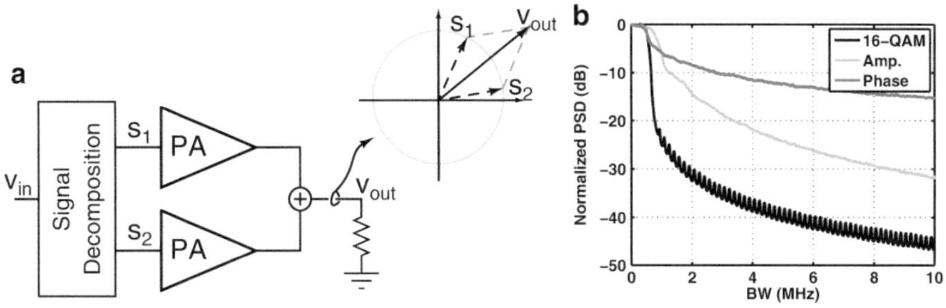

Fig. 8 (**a**) Principle and simplified schematic of a LINC/outphasing power amplifier and (**b**) a decomposition of a 1 MHz 16-QAM modulated signal in its polar form

This technique is also known as *LI*near amplification using *N*onlinear *C*omponents (LINC) and was introduced by Cox in 1974 [15], as a method to achieve linear amplification at microwave frequencies, based on the original outphasing technique proposed by Chireix in 1930 [16]. Outphasing combines the output of two nonlinear amplifiers exhibiting a constant envelope output signal.

The principle of the outphasing architecture is as follows: as shown in Fig. 8a, the input signal is

$$v_{in}(t) = V(t)cos(\omega t + \phi(t)) \tag{9}$$

where V(t) represents the variable envelope signal present on the carrier signal at frequency $\frac{\omega}{2\pi}$. Then the signal decomposition block splits the input signal into two constant envelope but phase modulated signals as

$$s_1(t) = V_{MAX}cos(\omega t + \chi(t)) \tag{10}$$

$$s_2(t) = V_{MAX}cos(\omega t + \theta(t)) \tag{11}$$

where the phases relate as

$$\chi(t) = \phi(t) - \psi(t) \tag{12}$$

$$\theta(t) = \phi(t) + \psi(t) \tag{13}$$

Then by recombining these constant envelope signals, $s_1(t)$ and $s_2(t)$, a correct linearly amplified version of the input signal is obtained with the following relationship:

$$V_{out}(t) = s_1(t) + s_2(t) \tag{14}$$

$$\psi(t) = bgcos\left[\frac{V_{out}(t)}{V_{MAX}}\right] \tag{15}$$

This LINC or outphasing technique has several advantages. It makes use of nonlinear amplifiers to realize an efficient linear amplification at RF and even at mm-wave frequencies [17]. The use of these highly efficient amplifiers, such as class D, E, F, results in a very efficient amplification. Unlike all conventional linear amplification such as class A, this technique is theoretical capable of achieving 100 % efficiency at all envelope levels of the output RF signal [5]. In addition, to achieve this high efficient performance, outphasing does not make use of supply modulation, saving some voltage headroom and efficiency lost in the supply modulator, like in polar modulation, ET or EER as will be discussed in the next sections.

However, the above signals, $s_1(t)$ and $s_2(t)$, needs to be successfully and accurately generated in order to guarantee these benefits of the outphasing techniques. This is known as the major challenge in the outphasing architecture: achieve accurate gain and phase matching between the two RF paths [18]. This is a typical challenge for all power amplifiers consisting of multiple RF paths. For this reason, the technique is not that popular and is not found in commercial applications for mobile devices so far. Errors in gain or phase will result in incomplete cancellation while summing both signal paths resulting in a number of unwanted spurious products or spectral regrowth in the output spectrum [19]. The outphasing amplifiers consequently suffer from limited operation range (DR). Indeed, the lowest output amplitude that can be represented is limited by the accuracy of the outphasing angle and the generation of two identical constant envelope signals from two different power amplifiers [20]: e.g. to represent a signal with a small amplitude, the outphasing angle $\psi(t)$ approaches 180° of two signals with constant peak amplitude. An approximation to quantify this limited dynamic range (DR) [21]:

$$DR = 20log_{10}\left(\frac{G_1 + G_2}{G_1 - G_2}\right) \tag{16}$$

where G_1 and G_2 represent the gain in signal paths of $s_1(t)$ and $s_2(t)$ respectively. A more detailed study on the limited dynamic range can be found in [21].

Another challenging issue in the outphasing architecture is the large bandwidth requirement for both signal path $s_1(t)$ and $s_2(t)$, since both signals experience large phase excursions [19]: $s_1(t)$ and $s_2(t)$ contain both phase variations of $\phi(t) \pm \psi(t)$. Typically, the bandwidth of a signal of the form $cos(\omega t + \phi)$ (denoted as Phase in Fig. 8b) occupies a larger bandwidth than its composite signal (i.e. black solid line in Fig. 8b), such as an modulated LTE signal [19] (i.e. 16-QAM signal with 1 MHz bandwidth in Fig. 8b). This effect is even more exacerbated in the outphasing topology and is therefore a major drawback in outphasing transmitters.

Finally, to realize the summation operator as shown in Fig. 8a, the design difficulty is pushed towards the summing device. Indeed, the signal originating from one PA may affect the performance of the other PA, resulting in performance degradation, resulting in spectral regrowth [19] or even corrupted amplification. If the outphasing architecture is implemented with combiners having isolated input ports, then amplifier choice is relatively unrestricted [22]. However, if lossless power combiners are chosen and thus the overall output impedance of each component amplifier is established by the outputs of both amplifiers, then the output interactions must be carefully designed.

5 PA Architectures with a RF Path and an Envelope Path

Some power amplifier architectures exhibit an RF path and an envelope path. As always, the ultimate goal of the power amplifier topologies in this group is to create a more linear operation starting from a non-linear but highly efficient amplifier or vice versa. Therefore, these amplifiers are sometimes called efficiency enhancement techniques or linearization techniques.

5.1 *Class G*

A class G amplifier requires more than one supply voltage, as shown in Fig. 9a, and at least two pair of active devices [5]. One pair is connected to the lower supply voltage to deliver the low output power and similarly, the pair connected to the high supply voltage generates the power during the high power mode. In the low power mode, only one part of the circuitry is operational while the other is shut down. At the high power level, the other part is operating as a current source while the lower pair is not active. The efficiency improvement comes from the fact that for lower signal levels, the low power part operates twice as efficiently as its high power counterpart. Therefore, this is beneficial when signals are applied that typically don't often need the high output power [23].

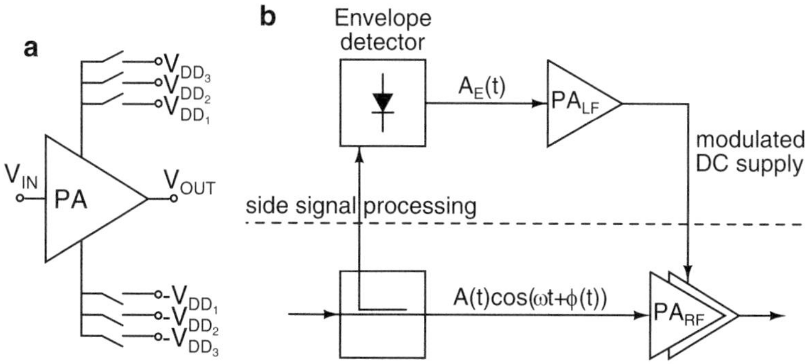

Fig. 9 (**a**) Simplified schematic of class G PA and (**b**) schematic of envelope tracking (ET) PA

5.2 Envelope Tracking

CMOS clearly has an advantage over other more exotic technologies when it comes down to signal processing. As such, efficiency enhancement and linearization techniques could be more easily realized on a CMOS RF PA. Since signal processing is a baseband operation, mainly those techniques with a strong baseband focus are successful in CMOS. For this reason, envelope modulation and restoration (EER), polar modulation and envelope tracking (ET) CMOS PAs in CMOS have seen a large success in the past few years [8, 24].

Envelope Tracking (ET) has been introduced to increase the efficiency of a linear PA. Envelope tracking adapts the supply voltage of the RF PA to the amplitude of the envelope signal, $A_E(t)$, as illustrated in Fig. 9b, and hence the DC current of the linear PA is adjusted to the amplitude of the RF signal. ET has emerged during the last decade because it is a successful mechanism to boost the efficiency at power back-off and even reach almost a flat efficiency-versus-output power graph when combined with a Class B RF PA. However, this technique also has its limitations: the power amplifier still needs to be sufficiently linear to cope with the specifications for the spectral mask and the EVM. In fact, changing the supply of a PA makes it much more non-linear so that the ET efficiency enhancement technique quite often needs to be combined with an additional linearization technique [25].

Additionally, to achieve an overall high average efficiency, the supply modulator needs to operate very efficiently. Thus, the design of an ET PA pushes some of its design difficulties into the supply modulator. However, some successful attempts have been published already of fully integrated envelope tracking PAs. The supply modulator is typically a combination of a linear supply modulator and a switching supply modulator [24, 26], which requires an additional large inductor off-chip, to achieve the high modulator efficiency. In practice, due to limited supply rejection ratio of the power amplifier, this switching modulator increases the spectral regrowth, which poses a severe threat to obey the stringent spectral purity requirements defined by the communication standards. In addition, substrate noise from the switching power supply modulator might affect the RF PA. To minimize the substrate noise which origins from the supply modulator in practice, both the supply modulator and the power amplifier are realized on a separate die [24, 26].

To conclude, due to its favourable efficiency behaviour, the envelope tracking technique gained a lot of popularity recently in industry. Indeed, in 2013, Qualcomm became the first company to realize an LTE compliant ET PA for mobile systems. As we speak today, the ET PA is present already in many of the high-end user smartphones such as Samsung Galaxy 3, Nexus 5... But, going to LTE-advanced, a limitation of ET is exposed: so far, no integrated envelope tracking RF PA has been reported that is able to support signal bandwidths between 20 and 40 MHz for mobile applications, whereas LTE-advanced targets uplink channel bandwidths up to 100 MHz.

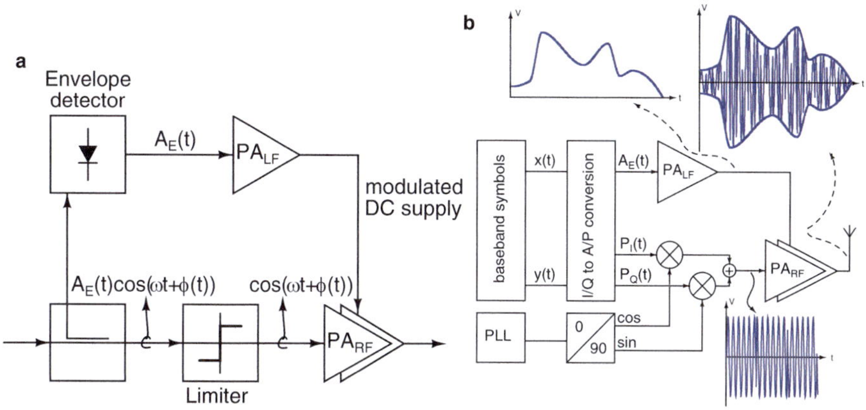

Fig. 10 (**a**) Block diagram of envelope elimination and restoration (EER) and (**b**) a basic schematic of a polar transmitter

5.3 *Envelope Elimination and Restoration*

Envelope elimination and restoration (EER) was proposed by Kahn [27] as an efficient and linear technique for power amplification. As shown in Fig. 10a, a limiter converts the RF signal consisting of both amplitude and phase modulation into a signal with constant envelope but varying phase. This process is also known as envelope elimination. Simultaneously, an envelope detector decomposes the signal and only the envelope signal remains. In the RF PA itself, both paths are recombined. By this construction, the amplitude modulation is again added to the output. This clearly explains why the EER architecture is categorized in the group of the RF path with envelope path.

The main advantage of EER is that the RF amplifier can be a saturated or switching PA which inherently has a high efficiency. This allows to achieve high efficiency even for amplitude modulated signals, provided that the envelope amplifier, which provides the supply voltage to the RF PA, can also be made power efficient.

At first sight, the only difference between EER and ET is the absence of the limiter as such that the amplitude and phase modulated carrier is amplified by the ET PA. This means that the envelope path in ET is not contributing to the signal accuracy, as in a polar architecture like EER or polar modulation, but its sole purpose exists in power reduction.

Thus in EER, the RF signal containing the phase and amplitude modulation is first split up into both the phase and the envelope signals using two different circuit blocks. In other words, the RF PA in EER is only amplifying a phase modulated carrier signal. As discussed before, a switching amplifier such as class D and class E, are excellent candidates, whereas amplitude linearity in the RF PA is still required in ET and hence the transistor in an ET power amplifier behaves as a current source.

All this becomes unnecessary when a fully integrated transmitter is realized [8, 28]. In this case, the digital signal processing block can provide the phase and the envelope signals separately which is often referred to as Polar Modulation which is discussed in the next section.

5.4 Polar Modulation

Starting from the EER system, the amplitude or envelope signal and phase modulated RF carrier can be directly generated from the I/Q signals. Figure 10b shows the basic system block of this technique, referred to as polar modulation. In polar modulation, the amplitude of the RF output signal of the switching PA is dependent of the modulated supply voltage and therefore the efficiency and linearity of polar modulation relies on great extend on the supply modulator.

One way to implement the envelope supply modulator is using linear regulator based on a low dropout regulator [29]. The feedback makes the output impedance very low and this circuit behaves as a voltage supply. This approach has the advantage of its wideband behaviour and circuit simplicity, but lower efficiency. When a switching RF PA is used, which maintains its high efficiency even when the supply is reduced, the overall efficiency versus output power curve becomes a square root behaviour as in the class B PA topology [29].

However, in contrast with a conventional class B amplifier, the linearity requirements of this design are handled by baseband circuitry. As such, a major improvement in linearity compared to the original RF class B amplifier is achieved. All this results in an overall better average efficiency and a higher peak power inasmuch the RF PA can now operate closer to the maximum output power as illustrated in [29].

Another approach to realize the amplitude modulator could be a switching-type modulator based on the principle of the buck or boost DC-DC converters [30]. The disadvantage of this technique is the quest for the necessary LC filter. This LC filter reconstructs the baseband signal from the switching square-wave waveform. These filters contain typically a rather large inductor. The quality factor of the inductor needs to be very high to have low-loss high current capability. Also the self-resonance frequency should be as high as possible. It becomes clear that the inductor is the limiting factor of this design and becomes a very expensive component.

The major drawback of polar modulation is in fact the decomposition of the AM and PM modulated RF signal into a separate amplitude or envelope signal and a phase modulated carrier. This decomposition is highly non-linear which will expand the bandwidth, typically by a factor 3…6. For the RF power amplifier, this is normally not an issue. But for the envelope amplifier, or supply modulator, this will create a real bottleneck. Indeed, with today's wideband systems of several tens of megahertz, this means that the envelope modulator needs to have a 50…100 MHz bandwidth. Clearly, such high bandwidth will somehow result in a reduced efficiency. On the other hand, predistortion can allow the system to have a lower envelope bandwidth.

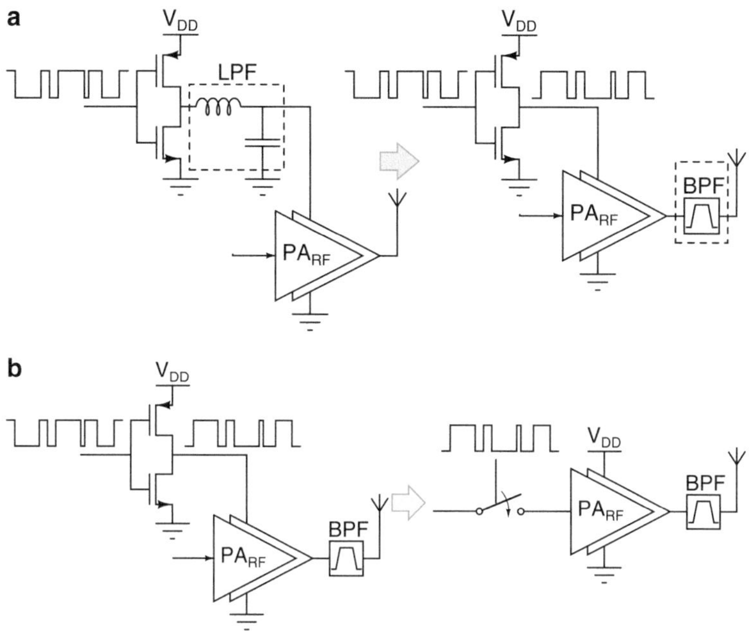

Fig. 11 (**a**) From low-pass to band-pass RF PA architecture and (**b**) from band-pass to switched carrier RF PA architecture

5.5 *Digital Polar Modulation and Burst Mode Operation*

As mentioned before, when a switching envelope modulator is used, the efficiency of the polar modulation scheme can be really high. The major drawback however is the low-loss filter needed between the switching envelope modulator and the RF power amplifier. This filter is required to filter the switching noise of the envelope modulator [31].

Repositioning the low pass filter to the output of the RF PA, would transform the low pass filter into a band pass filter [32], since the switching RF PA behaves like a mixer. This upconversion to a bandpass filter is depicted in Fig. 11a which illustrates this new architecture with the bandpass filter after the RF PA and the load.

The digital signal processing circuit controls the pulses which are delivered to the gate of the switching Class D or Class S PA. So this amplifier is constantly turned on and off at low frequency. The "bursty" signal now being produced at the output of the RF PA, is the result of the multiplication of the baseband low frequent envelope signal generated by the PWM or $\Sigma\Delta$ modulator with the high frequency phase modulated RF signal. Due to settling and start-up effects as a result of turning the amplifier on and off, this approach is not completely efficient. Introducing a switch at the gate to turn on and off the signal at the gate, could be a more favorable solution, which is clarified in Fig. 11b. This switch could then be incorporated in the modulator design. This PA topology is also known as a burst mode power amplifier (BMPA).

Let's have a closer look at some aspects of this bandpass filter at the output node of the RF PA. Such a filter is typically introduced to reduce the transmitted noise receive band. Consequently, this filter requires low loss components and could cause direct loss in performance of the PA itself. The low frequency modulator generates also higher harmonic components at multiples of the desired low envelope frequency. Obviously, the purpose of connecting a narrowband filter to the drain node of a PA, is broadcasting only in-band power. Assume an ideal brick-wall-filter is used, in this way the out-of-band impedance would be infinite. In this case, no input power contributes to out-of-band power.

In practice, the impedance seen by the power amplifier at out-of-band frequencies is not infinite, but needs to be as high as possible not to dissipate any out-of-band power. This out-of-band energy possibly violates spectral mask requirements of the wireless standard. Dissipating the out-of-band energy, reduces the efficiency and a similar efficiency-versus-output power curve as the square root behaviour in class B condition would be the result [33–36]. These extra requirements to improve efficiency, out-of-band impedance and the spectral mask requirements, result in an increase of the cost of the bandpass filter at the output.

To relax the filter requirements, the envelope frequency of the bursts is selected high enough such that the harmonic spectral components are pushed further away from the desired band. Also coding can be helpful to reduce the generated out-of-band power. As proof of concept, this has been implemented in [31, 33, 35–38].

Unlike polar, ET and EER, the burst mode amplifier doesn't suffer any more from alignment of the envelope path and the RF path to accurately create a variable envelope signal. A more detailed study on the trade-off between efficiency and the filter requirements in the burst mode power amplifier is presented in [39].

6 Conclusions

As we all know, CMOS offers the advantage of dealing more easily with complex systems. This paper focused on some recent design techniques to deal with the efficiency-linearity trade-off prominent in CMOS Power Amplifier design. In addition, this paper showed how power combining mitigates the risk to generate the required output power efficiently despite the low breakdown voltage in nanoscale CMOS processes but simultaneously, provide the flexibility that helps improving the flexibility at power back-off. As presented in this paper, a fully-integrated Doherty RF PA with on-chip power combiner is one option. But several techniques are described in this paper how to achieve efficiency enhancement at power back-off. Each of the techniques have their advantages and drawbacks or practical pitfalls. But due to a powerful processing platform with its unparalleled integration level and extensive digital processing capability offered by the CMOS technology for realizing a fully integrated radio System-on-Chip (SoC), a flexible reconfigurable RF PA comprising several enhancement techniques will certainly take the lead in future power amplifier design.

References

1. Raab F, Asbeck P, Cripps S, Kenington P, Popovic Z, Pothecary N, Sevic J, Sokal N (2002) Power amplifiers and transmitters for RF and microwave. IEEE Trans Microwave Theory Tech 50(3):814–826
2. Aoki I, Kee S, Rutledge D, Hajimiri A (2002) Distributed active transformer—a new power-combining and impedance-transformation technique. IEEE Trans Microwave Theory Tech 50(1):316–331
3. François B, Reynaert P (2011) A fully integrated CMOS power amplifier for LTE-applications using clover shaped DAT. In: 2011 proceedings of the ESSCIRC, pp 303–306
4. Cripps S (1999) RF power ampliers for wireless communication. Artech House, Norwood
5. Kenington P (2000) High-linearity RF amplier design. Artech House, Norwood
6. François B, Reynaert P (2012) A fully integrated Watt-level linear 900 MHz CMOS RF power amplifier for LTE-applications. IEEE Trans Microwave Theory Tech 60(6):1878–1885
7. Kaymaksut E, François B, Reynaert P (2013) Analysis and optimization of transformer-based power combining for back-off efficiency enhancement. IEEE Trans Circuits Syst Regul Pap 60(4):825–835
8. Reynaert P, François B, Kaymaksüt E (2009) CMOS RF PA design: using complexity to solve the linearity and efficiency trade-off. In: IEEE international symposium on radio frequency integration technology (RFIT), pp 207–212
9. Doherty WH (1936) A new high efficiency power amplifier for modulated waves. Proc Inst Radio Eng 24(9):1163–1182
10. Cripps S (2004) RF power ampliers for wireless communication, 2nd edn. Artech House, Norwood
11. Kaymaksut E, Reynaert P (2014) 3.4 A dual-mode transformer-based Doherty LTE power amplifier in 40 nm CMOS. In: 2014 IEEE international solid-state circuits conference digest of technical papers (ISSCC), pp 64–65
12. Kaymaksut E, Reynaert P (2012) Transformer-based uneven doherty power amplifier in 90 nm cmos for wlan applications. IEEE J Solid-State Circuits 47(7):1659–1671
13. Choi J, Kang D, Kim D, Kim B (2009) Optimized envelope tracking operation of Doherty power amplifier for high efficiency over an extended dynamic range. IEEE Trans Microwave Theory Tech 57(6):1508–1515
14. Kaymaksut E, Zhao D, Reynaert P (2014) E-band transformer-based Doherty power amplifier in 40 nm cmos. In: 2014 IEEE radio frequency integrated circuits symposium, pp 167–170
15. Cox D (1974) Linear amplification with nonlinear components. IEEE Trans Commun 22(12):1942–1945
16. Chireix H (1935) High power outphasing modulation. Proc Inst Radio Eng 23(11):1370–1392
17. Zhao D, Kulkarni S, Reynaert P (2012) A 60-GHz outphasing transmitter in 40-nm CMOS. IEEE J Solid-State Circuits 47(12):3172–3183
18. Sundstrom L (1995) Automatic adjustment of gain and phase imbalances in linc transmitters. Electron Lett 31(3):155–156
19. Razavi B (2011) RF microelectronics, 2nd edn. Prentice-Hall, Upper Saddle River
20. Fritzin J, Jung Y, Landin P, Handel P, Enqvist M, Alvandpour A (2011) Phase predistortion of a class-D outphasing RF amplifier in 90 nm CMOS. IEEE Trans Circuits Syst Express Briefs 58(10):642–646

21. Casadevall F, Olmos J (1990) On the behavior of the LINC transmitter. In: 1990 IEEE 40th vehicular technology conference, pp 29–34
22. Zhang X, Larson EL, Asbeck P (2003) Design of linear RF outphasing power amplifiers, no. 1-58053-374-4. Artech House, Norwood
23. Yoo S-M, Jann B, Degani O, Rudell J, Sadhwani R, Walling J, Allstot D (2012) A class-G dual-supply switched-capacitor power amplifier in 65 nm CMOS. In: 2012 IEEE radio frequency integrated circuits symposium (RFIC), pp 233–236
24. Kang D, Park B, Kim D, Kim J, Cho Y, Kim B (2013) Envelope-tracking CMOS power amplifier module for LTE applications. IEEE Trans Microwave Theory Tech 61(10): 3763–3773
25. Wang F, Ojo A, Kimball D, Asbeck P, Larson L (2004) Envelope tracking power amplifier with pre-distortion linearization for WLAN 802.11g. In: 2004 IEEE MTT-S international microwave symposium digest, vol. 3, pp 1543–1546
26. Li Y, Lopez J, Wu P-H, Hu W, Wu R, Lie D (2011) A SiGe envelope-tracking power amplifier with an integrated CMOS envelope modulator for mobile WiMAX/3GPP LTE transmitters. IEEE Trans Microwave Theory Tech 59(10):2525–2536
27. Kahn L (1952) Single-sideband transmission by envelope elimination and restoration. Proc IRE 40(7):803–806
28. Reynaert P, François B, Kaymaksüt E (2009) Challenges for mobile terminal CMOS power amplifiers. In: Proceedings of advances in analog circuit design (AACD), pp 295–303
29. Reynaert P, Steyaert M (2005) A 1.75-GHz polar modulated CMOS RF power amplifier for GSM-EDGE. IEEE J Solid-State Circuits 40(12):2598–2608
30. Cantrell W, Davis W (2003) Amplitude modulator utilizing a high-Q Class-E DC-DC converter. In: 2003 IEEE MTT-S international microwave symposium digest, vol 3, pp 1721–1724
31. Stauth JT, Sanders SR (2008) A 2.4 GHz, 20 dBm class-D PA with single-bit digital polar modulation in 90 nm CMOS. In: IEEE custom integrated circuits conference (CICC), pp 737–740
32. Reynaert P, Steyaert M (2006) RF power amplifiers for mobile communications, no. 978-1-4020-5116-6. Springer, Dordrecht
33. Laflere W, Steyaert M, Craninckx J (2008) A polar modulator using self-oscillating amplifiers and an injection-locked upconversion mixer. IEEE J Solid-State Circuits 43(2):460–467
34. Reynaert P, Laflere W, Steyaert MSJ, Craninckx J (2008) Self-oscillating RF amplifiers. In: 2008 Gigahertz symposium, Goeteberg, p 6
35. François B, Reynaert P, Wiesbauer A, Singerl P (2010) Analysis of burst-mode RF PA with direct filter connection. In: 2010 European microwave conference (EuMC), pp 974–977
36. François B, Kaymaksut E, Reynaert P (2011) Burst mode operation as an efficiency enhancement technique for RF power amplifiers. In: 2011 XXXth URSI general assembly and scientific symposium, pp 1–4
37. Nuyts P, Singerl P, Dielacher F, Reynaert P, Dehaene W (2012) A fully digital delay line based GHz range multimode transmitter front-end in 65-nm CMOS. IEEE J Solid-State Circuits 47(7):1681–1692
38. Nuyts P, François B, Dehaene W, Reynaert P (2012) A CMOS burst-mode transmitter with Watt-Level RF PA and flexible fully digital front-end. IEEE Trans Circuits Syst Express Briefs 59(10):613–617
39. François B, Singerl P, Wiesbauer A, Reynaert P (2011) Efficiency and linearity analysis of a burst mode RF PA with direct filter connection. Int J Microwave Wireless Technol 3(3): 329–338

Energy-Efficient Phase-Domain RF Receivers for Internet-of-Things (IOT) Applications

Yao-Hong Liu

Abstract This paper presents an ultra-low power 2.4 GHz FSK/PSK RX for wireless personal/body area networks. An energy-efficient single-channel phase-domain receivers based on a sliding-IF phase-to-digital conversion (SIF-PDC) loop is presented. It equivalently transforms the signal processing from analog I/Q domain to the digital phase domain, which save almost 40 % power consumption. A phase rotator is employed in the phase-tracking loop to decouple the carrier generation and frequency demodulation, which guarantees the frequency stability and enables wideband operation. The analog multiplier and the single-bit quantization implementation improves the interference rejection. Fabricated in a 90 nm CMOS technology, the presented RX consumes 2.4 mW at 2 Mbps data rate, i.e., 1.2 nJ/b efficiency, and achieves a sensitivity of −92 dBm.

1 Introduction

This paper presents an ultra-low power (ULP) 2.4 GHz RX for short-range wireless personal and body area networks. In such applications, the RF transceiver typically consumes more than 90 % of the total battery energy in a remote sensor node. To achieve a long operation lifetime, improving the energy efficiency (i.e., power consumption/data rate) of the radios to below 1 nJ/bit is a primary design goal. Although energy-detection or super-regenerative ASK RXs [1] are very efficient, they are vulnerable to interference, which lead to a poor quality of the wireless link in a crowded 2.4 GHz ISM band. Figure 1 shows the Bluetooth Low Energy 2.4 GHz channel plan as an example. Up to 40 Bluetooth Low Energy radio devices can operate at the same time, so the RXs operating at this band should have sufficient selectivity and interference rejection.

Y.-H. Liu (✉)
Holst Centre/IMEC, Eindhoven, The Netherlands
e-mail: Yao-hong.Liu@imec-nl.nl

K.A.A. Makinwa et al. (eds.), *Efficient Sensor Interfaces, Advanced Amplifiers and Low Power RF Systems*, DOI 10.1007/978-3-319-21185-5_17

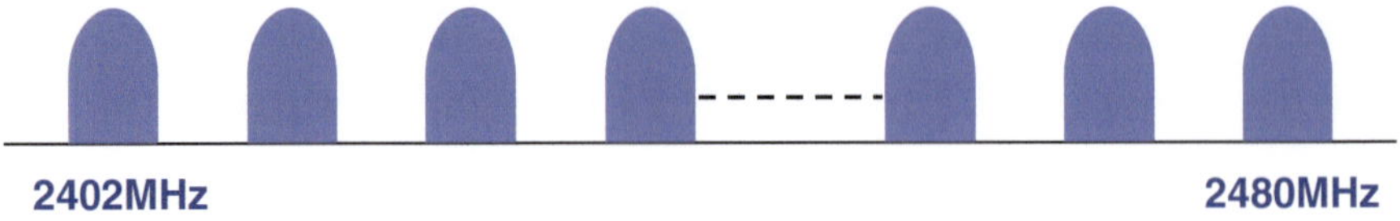

Fig. 1 Channel plan of bluetooth low energy standard

2 FSK RXs with Frequency Demodulators

FSK/PSK-type modulations are popular in the target applications because of their power-efficient hardware and higher immunity to interference. They are also widely adopted in many short-range wireless standards (e.g., IEEE802.15.4, Bluetooth Smart, etc.).

Thanks to the constant-envelope nature of FSK-type modulations (e.g., GFSK, MSK, HS-OQPSK), they only modulate data on the carrier frequency or phase, so the TX hardware can be simplified, while the efficiency can be enhanced by driving the circuits into a saturated mode (e.g., PLL-based FSK TXs [2, 3]). Similarly in the RX counterpart, instead of demodulating the frequency-modulated signal in the complex I/Q domain, it can be demodulated in the phase domain. The FSK RXs can be categorized into several categories: Quadrature-correlation based, zero-crossing detector based, phase ADC based and phase-tracking RXs.

2.1 Correlator-Based FSK RXs

Quadrature correlation is one of the most popular FSK demodulators [4] because of its low complexity. They measure the frequency deviation by multiplying with their own delayed or differentiated version of the signals, as shown in Fig. 2. The quadrature correlation typically uses a low-IF RX architecture, but it can also be implemented in zero-IF RXs [5]. The quadrature correlator can be implemented fully in digital domain after digitization by two medium to high resolution I/Q ADCs [6].

2.2 Zero-Crossing Detector (ZCD) Based FSK RXs

Another popular frequency demodulator is based on zero-crossing detection [7, 8], as shown in Fig. 3. These RXs typically employ the low-IF architecture, since zero-crossing detection requires a non-zero IF.

In the low-IF implementation, the image signal is removed by the complex image reject band-pass filter (BPF). After the filtering, a hard limiter boost the signal to rail-to-rail before outputting to the zero-crossing detector. The concept of these

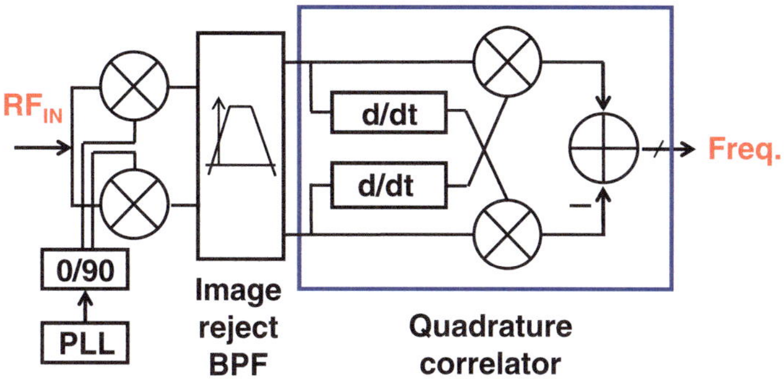

Fig. 2 FSK RX with a quadrature correlator

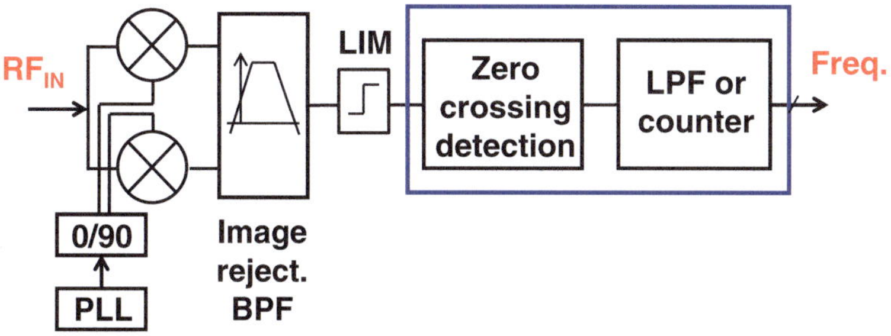

Fig. 3 FSK RX with a zero-crossing detector

demodulators is to produce a pulse once the zero crossing is detected. Then the demodulators use either a low-pass filter or a counter to extract the information of the frequency deviation.

2.3 Phase-ADC Based FSK RXs

As mentioned in the previous section, the digital demodulator can be implemented in digital domain, which can gain many benefits of having powerful digital algorithms to improve the robustness of the demodulation, e.g., frequency offset removal. However, it requires a medium to high resolution ADC. As shown in Fig. 4, a low-resolution phase ADC can replace I and Q amplitude domain ADCs, and it directly measures and quantizes the signal phase based on the relation between I and Q signals [9, 10]. This reduces the power consumption of quantization and simplifies the digital baseband. The frequency deviation is measured by differentiating the digitized phase signal. The FSK RXs with phase ADC typically employ zero-IF architecture.

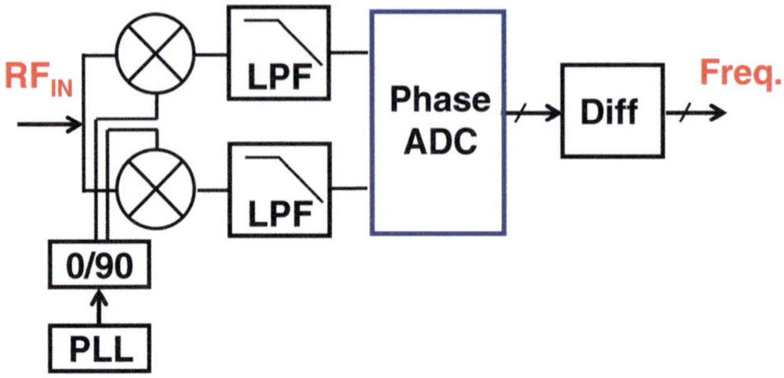

Fig. 4 FSK RX with a phase ADC

2.4 Phase-Tracking FSK RXs

The FSK RXs mentioned above all require a power-hungry high-frequency quadrature-phase LO generation and "2-dimensional" down-conversion and filtering circuits. In the single-channel phase-tracking RX of [11], a free-running VCO is part of the RX carrier recovery and frequency-demodulation loop, as shown in Fig. 5a. The RX power consumption can be significantly reduced, since it only needs one channel and does not require high-frequency quadrature LO generation. However, if it is used as a front-end to directly receive RF modulated signals, the center frequency of a free-running VCO can be easily "pulled away" by an interferer that is stronger than the desired signal during the carrier recovery. This then leads to a poor RX selectivity. Moreover, the RX sensitivity is also degraded due to the slow frequency drift of a free-running VCO which translates to low-frequency noise at the RX output. Finally, an ADC that digitizes the VCO tuning voltage contributes quantization noise and limits the accuracy of the frequency digitization.

A $\Delta\Sigma$-frequency-to-digital converter ($\Delta\Sigma$-FDC) as a frequency demodulator [12], as shown in Fig. 5b, uses a counter as a precise LO phase controller to improve carrier frequency stability and a quantization noise shaping technique to enhance the resolution of the frequency digitization. The counter, sampled by a clock (f_{CNT}) eight times higher than carrier frequency f_C, provides its "Carry" as a time-domain feedback to the phase detector. This approach has a stable center frequency which avoids the issues of a free-running VCO as mentioned previously, i.e., slow frequency drift and center frequency pulling by interference. However, the ADC needs to be sampled at the signal carrier frequency f_C, making it inefficient to directly operate at the RF input frequency above few-hundred MHz. In addition, the ADC sampling time (Carry + 1) changes along with the modulated data and leads to a non-uniformly-sampled digital output, which further complicates the hardware of the following digital signal processing.

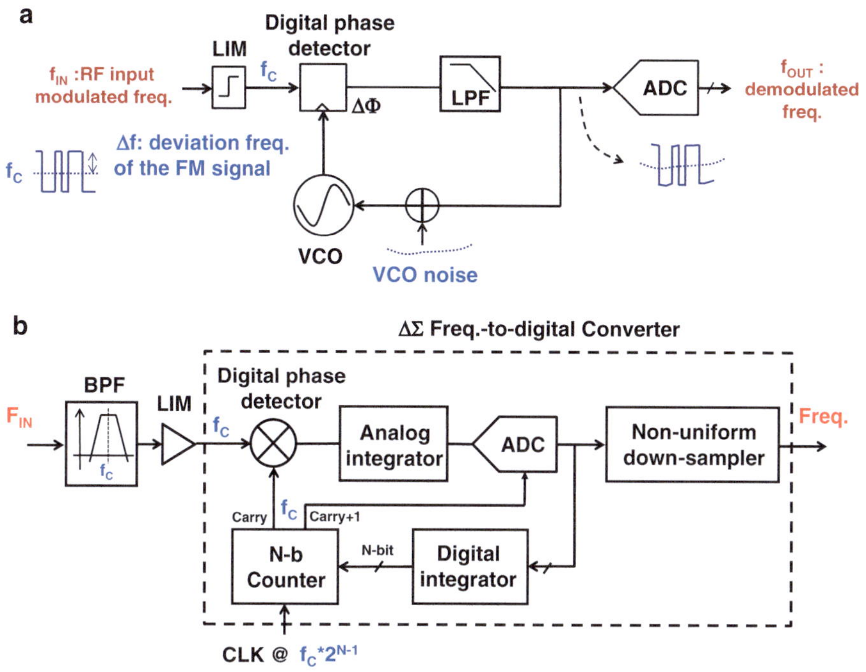

Fig. 5 (**a**) Phase-tracking FSK RX with a free-running VCO and (**b**) phase-tracking FSK RX with a ΔΣ-frequency-to-digital converter

3 Phase-Tracking RXs with Phase Rotator

As shown in Fig. 6a, rather than using a counter as the LO phase controller, a phase rotator [13, 14] (consisted of a phase selector and a phase integrator) is employed in [15] to decouple the input center frequency (f_C) and the sampling frequency (f_{CLK}) of the ADC. Hence, the selection of the oversampling clock f_{CLK} becomes flexible and the non-uniform sampling as in [12] is avoided.

Moreover, instead of implementing the phase integration in the analog domain (first integration then quantization [12, 13]), it is moved to the digital domain (first quantization then accumulation) to avoid an issue of limited phase integration range, which is critical when demodulating FSK/PSK signals with random data. For instance, when an FSK RX with an analog phase integrator receives several consecutive "1"s, the integrator output will increase continuously and eventually result in clipping due to the saturation point (e.g., supply voltage V_{DD}) of the analog integrator or the ADC. On the other hand, the range of digital phase integration is not limited because its output codes (i.e., 0 to $2^N - 1$) can be designed to match one cycle of the LO phase range (i.e., $-\pi$ to π). Therefore, even a digital integrator/accumulator with only few bits of length (e.g., 4 bits in this work) will not reach a saturation point as it digitally wraps back after reaching full scale, as

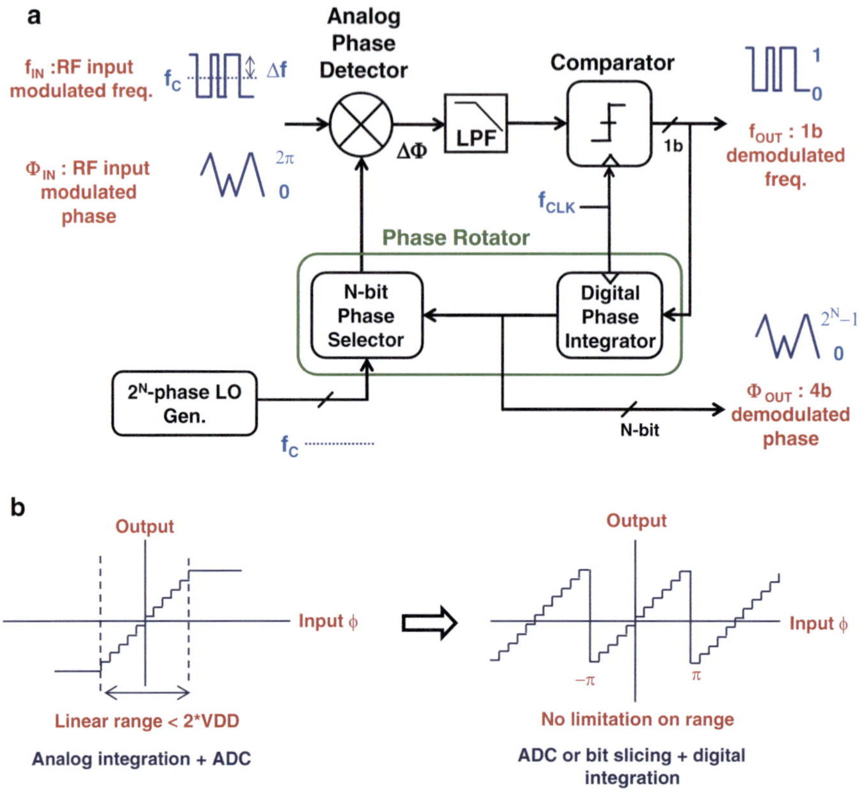

Fig. 6 (**a**) The proposed phase-to-digital conversion loop with a phase rotator and (**b**) the output phase range of the analog phase integration/quantization and digital phase integration

illustrated in Fig. 6b. Furthermore, implementing the phase integration in the digital domain allows the RX to directly readout the digitized phase information and to perform a direct RF phase-to-digital conversion (PDC).

One of the biggest challenges of the phase-tracking RXs is the selectivity. Rather than using a digital phase detector [12], an analog phase detector based on a Gilbert-cell mixer is adopted in the proposed PDC loop because it can accommodate a stronger interference. A limiting amplifier that is required before the digital phase detection has an extremely non-linear characteristic, which can seriously distort the phase information of the desired signal when a strong interference is present. Although a band-pass filter (BPF) can be added before the limiting amplifier to suppress the interference, it needs to have a high quality factor ($Q > 100$) when the input frequency is above few-hundred MHz. On the other hand, the mixer-based analog phase detector has a relatively linear transfer function before the 1-dB compression point, which allows the RX to accommodate the interference to a certain level without requiring a high-Q BPF. The interference is filtered by the LPF after the mixing.

The analog phase detector has an amplitude-dependent phase-to-voltage transfer function, which makes the proposed PDC loop sensitive to amplitude fluctuation, e.g., channel fading or a change of link distance. Since the targeted reception signals (i.e., FSK, GFSK, HS-OQPSK) contain only binary frequency-modulated information, a 1-bit comparator is used in the PDC. The 1-bit quantization not only gets rid of amplitude information, it also acts as a limiter after the LPF, which allows the PDC to benefit from the "capture effect" [13]. The combination of the analog phase detector, the LPF, and the capture effect from the 1-b quantization improve the overall interference robustness of the PDC loop.

Although zero-IF implementations do not require an extra image rejection filter, the multiphase generation directly at high frequency is very power consuming. A low-IF approach reduce the power consumption of multiphase generation, but the 2-dimensional I and Q signal branches are still needed to implement a complex band-pass image rejection filter.

In this work, a single-channel sliding-IF (at 1/9 of input frequency) RX with a phase-to-digital converter (SIF-PDC) is proposed, as illustrated in Fig. 7. The sliding-IF architecture is adopted because it can effectively reduce the power consumption of multi-phase LO generation [2, 3], while the image frequency is around 500 MHz away from the input frequency, which can be easily filtered out by the intrinsic band-pass characteristic of the LNA.

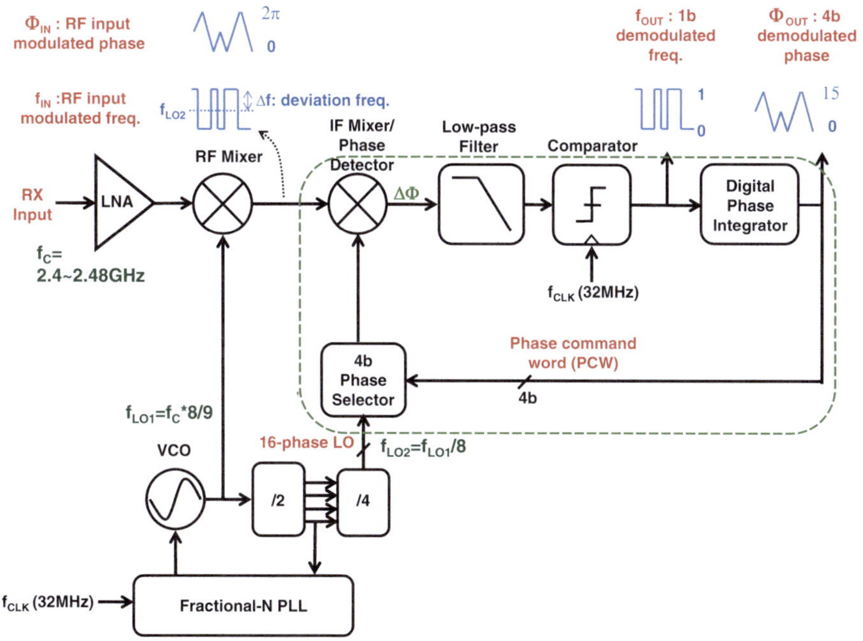

Fig. 7 The simplified block diagram of the proposed RX with SIF-PDC

As shown in Fig. 7, a fractional-N PLL locked to a 32-MHz reference clock is employed to generate a stable carrier, which guarantees the carrier frequency stability. The VCO is locked at 8/9 of the input center frequency (e.g., $f_C = 2.475$ GHz) as the LO of the RF mixer (e.g., $f_{LO1} = 2.2$ GHz). Then it is further divided by 8 (e.g., $f_{LO2} = 275$ MHz)) to provide a 16-phase LO for the IF mixer which also performs the phase detection. The detected phase difference ($\Delta\Phi$) is then filtered by the LPF and digitized by the comparator. The 1-bit comparator output represents the binary frequency demodulated data (f_{OUT}), and a 4-bit digital phase integrator output represents the demodulated phase (Φ_{OUT}). The phase selector picks one of 16 phases according to Φ_{OUT}. With the current frequency plan, the image signal is approximately 550 MHz away, and the input impedance matching along with the LC-tank of the LNA provides approximately 25-dB image rejection.

SIF-PDC forces the selected phase to be synchronized with the input modulated phase (Φ_{IN}), and it is independent from the carrier generation loop, i.e., the fractional-N PLL. Hence, the signal demodulation bandwidth and the PLL bandwidth can be separately optimized. While the PLL bandwidth is preferably low (around 100 kHz) to insure optimum phase noise and a stable carrier frequency, the demodulation bandwidth of the SIF-PDC can be higher to allow the RX to demodulate the targeted FSK/PSK signals up to few Mbps.

4 Simplified Mathematical Model

A phase-domain approximated linear model of the SIF-PDC is presented in Fig. 8. The phase detector implements a phase subtraction and has an amplitude-dependent phase-difference-to-voltage gain of "$A * K_{PD}$," where "A" is the input amplitude of the phase detector [16]. The LPF has a low-pass transfer function of $K_{LPF}(s)$. The comparator functions as a "bit slicer" in the SIF-PDC, detecting only the polarity of the LPF output and boosting it to a rail-to-rail signal regardless of its input amplitude. The comparator has a non-linear signal-dependent gain K_{CMP}, and introduces frequency quantization noise, "f_{QN}."

The phase rotator, which functions as a digitally-controlled oscillator (DCO), has an equivalent continuous-time transfer function of "KDCO/s," and adds phase quantization noise "Φ_{QN}" due to the discrete phases. The phase rotator gain (K_{DCO}) is determined by the accumulation rate of the digital phase integrator. This parameter also represents the full scale of the frequency digitization range of the SIF-PDC.

Similar to a $\Delta\Sigma$ ADC, the SIF-PDC performs a 1st-order $\Delta\Sigma$ 1-bit frequency-to-digital conversion.

Since the comparator gain K_{CMP} is approximately inversely proportional to its input amplitude, it can be approximated as

$$K_{CMP} \approx \frac{1}{A \cdot K_{PD} \cdot K_{LPF}} \tag{1}$$

The LPF is simplified as a constant gain in (1), i.e., K_{LPF}, for the sake of simplicity. Note the comparator is an extremely non-linear component and its gain is strong signal dependent, so it is difficult to model a comparator as a linear component. Therefore, the actual K_{CMP} should be found from extensive numerical simulations to avoid a misleading result predicted by the linear model [16].

The loop bandwidth of the SIF-PDC can be derived from the linear model as

$$f_{PDC} = A \cdot K_{PD} \cdot K_{LPF} \cdot K_{CMP} \cdot K_{DCO}/2\pi \approx K_{DCO}/2\pi \tag{2}$$

Equation (2) indicates the loop bandwidth of the SIF-PDC can be approximated to the accumulation rate of digital phase integrator, K_{DCO}. In addition, the dependency to the input signal amplitude "A" of the phase detector is compensated by the comparator, so the overall SIF-PDC transfer function is now approximately amplitude independent.

The loop bandwidth selection of the SIF-PDC is determined by the requirements of the interference rejection, quantization noise suppression, and the frequency digitization range. In-band quantization noise of the SIF-PDC is suppressed by a high-pass NTF, so ω_{PDC} should not be too low. In the meantime, the low-pass STF helps to suppress out-band unwanted components, so ω_{PDC} is preferably low. Moreover, the loop bandwidth is approximately equal to the full scale of the frequency digitization range, so it has to be at least equal or larger than the deviation frequency (Δf) of the targeted frequency-modulated signals.

A LPF with a cut-off frequency of "f_{LPF}" can be employed in the SIF-PDC to further assist the suppression of out-band interference, so the interference rejection requirement is decoupled from the loop bandwidth selection. Assuming the LPF is a first-order filter to simplify the analysis, and the LPF bandwidth f_{LPF} is significant larger than the loop bandwidth f_{PDC} and the signal bandwidth, so the assumption in (1) and (2) are still valid. The open-loop transfer function ($TF_{OpenLoop}$), frequency responses of the signal transfer function (STF), and the noise transfer function of both frequency quantization noise (NTF_f) are derived as

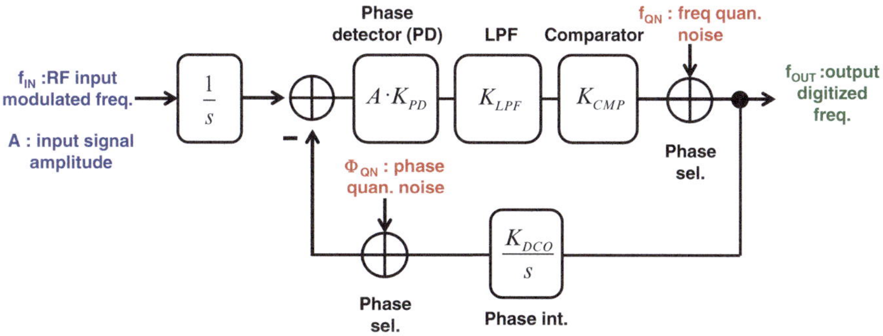

Fig. 8 An equivalent frequency-domain linear model of SIF-PDC

$$TF_{OpenLoop}(s) = \frac{\omega_{PDC}}{s} \cdot \frac{\omega_{LPF}}{s + \omega_{LPF}} \tag{3a}$$

$$STF = \frac{f_{OUT}(s)}{f_{IN}(s)} = \frac{\omega_{LPF}}{s^2 + s \cdot \omega_{LPF} + \omega_{LPF} \cdot \omega_{PDC}} \tag{3b}$$

$$NTF_f = \frac{f_{OUT}(s)}{f_{QN}(s)} = \frac{s^2 + s \cdot \omega_{LPF}}{s^2 + s \cdot \omega_{LPF} + \omega_{LPF} \cdot \omega_{PDC}} \tag{3c}$$

5 Circuit Implementation

A single-ended LNA, also functioning as an image rejection filter, is used because of the low power consumption and to avoid a lossy Balun. One of the main drawbacks of a single-ended LNA is its higher LO leakage to the antenna port, which may lead to a self-mixing DC offset. Fortunately, it is not an issue in the presented sliding-IF architecture, since it will be up-converted to IF frequency by the IF mixer.

While the single-balanced mixers are not favored due to the large LO feed-through, a pseudo-differential double-balanced mixer [2] with one of the inputs being AC-grounded is typically used to interface with the single-ended LNA. The drawback of this conventional solution is the 50 % waste of both power consumption and transconductance (gm), thus resulting in a serious degradation of the noise performance and the efficiency. In order to address this issue, a push-pull mixer structure, as shown in Fig. 9, is proposed to improve both the gm efficiency and noise performance. Moreover, this structure inherently provides a lossless single-ended to differential conversion. The LO feed-through is first-order cancelled by the proposed structure. A common-mode feedback (CMFB) further reduces the LO feed-through by minimizing the bias current mismatch between N- and P-MOS gm (GM_N and GM_P). Simulations shows that 1.9 GHz LO feed-through can be suppressed to 10 mV, which can be easily accommodated by the following stage with a 1 dB compression of 200 mV until it is safely filtered out by the low-pass filters (LPFs).

Figure 10 shows the implementation of the IF mixer and the analog baseband. The IF signal is further down-converted by a passive mixer. A 25 % duty-cycled LO for minimizing noise leakage is not required because of the single-channel implementation. A third-order Butterworth LPF with a programmable bandwidth suppresses adjacent channel interference. The LPF also functioned as a PGA, and provides a maximum gain of 36 dB. Finally, a dynamic comparator sampled at 32 MHz provides a 1b digital output that represents a demodulated frequency. The simulated comparator delay is less than 2 ns, so the output data will be ready before the rising edge of the next clock which triggers the digital phase integrator.

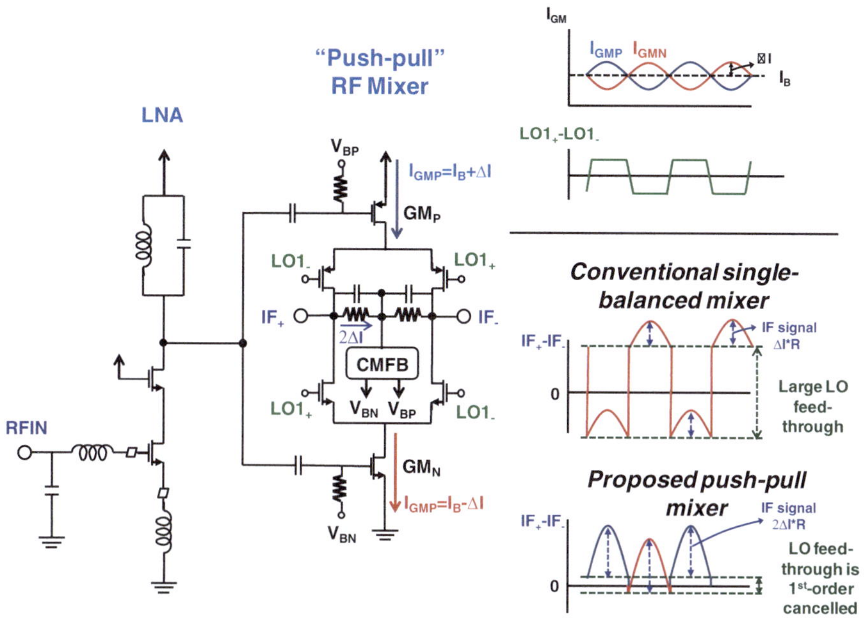

Fig. 9 A single-ended LNA and a push-pull mixer

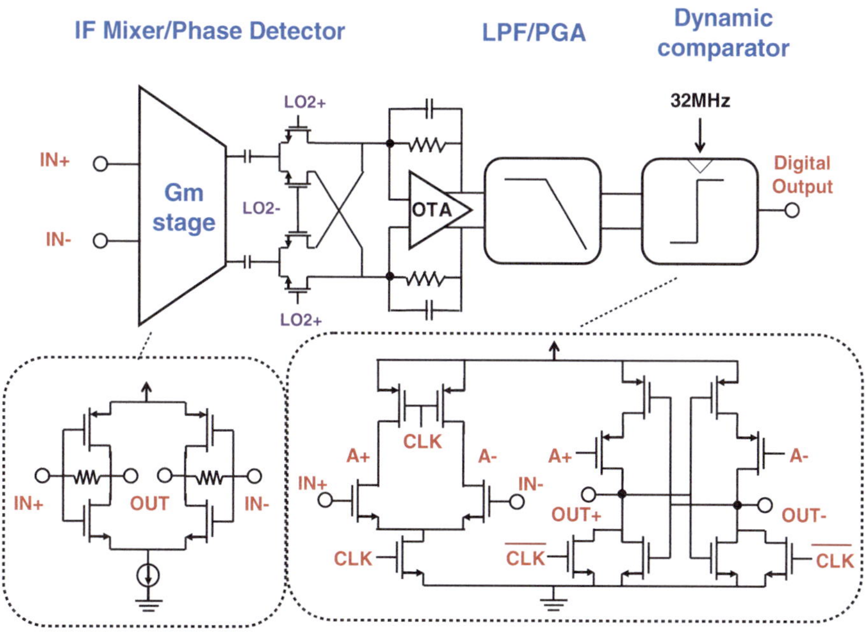

Fig. 10 The IF mixer and the analog baseband

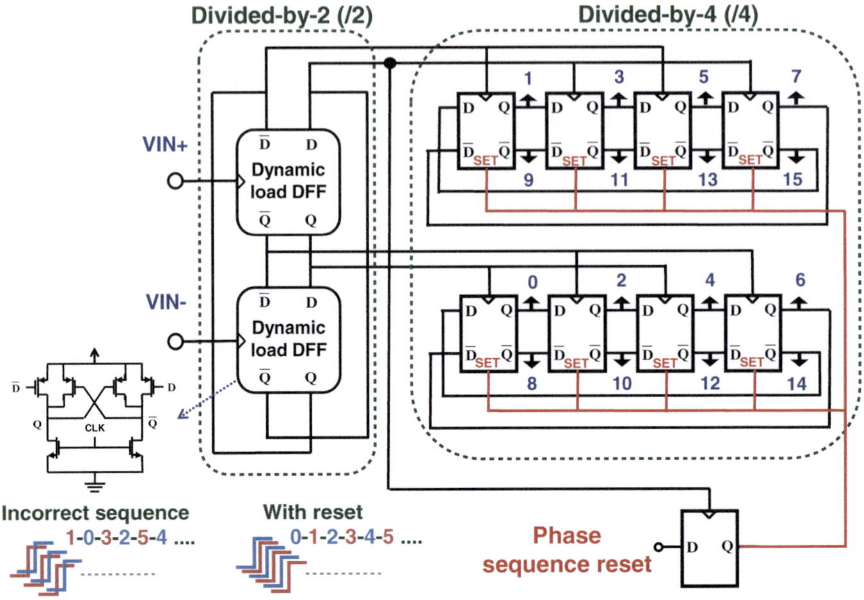

Fig. 11 Simplified schematic of the multi-phase generation

The implementation of the multi-phase LO generator and the phase selector are shown in Fig. 11. The 16 LO phases are generated through a frequency division of 8 that is performed in two cascaded stages (/2 and /4). In comparison to a single stage implementation with eight D-Flip-Flops (DFFs) connected as a ring, this approach relaxes the timing requirements and reduces power consumption by half. The first divide-by-2 stage (/2) is realized with high-speed, low-power, and low-voltage dynamic-load DFFs [17]. Its 4-phase output triggers a parallel divide-by-4 stages (/4) to further produce 16 phases.

A phase sequence reset eliminates the ambiguous state by guaranteeing the even-phase sequence $\{0, 2 \ldots 14\}$ is always triggered before the odd-phase sequence $\{1, 3, \ldots 15\}$ during start up. Finally, the phase selection is implemented by a 4b digital multiplexer.

6 Measurement Results

This RX is implemented in a 90 nm CMOS technology. The total core chip area is around 0.9 mm^2. The measured time-domain waveform is shown in Fig. 12. A 2 Mbps MSK modulation (IEEE802.15.4 equivalent) and a 1 Mbps GFSK modulation (Bluetooth Smart compliant) with a fixed-pattern of "1111-0000-1010" are provided from a 2.4 GHz TX. The proposed SIF-PDC can track the frequency/phase modulations and directly provide demodulated digital outputs.

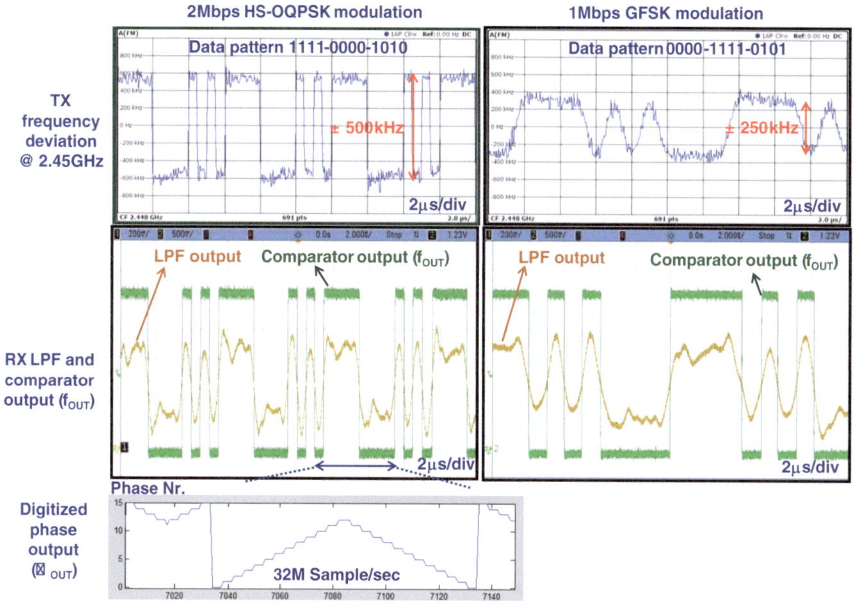

Fig. 12 Measured time-domain demodulated waveform of a 2 Mbps HS-OQPSK and a 1 Mbps GFSK with the proposed SIF-PDC

The measured noise figure is 6 dB. Figure 13 shows the measured raw bit error rate of 2 Mbps HS-OQPSK signal with a pseudo-random bit stream. The proposed RX achieves a sensitivity level of −92 dBm. The sensitivity is defined as the input power corresponding to a BER of 10^{-3}.

The adjacent channel interference rejection (ACR) is measured with an unmodulated CW tone as interference and a modulated desired signal with a level 3 dB higher than the sensitivity. The co-channel interference rejection is mainly determined by the SNR requirement of the demodulation (of FSK or ASK), so these RXs have around the same performance. The measured ACR is −3/12/17 dB at the offset frequency of 2/4/6 MHz, respectively. The second and third ACR are around 8–10 dB lower than conventional I/Q RXs [2] but 20–30 dB better than super-regenerative RXs [1], which offers sufficient selectivity for the target applications.

Figure 14 also shows the power consumption breakdown of the SIF-PDC. The proposed RX architecture leverages the advantages in nanometer technologies, as the power consumption and precision of the multi-phase generation can be further improved by the fast switching speed of the nanoscale CMOS devices. As a result, the continuous technology scaling will benefit the proposed phase-domain RX architecture.

Table 1 summarizes and compares the performance of the presented RX with the state-of-the-art low-power 2.4-GHz RXs. The proposed RX further reduces the power consumption by up to nearly 40 % compared to the previous work [2] at 2 Mbps, thus leading to an excellent energy efficiency of 1.2 nJ/bit, but without

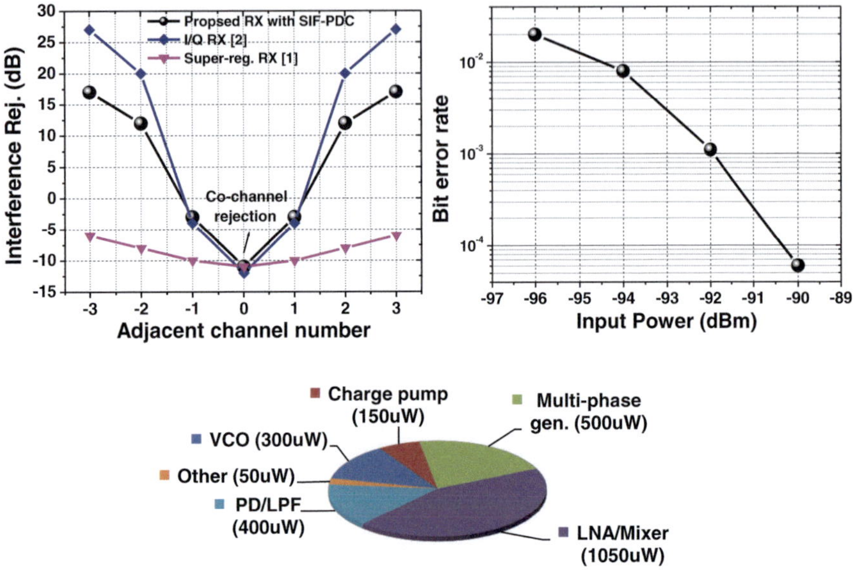

Fig. 13 Measured bit error rate with 2 Mbps HS-OQPSK signal, adjacent channel rejection and power breakdown of the proposed RX

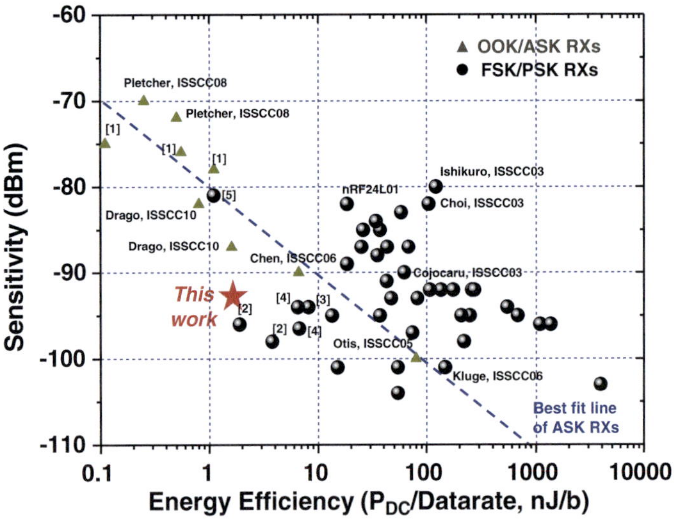

Fig. 14 Benchmark of low-power 2.4-GHz RXs

dramatically degrading its sensitivity as in [9] or selectivity as in [1]. A RX figure of merit (FoM) for low-power applications is defined as the inverse of energy efficiency multiply with RX sensitivity, as shown in (4),

Table 1 Comparison table

	This work	[1] ISSCC'11	[2] ISSCC'13	[7] JSSC'03	[3] ISSCC'12	[9] MTT'13
Data rate and modulation	2-Mbps HS-OQPSK	5-Mbps OOK	2-Mbps HS-OQPSK	1-Mbps GFSK	1-Mbps GFSK	1-Mbps GFSK
Technology	90 nm	90 nm	90 nm	0.35 μm	130 nm	130 nm
Architecture	Sliding-IF	Super reg.	Sliding-IF	Low-IF	Sliding-IF	Zero-IF
Demodulator	Phase tracking	NA	Digital Quad. Corr.	Zero-crossing	NA	Phase ADC
Supply voltage (V)	1	1/1.2	1.2	3	1/1.5	1
Power consumption (mW)	2.4	0.53	3.8	180	6.5	1.1
Sensitivity[a] (dBm)	−92	−75	−96	−82	−94	−81
ACR (second/third)[b] (dB)	12/17	−8/−6	20/27	NA	NA	NA
RX energy eff. (nJ/b)	1.2	0.1	1.9	180	6.5	1.1
FOM[c]	181	175	183	150	175	171

[a]Based on BER of 10^{-3} without error corrections
[b]ETSI EN 300 440-1V1.3.1 (2001–09) page 27
[c]RX FoM = −sensitivity − 10 * log(P_{DC}/data rate)

$$FoM_{ULP\_RX} = -Sensitivity - 10 * \log\left(P_{DC}/Data\_Rate\right) \tag{4}$$

which implies the RX sensitivity can be improved by either reducing data rate or by increasing power consumption. The presented phase-domain RX improves the energy efficiency while maintaining excellent sensitivity, thus resulting in an excellent FoM of 181 dB.

Figure 14 benchmarks the state-of-the-art 2.4 GHz ultra-low power RXs. Similar to the RX FoM defined in Table 1, a general trend shows that the RX sensitivity is inversely proportional to its energy efficiency. There are two RX groups benchmarked in Fig. 9: ASK energy-detection RXs and FSK/PSK RXs. The energy-detection ASK RXs typically possess excellent energy efficiency (typically in the range of 0.1–1 nJ/bit) mainly because they only detect the on/off energy of the desired signals. These ASK RXs typically have inferior sensitivity due to the lack of a precise and stable LO that provides selectivity. Hence, a wider filter bandwidth is required, leading to a poorer noise performance. On the other hand, the FSK/PSK RXs typically hold a better sensitivity at the expense of higher power consumption (i.e., worse energy efficiency), which is mainly consumed by the LO generation (e.g., PLL). The presented phase-domain RX achieves comparable energy efficiency (1.2 nJ/b) as energy detection ASK RXs, while maintains a similar sensitivity level as the FSK/PSK RXs.

7 Conclusions

A new phase-domain RX based on a sliding-IF phase-to-digital converter (SIF-PDC) for directly demodulating and digitization FSK/PSK signals is presented in this paper. In comparison to the conventional Cartesian I/Q RXs, the proposed SIF-PDC saves RX power consumption without compromising its sensitivity and selectivity performance. The proposed SIF-PDC transforms the signal processing from the analog amplitude domain to the digital phase domain, making it favorable for low-voltage operation and technology scaling.

References

1. Vidojkovic M et al (2011) A 2.4 GHz ULP OOK single-chip transceiver for healthcare applications. In: ISSCC digest of technical papers, pp 458–459
2. Liu Y-H et al (2013) A 1.9 nJ/bit 2.4 GHz multistandard (bluetooth low energy/zigbee/IEEE802.15.6) transceiver for personal/body area networks. In: ISSCC digest of technical papers, pp 446–447
3. Wang A et al (2012) A 1 V 5 mA multimode IEEE 802.15.6/bluetooth low-energy WBAN transceiver for biotelemetry applications. In: ISSCC digest of technical papers, pp 300–301
4. Quilan P et al (2004) A multi-mode 0.3–128 kb/s transceiver for the 433/868/915 MHz ISM bands in 0.25 μm CMOS. In: ISSCC digest technical papers, pp 274–528
5. Saitou M et al (1995) Direct conversion receiver for 2- and 4-level FSK signals. In: IEEE universal personal communications, pp 392–396
6. Park J et al (1999) A 5-MHz IF digital fm demodulator. IEEE J Solid State Circuits 34:3–11
7. Sheng W et al (2003) A 3-V 0.35 μm CMOS bluetooth receiver IC. IEEE J Solid State Circuits 38:30–42
8. Wilson JF et al (1991) A single-chip VHF and UHF receiver for radio paging. IEEE J Solid State Circuits 26:1944–1950
9. Masuch J et al (2013) A 1.1-mW-RX -81.4-dBm sensitivity CMOS transceiver for bluetooth low energy. IEEE Trans Microwave Theory Tech 61:1660–1673
10. Samadian S et al (2003) Demodulators for a zero-IF bluetooth receiver. IEEE J Solid State Circuits 38:1393–1396
11. Gustat H et al (2003) Integrated FSK demodulator with very high sensitivity. IEEE J Solid State Circuits 38:357–360
12. Galton I et al (1998) A delta-sigma PLL for 14-b, 50k sample/s frequency-to-digital conversion of a 10 MHz FM signal. IEEE J Solid State Circuits 33:2042–2053
13. Kashmiri S et al (2009) A temperature-to-digital converter based on an optimized electrothermal filter. IEEE J Solid State Circuits 40:2026–2035
14. Liu Y-H et al (2009) A wideband PLL-based G/FSK transmitter in 0.18 μm CMOS. IEEE J Solid State Circuits 44:2452–2462

15. Liu Y-H et al (2014) A 1.2 nJ/bit 2.4 GHz receiver with a sliding-IF phase-to-digital converter for wireless personal/body area networks. IEEE J Solid State Circuits 49:3005–3017
16. Best RE (2003) Phase-locked loops, 2nd edn. McGraw Hill, New York
17. Razavi B et al (1995) Design of high-speed, low-power frequency dividers and phase-locked loops in deep submicron CMOS. IEEE J Solid State Circuits 30:101–109

A Low-Power Versatile CMOS Transceiver for Automotive Applications

Jérémie Chabloz, Andreas Ott, Denis Ruffieux, Peter Teichmann, Frédéric Sacksteder, Nicolas Raemy, Nicola Scolari, Alexandre Vouilloz, Pascal Persechini, and Wouter Couzijn

Abstract In this work, we will present a wake-up controller system complete with UHF and LF transceivers. Typical targeted applications are automotive remote/passive keyless entry key-fob and central unit solutions. Details for the implementation of the UHF transceiver front-end and signal processing are given that demonstrate the desired versatility in frequency, data rate and output power configurations.

1 Introduction

Wireless communications in an automotive environment are used for many different applications. Amongst those, the most pervasive and well known are keyless entry and tire pressure monitoring.

In remote keyless entry (RKE), the wireless transmission is used as a mean to transmit authentication information in order to authorize the unlocking of the vehicle's doors. The system functionality can also be completed with remote keyless ignition (RKI) and *immobilizer* functions, which are used to authorize the ignition of the vehicle's motor and fuel pump operation. Classical RKE requires an action of the user, such as pushing a button on a key-fob, while so-called *passive* keyless entry (PKE) acts automatically when the user gets sufficiently close to the car or when he pulls the door handle.

A direct tire pressure monitoring system (TPMS) measures the pressure and temperature from within the tire and transmits the information to the car electronic

J. Chabloz (✉) • A. Ott • D. Ruffieux • P. Teichmann • F. Sacksteder
Melexis Technologies, Bevaix, Switzerland
e-mail: jch@melexis.com

N. Raemy • N. Scolari • A. Vouilloz • P. Persechini
Swiss Center for Electronics and Microtechnology, Neuchâtel, Switzerland

W. Couzijn
PosEdge, Prinsenbeek, Netherlands

K.A.A. Makinwa et al. (eds.), *Efficient Sensor Interfaces, Advanced Amplifiers and Low Power RF Systems*, DOI 10.1007/978-3-319-21185-5_18

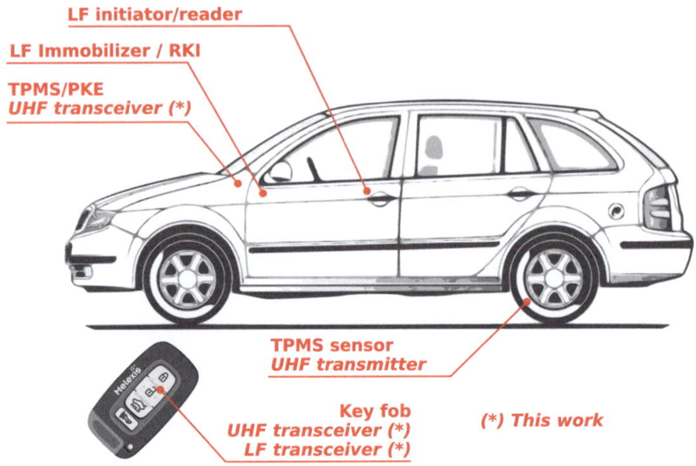

Fig. 1 Overview of wireless devices in a car

control unit (ECU). Future generations of such sensor systems will most probably evolve into measuring and transmitting even more complex data, such as multiple-axes acceleration or information about the estimated tire tread depth or overloading of the vehicle.

Figure 1 gives an overall picture of the different devices required to implement the applications described above and their approximative locations.

Usually located close to the car electronic control unit (ECU), a UHF RF transceiver is shared (in most recent systems) to receive signals both from the TPMS sensor, located in the tires, or receive and transmit authentication information from and to the key fob for keyless entry. The TPMS sensors are fitted with corresponding UHF transmitters and the key fob also comprises a complete transceiver. To fully implement *passive* keyless entry functionality and/or RKI, near-field LF-band communication is typically used with a receiver in the key fob and a reader in the car with antennas close to or even in the door handles.

From this pictorial overview, it is clear that the amount of different wireless communication devices that can be found in or around the car has been increasing significantly with the number of features implemented. Moreover, the required compliance with diverse regulations and standards, more specifically in terms of frequency bands and emitted powers, further complicates the choice of all the required pieces of hardware. With the goal to propose a global and simple solution with a minimized number of components, this work presents a generic UHF & LF transceiver that could be used indifferently in the key fob or in the car as a shared TPMS/RKE receiver unit, as indicated in Fig. 1.

This work will describe the proposed solution both at system and circuit levels. It will especially focus on details of the implementation of the UHF transceiver, from the analog front-end to the digital signal processing. It is further organized as follows; in Sect. 2 we will define concept and the expected high-level specifications;

in Sect. 3 we will present the system architecture of the proposed transceiver and of the main building blocks; in Sect. 4 we are going to focus on specific implementation details of the UHF transceiver. We will conclude with a short summary in Sect. 5.

2 Concept

Essentially, all that does a key fob, is waiting for *something* to happen to wake up and proceed to transmit the required information to the central unit. The same is also true for the central unit which is waiting to receive a relevant message from a key fob or a TPMS sensor. The proposed concept is therefore to develop a *wake-up controller* and UHF/LF communication peripheral for a passive RKE/RKI key-fob controller that can double as a classical RF transceiver for a common RKE/TPMS central unit.

This system is basically a peripheral that can be used in order to wake-up an external microcontroller according to pre-defined conditions. This external controller is therefore responsible to configure the wake-up behaviour of the system prior to getting into sleep mode. From this point on, the wake-up controller has to be completely autonomous and could even be used to switch off the external controller supply for optimum sleep mode current consumption. Figure 2 illustrates this principle; a wake-up event may be triggered by the correct reception of an expected RF or LF communication or by an internal programmable timer. Programmable general-purpose IO (GPIO) pins can be used to wake-up the main controller and optionally control a discrete power switch.

2.1 *High-Level Specifications*

The communication interfaces must have all the required flexibility to provide back-compatibility with communication protocols already in use, as well as support

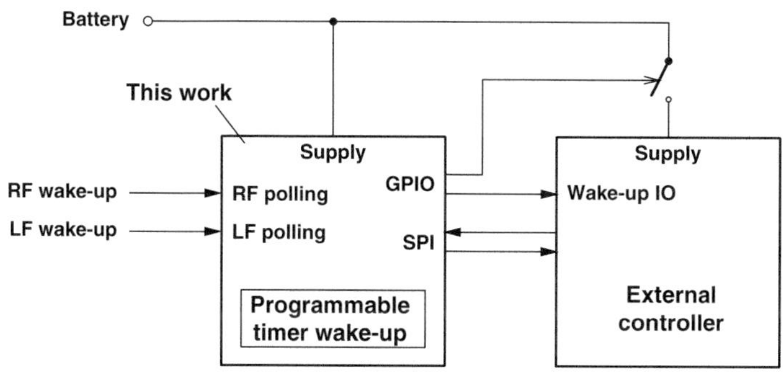

Fig. 2 Wake-up controller principle

Table 1 Specifications summary

Parameter	Min	Typ	Max	Unit	Comments
Supply voltage	2.1	3.0	3.6	V	
Temperature	−40	27	125	° C	
RF frequency	300	315	330	MHz	Band 1
	426	434	447	MHz	Band 2
	860	868	960	MHz	Band 3
RF data rate	1.2		200	kbps	FSK
RF output power	−10		13	dBm	<1dB steps
LF carrier frequency	115	125	140	kHz	
LF data rate	3.6	3.9	4.4	kbps	Manchester

future evolutions. The primary targeted frequency bands are mainly the *industrial, scientific and medical* (ISM) bands, which allow unlicensed operation and are typically used for the targeted applications. The two sub-GHz ISM bands are the 433.92 and 915 MHz bands, depending on region. Other non-ISM bands allowing unlicensed operation for short-range devices (SRD) are the 315 MHz band in the US and Japan, 868 MHz band in Europe as well as the band going from 950 to 956 MHz as defined by Japan's ARIB association. Allowed maximal RF output powers are also band- and country-dependent. Output power is additionally an optimization factor that can be chosen according to the tradeoff between link budget and current consumption. It is therefore required to have a wide configurability for this parameter as well. The transceiver has to provide both frequency-shift keying (FSK) and on-off keying (OOK) modulation schemes.

For the battery-powered key-fob, the supply voltage range shall be aligned onto usual batteries voltage ranges (up to a nominal 3.6 V for a Li-Ion battery). Due to the automotive environment, the targeted temperature range corresponds to the most usual AEC range of −40 to 125° C (grade 1) [1].

Based on previous considerations, the proposed system specifications, with a focus on the UHF and LF interfaces, are summarized in Table 1.

3 System Overview

Figure 3 shows a simplified block diagram of the proposed system, laying out all the building blocks necessary to implement the requirements described above. The sections that follow describe more in detail the function and architecture for each of these building blocks.

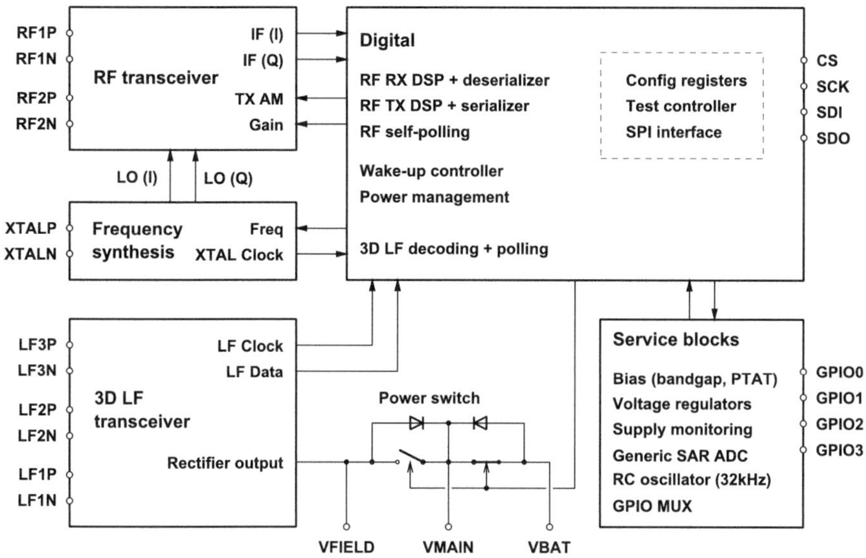

Fig. 3 Simplified block diagram

3.1 *RF Transceiver*

The RF transceiver contains all the required analog blocks for the RX and TX paths of the UHF link. The RF front-end is doubled so that two different sets of RF IO pins can be used, as shown in the diagram by RF1 and RF2 pins. This allows to use antenna and/or frequency diversity to improve the communication link reliability, especially for central unit designs. This could also potentially be used to split RX and TX paths in order to apply external amplifiers on RX and/or TX, depending on applicative constraints. On the RX path, the received signal is downconverted to a intermediate frequency (IF) of 500 kHz and converted into digital signals that are then fed to the demodulator digital signal processing (DSP). Both the gain of the RX path and the transmitted output power of the TX path are controlled from the digital part.

Further details of the RF transceiver analog front-end architecture are shown in Fig. 4. On the TX path, the carrier signals provided by the frequency synthesis are amplified by several preamplifier stages and finally the power amplifier last stage drives the RF output load. The transmitter features a coarse and fine control of the power amplifier output current (see Sect. 4.2) that can be used to set the nominal output power as required in Table 1, and also modulate the RF signal amplitude. It is symbolized in the block diagram by a digital-to-analog converter (DAC). The same IO pads are shared by the receive (RX) path, which amplifies the received signal and translates it down to the IF via a quadrature downconversion mixer. This topology is commonly described as a *low-IF receiver* [2]. With this topology, the rejection

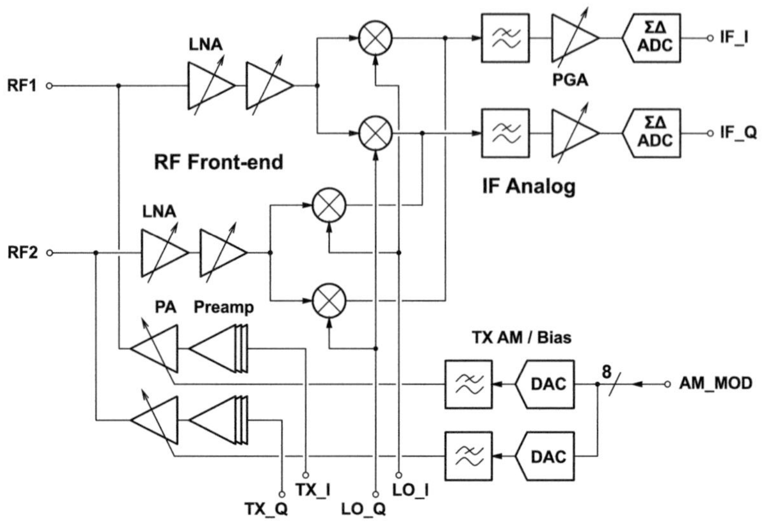

Fig. 4 RF transceiver block diagram

of the *image signal* requires a well defined phase-shift of 90° and good amplitude matching between the two in-phase (I) and in-quadrature (Q) paths [2, 3]. The receiver digital signal processing provides a mean to compensate for this imbalance (Sect. 3.6). The low-noise amplifier (LNA) implementation will be described in detail in Sect. 4.3. As previously explained, the complete RF front-end is doubled. On the RX path, signals of both front-ends are combined at the mixer load, where it is easy to sum up currents together and where the frequency of signals is much lower. Both TX paths are however completely separated.

Once downconverted to the intermediate frequency the signal is filtered and further amplified by a programmable-gain amplifier (PGA), which gain can be controlled between 0 and 64 dB in 1 dB steps. It is then converted in digital by means of a second-order continuous-time loop filter ΣΔ ADC which yields a 32 Mbps single bit stream. The noise transfer function is designed so that it shows a notch at the center of the IF band (500 kHz). For optimal signal amplitude conditions, the ADC provides a SNDR of approximately 60 dB for a IF bandwidth of 100 kHz.

Note also that, even though they have been simply represented by single paths in the block diagram, all RF and IF signals are differential.

3.2 *Frequency Synthesis*

The frequency synthesis is essentially a fractional phase-locked loop (PLL) used to synthesize the signal for the RF transmitter or local oscillator quadrature signals for the receiver. The crystal oscillator uses the resonance of an external quartz as

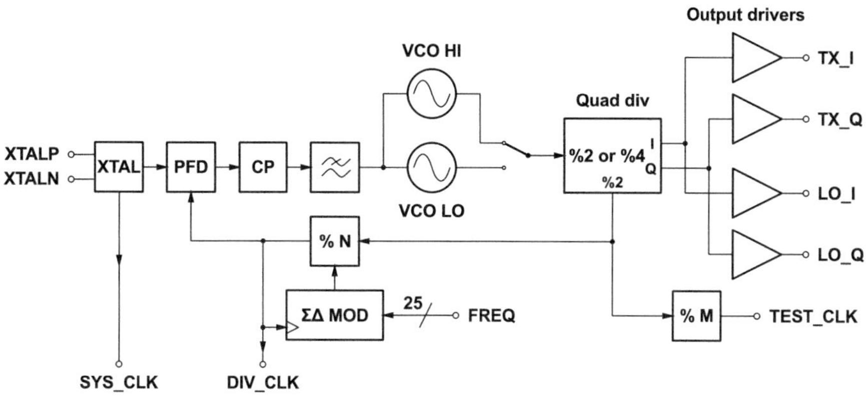

Fig. 5 Frequency synthesis block diagram

a frequency reference, nominally 32 MHz. It also provides a clock to the digital in order to clock mainly the RF DSP and RF serializer/de-serializer part (Sect. 3.6). The frequency, as set by the modulated PLL prescaling ratio, is digitally controlled and this is used for direct frequency modulation as well as for setting the nominal carrier frequency.

Figure 5 shows a more detailed block diagram of the frequency synthesis. In order to cover the required frequencies (see Table 1), the voltage-controlled oscillator (VCO) is actually implemented as two separate oscillators. One for the highest frequencies and one for the lowest ones. The frequency of the oscillator is then further divided by a factor of 2 or 4 before being used. The signal injected into the prescaler is always divided by a factor 2. This has several advantages, since it allows to cover a wider frequency range, reduces the required integrated inductors area, easily provides quadrature signals and helps avoiding VCO pulling effects which would lead to transmitted spectrum degradation.

The prescaler by N is implemented as a programmable divider using a modular approach and dynamic logic based dual-modulus cells [4]. The wide range of programmable division ratio contributes to provide the required flexibility to the presented transceiver.

3.3 3D LF Transceiver

The 3D LF transceiver is responsible for amplifying and demodulating LF signals that can be received by inductive coupling on one or all of the three channels. Three similar channels are implemented in order to be able to use a three-dimensional antenna with a different orientation for each channel in order to minimize reception blind spots. Information is transmitted by using a 100 % amplitude shift keying (ASK) modulation of the LF carrier and with a Manchester encoding applied.

The receiver is made in such way that the strongest of the three signals is automatically the one chosen for data extraction. A logarithmic-scale received-signal strength indicator (RSSI) is implemented that can later be used to determine the relative distance of the key-fob from the LF reader for the immobilizer functionality. Information can be transmitted from the 3D LF transceiver back to the reader by using impedance back-modulation, similar to what is described in [5], even though the application is very different.

3.4 Power Switch and Battery-Backup Mode

Another purpose of the LF transceiver is to provide a *battery backup* feature; when the battery is empty or too weak and when the inductively coupled LF field is able to provide enough energy, it is possible to supply the 3D LF transceiver and digital solely with the harvested energy from the LF field. The main reason for this is that the immobilizer feature has to work even in case of an empty battery. In this use case, the external controller obviously needs to be powered in the same way. In the diagram of Fig. 3, it is the purpose of the *power switch* to commute the main supply line V_{MAIN} between the battery V_{BAT} and rectified field voltage V_{FIELD}. The main supply line can be shared with an external controller.

The power switch is managed at power-on by a dedicated digital finite-state machine; it starts by dynamically selecting the highest voltage and once the power-on reset (POR) is released, the battery voltage is compared with a predefined threshold and if it is found to be above it, the power switch is forced to choose the battery voltage. Otherwise, the power switch is forced to the other position, favoring the rectified LF field voltage.

3.5 Service Blocks

Several so-called *service blocks* are also included in the circuit. These entail all the usual biasing circuits, such as a bandgap voltage reference, PTAT current generator and bias currents distribution. Also comprised in this category are the different low drop-out (LDO) voltage regulators that are providing well defined voltages to the high-performance analog (mostly RF and PLL) and to the high-speed digital parts (mostly the RF DSP). Supply monitoring circuits are used to verify that the battery voltage and all regulated supply voltages are above pre-defined thresholds, constructed as ratios of the bandgap voltage.

A general-purpose 10bits successive-approximation register analog-to-digital converter (SAR ADC) can be used to convert internal or external analog voltages into a digital quantity. Different references for the ADC can be selected, such as ratios of the supply voltages or the internal bandgap voltage.

A 32 kHz *RC* oscillator is used to clock the internal programmable wake-up timer as well as the power-on state machine and LF decoding. It was chosen not to trim the frequency of this oscillator, but rather to rely on self-calibration for critical functions linked to it. For example, the programmable wake-up timer can rely on the crystal clock frequency as a self-calibration reference, while the LF Manchester decoding uses the LF carrier own frequency as a self-calibration reference.

Programmable general-purpose input/output (GPIO) pins can be used as a communication interface between the system and an external controller (as illustrated in Fig. 2), for test or for any applicative functionality. The numerous possibilities provided by the GPIO pins are too many to be described here in detail.

3.6 Digital

The digital part of the system can be split into two categories; everything related to SPI interface, configuration, power management and wake-up state machines as well as interface with the LF transceiver goes into the first category while all the digital signal processing associated with RF reception or transmission is into the second one. These categories distinguish themselves mainly from the fact that the first one always stays supplied (on V_{MAIN} supply) and consumes relatively low dynamic power, while the second one is much more power hungry and is supplied by a regulated lower supply voltage which is interruptible. The split is made in order to reduce both dynamic current consumption while receiving or transmitting and leakage current while "sleeping". Such a method, called *power shut-off* or *power gating*, is one of the easiest to implement and most effective to reduce leakage currents in digital [6] and thus reach the required low standby power consumption for systems using heavy duty cycling such as this one.

The transceiver digital signal processing is illustrated by the diagram of Fig. 6. Without entering into too much detailed descriptions, the processing of the signal on the RX path comprises the following steps:

- Input decimation filter of the ADC I and Q bit streams with a fixed decimation factor.
- Correction of I/Q imbalance in phase and amplitude in order to improve image rejection. The algorithm is designed in such a way that it does not require any specific training signal.
- Downconversion from the intermediate frequency to DC.
- Channel filtering with programmable bandwidth. This part is essential to provide the required data rate flexibility. The bandwidth can be programmed between 9 and 600 kHz.
- A CORDIC algorithm implementation is used to extract the signal complex envelope (amplitude and phase) out of the separate I and Q signals.
- ASK or FSK demodulation and carrier, clock and data recovery.

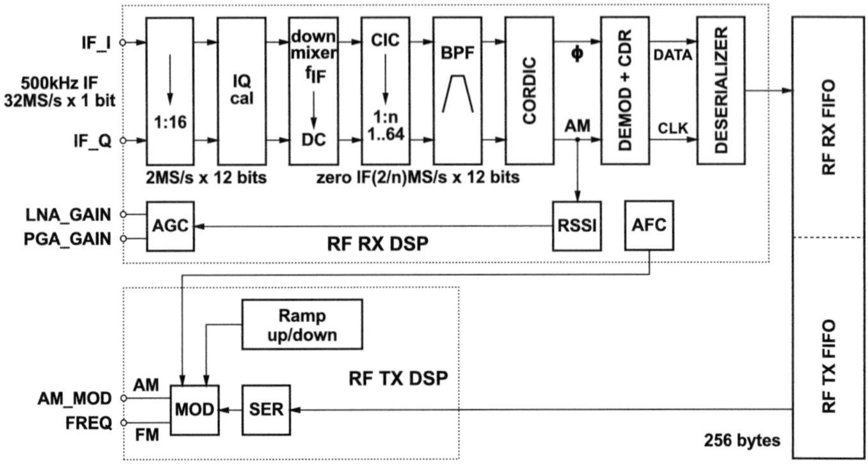

Fig. 6 RF transceiver digital signal processing overview

A received-signal strength indicator (RSSI) can be directly extracted from the complex envelope amplitude, which in turn can be used in the automatic gain control (AGC) mechanism. In this work, the chosen strategy was to implement an *SNR tracking* AGC; the gain of the RF front-end and PGA is adapted to the minimum gain required to guarantee a certain signal-to-noise ratio (SNR), thus always preserving the maximal possible headroom for a possible stronger out-of-band interfering signal. The carrier recovery error value can also be used to implement an automatic frequency compensation (AFC) mechanism.

On the transmitter side, the digital signal processing provides the required frequency values to the PLL $\Sigma\Delta$ modulator in case of an FSK modulation or the signal amplitude modulation values to the transmitter power control in case of an ASK modulation. The same power control mechanism can be used for controlling the ramp-up and ramp-down of the transmitted signal in order to avoid unwanted transient power in the adjacent bands [7].

Serializer and de-serializer are responsible to automatically handle the packets formatting, with many different options possible.

4 Implementation Details

In this section, we will focus on implementation details of the RF transceiver front-end, highlighting the chosen means to implement the required versatility of the transceiver in terms of output power and frequency.

4.1 RX/TX Combination

The first challenge for the implementation of the RF interface is to fulfill the requirement to couple both RX and TX paths onto the same I/O pins. The chosen solution is illustrated by the simplified schematic of Fig. 7.

Switches in the schematic represent the way to commute the front-end between TX and RX operating modes. Note that switches position in the drawing of Fig. 7 correspond to the transmit mode. In this mode, the transistor M0 acts as a common-source transconductance amplification stage and transistor M1 as a cascode transistor, useful both for improving the stability by increasing the isolation and acting as a protection against high peak voltages for M0 [3]. External DC-feed inductor L_0 connected to the power amplifier supply V_{PA} and external AC coupling capacitor C_0 complete the transmit path towards the RF load. Note that this is the simplest expression of a full matching network that would be required to translate the RF load impedance into the wanted impedance

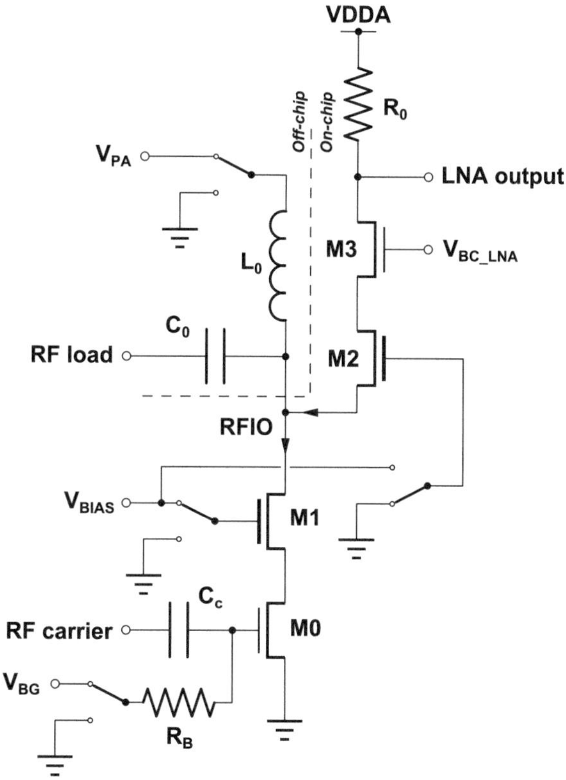

Fig. 7 Principle for shared RX and TX RF interface pins

In receive mode, consider that all switches in Fig. 7 have commuted. In this mode, M2 acts as a common-gate input stage and M3 as a cascode towards resistive load R_0. The DC-feed inductor L_0 is grounded in order to bias M2 with a positive V_{GS} voltage. Note also that the biasing circuitry providing V_{BIAS} to M1 and M2 can be shared.

Transistors M1 and M2 are thick-oxide transistors, allowing them to sustain higher voltages. Furthermore, they are laid out with an extension of the diffusion region on the side connected to the RF I/O, in order to improve their ESD behaviour; The series resistance of the extended diffusion region acts as a ballast, ensuring a good repartition of the ESD current over all fingers of the transistor. It also helps limiting the electrical field magnitude, reducing potential hot carriers issues. The resulting series resistance stays however sufficiently small to be neglected in the rest of the discussion. With this scheme, RX and TX paths are clearly separated from each other, sharing only the I/O net. One limitation of sharing this net, is that the load impedance as seen by the PA output in TX mode has to be the same as the LNA input impedance if impedance matching is a requisite;

$$Z_{\mathrm{load,PA}} = Z_{\mathrm{in,LNA}}. \tag{1}$$

Note however that exact impedance matching is usually not resulting in the best receiver sensitivity and, depending on the wanted maximum power and chosen impedance, it will usually be beneficial to voluntarily bias the LNA so that it yields a smaller input impedance.

4.2 Power Control and Bias

One of the required flexible parameters specified in Table 1 concerns the transmitted output power. Figure 8 illustrates the power control principle that was chosen. It is based on the cascode transistor biasing. The main idea is for the cascode transistor to act as a *current limiter*. If the biasing of the cascode gate is based on a current matching principle, there will be a direct relation between the maximum output current I_{OUT} and the biasing current I_{Bx}. In a first order approximation, it can even be considered that the equivalent conduction angle of the power amplifier output stage will not be affected.

This principle is directly applied to the power amplifier cascode via the biasing circuit shown in Fig. 9. It is actually proposed to extend the concept to not only allow to define the nominal output power, but also use it as a mean to modulate the amplitude of the transmitted carrier. The current of the biasing circuitry is provided by an 8-bits current digital-to-analog (DAC) converter. This current is further multiplied by a factor determined by two sets of switchable transistors. To get the required tuning range on a logarithmic scale, a "piecewise logarithmic" approximation is implemented by the multiplication of coarse and fine tuning factors. The coarse tuning implements a $\Theta_{\mathrm{oct}} = 2^n$ factor with $0 \leq n \leq 5$

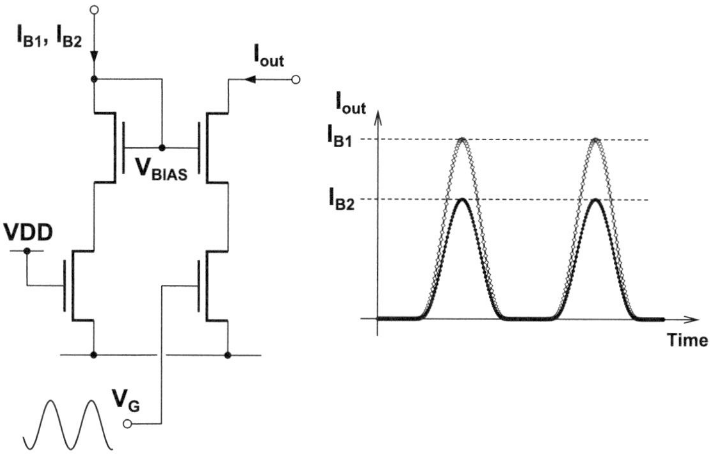

Fig. 8 Power control mechanism principle

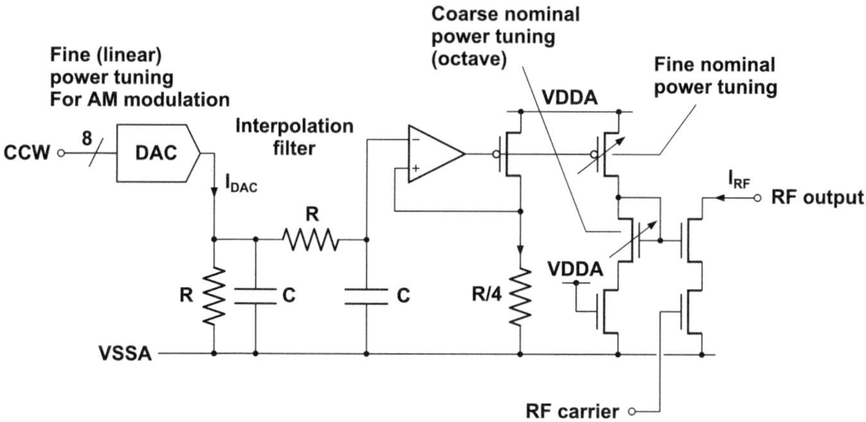

Fig. 9 Power control and amplitude modulation circuit

(octave tuning) while the fine tuning yields a linear factor between 1 and 2 such that $\Theta_{\text{lin}} = 1 + m/8$ with $0 \leq m \leq 7$. If we consider the DAC as a modulation factor Θ_{DAC} between 0 and 1 with 1 corresponding to the full-scale output current, the overall tuning factor Θ can be expressed as

$$\Theta = \Theta_{\text{DAC}} \cdot \Theta_{lin} \cdot \Theta_{oct}. \tag{2}$$

In practice, Θ_{oct} and Θ_{lin} are used to set the nominal output power (i.e. full-scale) while Θ_{DAC} is for fine amplitude modulation and output power ramp-up/down.

A nice advantage of this amplitude modulation scheme, is that it allows to easily implement pulse-shaped on-off keying modulation, yielding a tighter spectrum

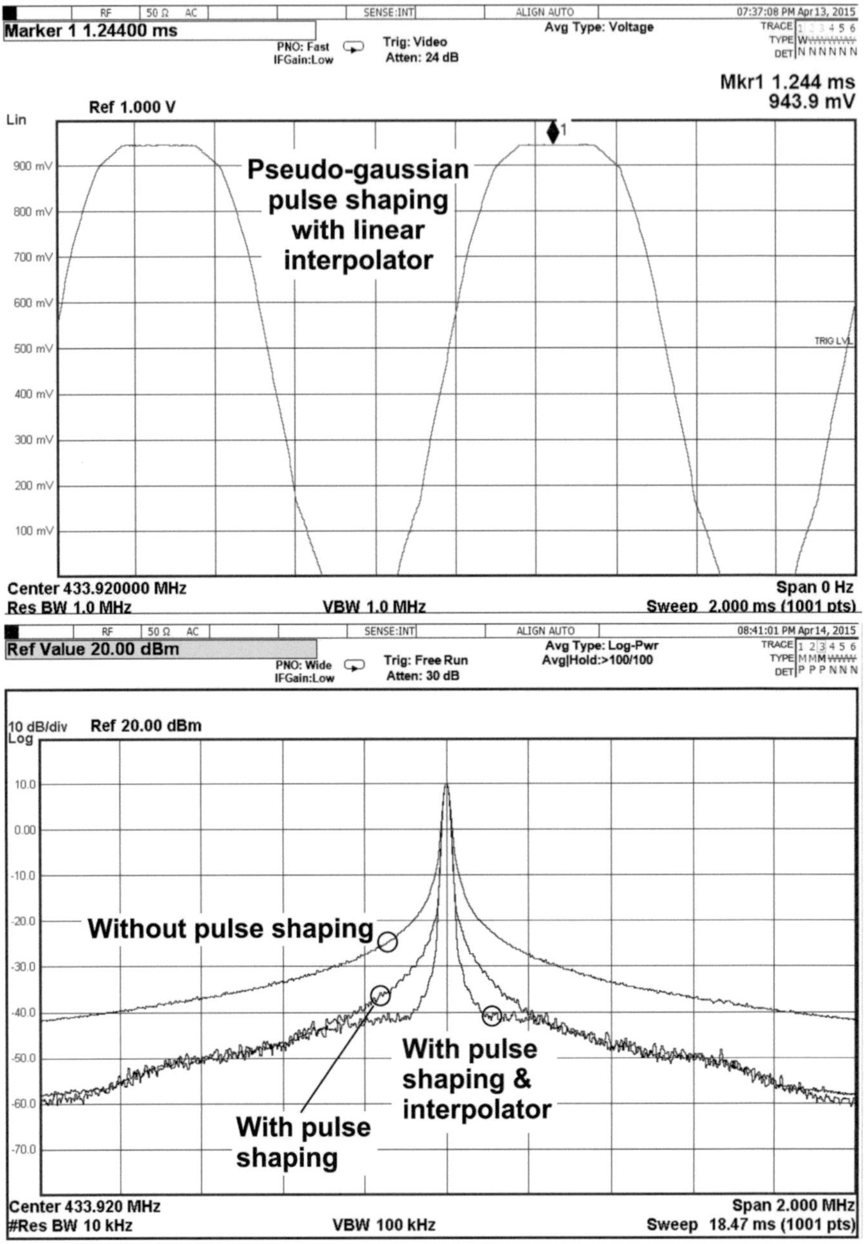

Fig. 10 Modulated OOK spectrum with and without pulse shaping applied

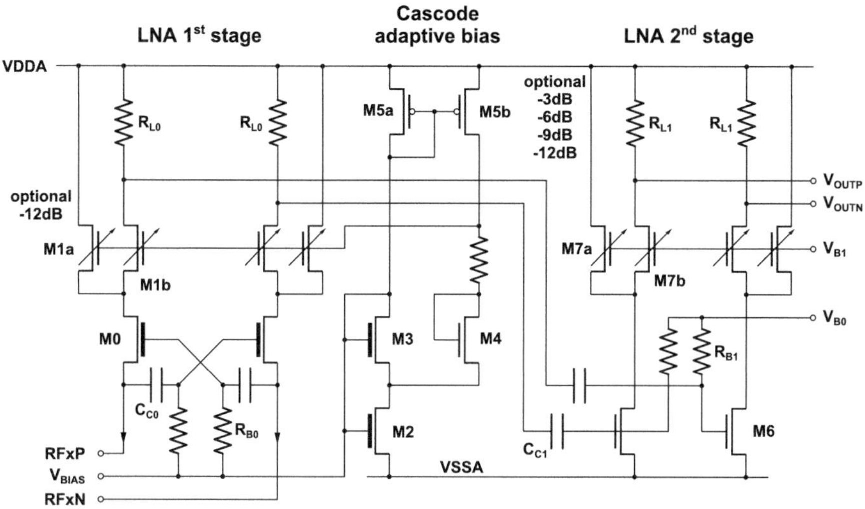

Fig. 11 Details of the low-noise amplifier implementation

and making it easier to pass the regulatory requirements concerning transmitted bandwidth (see [7]). This is illustrated by the measurement results shown in Fig. 10. It can be seen that approximately 10 dB can be gained on the spectrum skirts using an pseudo-gaussian pulse shaping.

4.3 *Low-Noise Amplifier*

A detailed view of the low-noise amplifier (LNA) implementation is given in Fig. 11. As partially shown already in Fig. 7, the input stage corresponds to a common-gate topology. As explained in Sect. 4.1, remember that the RF input nodes are DC-biased to ground via an external inductor when the front-end is used as a receiver. A "transconductance-boosting" topology is realized by using cross-coupled capacitors between gates and sources of the input transistors [8]. Less bias current is required for a given transconductance gain, improving current consumption and noise figure of the input stage. The small-signal input impedance $Z_{\text{in,LNA}}$ is inversely proportional to the input transistors source conductance. If they are biased in weak inversion, it means that the input impedance is inversely proportional to the biasing current

$$Z_{\text{in,LNA}} \propto \frac{1}{I_{\text{LNA}}}. \tag{3}$$

A side-benefit of the highly tunable biasing used for power control and amplitude modulation described in Sect. 4.2 is that, since the gate biasing voltage of the power

Table 2 Measured typical UHF receiver FSK sensitivity $f_{RF} = 434\,\text{MHz}$, BER $= 10^{-3}$

Data rate (kbps)	Δf (kHz)	Channel bandwidth (kHz)	Sensitivity (dBm)	Current (mA)
2.4	4.0	15.0	−120	13.9
5.0	1.3	9.0	−119	13.9
50.0	12.5	75.0	−109	13.9
100.0	25.0	150.0	−103	14.0
200.0	50.0	300.0	−102	14.5

amplifier cascode V_{BIAS} corresponds to the LNA common-gate input transistor gate biasing voltage as well (see Fig. 7), it can be used to finely control the current in the common-gate transistors over a very wide range as well. Taking (3) into account, this makes it easy to realize (1) over a large range of possible impedances.

Since the current in the input transistor is highly tunable, the cascode transistor bias, also shown in Fig. 11, needs to adapt itself. Transistors M0,M2 and M3 individual fingers are matched, M3 being much wider than M2. With the diode-mounted M4 matched to cascode M1, this arrangement allows to bias the cascode transistors M1 to the saturation limit of M0 [9], therefore always providing an optimal operating point.

A second stage follows to provide power gain and improve the noise masking ability of the LNA, since due to the output-to-input impedance ratio, the first stage provides voltage gain but almost no power gain. Both first and second stage feature a selectable attenuation that can be used in the AGC mechanism to improve the receiver's dynamic range when the received signal is strong.

As a measure of the complete receiver performance, Table 2 gives typical measured values for the complete receiver sensitivity with FSK modulation.

4.4 Power Amplifier

An overview of the complete power amplifier (PA) implementation, is given by Fig. 12. A first preamplifier input stage drives a combination of up to four parallel output stage. The reason for splitting the power amplifier output driver into several parallel stages is to be able to slightly improve the global efficiency if the transmitter has to be optimized for a lower output power (e.g. 0 dBm). In this case, it makes sense to try and spare some current in the preamplifier.

Figure 13 shows plots of the measured output power and output stage efficiency for a load impedance configuration optimized for a 13 dBm output power. For power settings > 55, the V_{BIAS} voltage is pulled-up to the analog regulated voltage instead of being determined by the biasing circuit as described by Fig. 8, which is why a significant step can be seen in the output power plots.

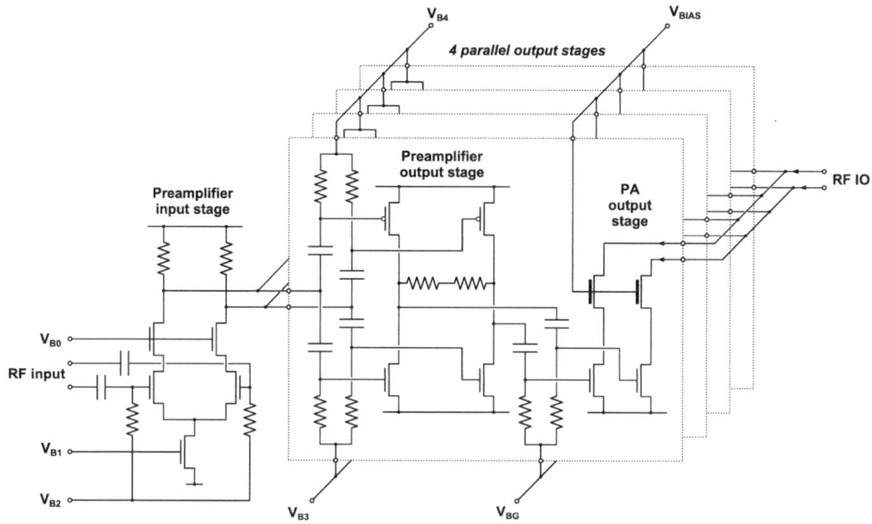

Fig. 12 Details of the power amplifier stages with pre-amplifier

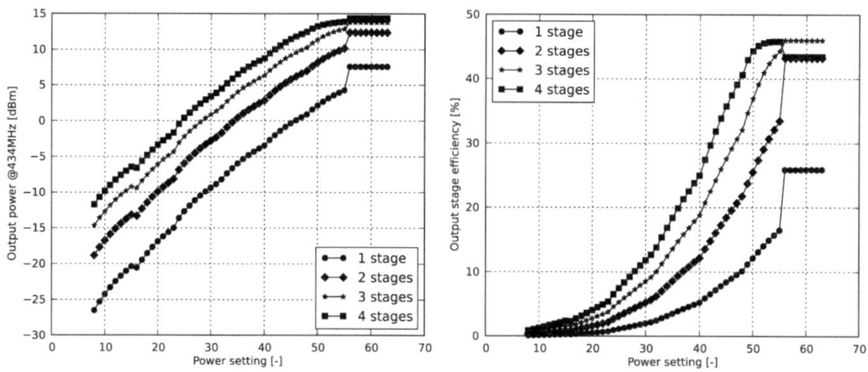

Fig. 13 Measured power amplifier output power and output stage efficiency

4.5 System Integration

Figure 14 shows a photograph of the integrated system. The size of the die is 2.56 mm × 2.65 mm. The technology used is a standard 0.18 m CMOS process without specific dedicated RF options, with 1 poly layer and five metal layers.

It has been assembled within a QFN32 5 × 5 mm package in two different bond-out versions; one with the double RF front-end bonded out but without 3D LF transceiver (aiming the central unit application), and one with only one RF front-end accessible but with the 3D LF transceiver three channels available (aiming the key-fob application).

Fig. 14 Photograph of the integrated system

5 Conclusion

In this work, we have presented a complete low-power wake-up controller system specifically suited for automotive keyless entry/ignition applications. The availability of both UHF and LF communication interfaces in the same device enables the realization of the applicative solution with a lower bill-of-material. We have more specifically shown the requirements, detailed system architecture and circuit implementation for the UHF transceiver front-end and digital signal processing enabling the versatility that is required for the back-compatibility with existing systems and to cope with the very diverse regional regulations.

It is believed that the automotive industry will follow steps with the trend that has already been going on for several years in the mobile phone and entertainment industries to combine most, if not all, of the wireless communication interfaces in single chips solutions. The easiness of reconfigurability and versatility that are seemingly contradictory with optimized low-power solutions, will probably become a more important part of the requirements in order to lower costs in the integration of more advanced features. The present work is seen as a step in this direction.

References

1. AEC-Q100 (2014) Failure mechanism based stress test qualification for integrated circuits. Automotive Electronics Council Std., Rev. H, Sept 2014
2. Crols J, Steyaert MSJ (1998) Low-IF topologies for high-performance analog front ends of fully integrated receivers. IEEE Trans Circuits Syst II 45(3):269–282
3. Razavi B (2012) RF microelectronics, 2nd edn. Prentice Hall, New York
4. Chabloz J, Ruffieux D, Enz C (2008) A low-power programmable dynamic frequency divider. In: 34th European solid-state circuits conference, ESSCIRC 2008, pp 370–373
5. Mandal S, Sarpeshkar R (2008) Power-efficient impedance-modulation wireless data links for biomedical implants. IEEE Trans Biomed Circuits Syst 2(4):301–315
6. A Practical Guide to Low Power Design (2009) Power forward initiative, Cadence design systems. Introduction to low power. Available: http://www.powerforward.org/DesignGuide.aspx (Online)
7. ETSI EN 300 220-1 (2012) Electromagnetic compatibility and radio spectrum matters (ERM); short range devices (SRD); radio equipment to be used in the 25 to 1000 MHz frequency range with power levels ranging up to 500 mW; Part 1: technical characteristics and test methods, ETSI Std., Rev. 2.4.1, May 2012
8. Li X, Shekhar S, Allstot D (2005) G_m-boosted common-gate LNA and differential colpitts VCO/QVCO in 0.18-μm CMOS. IEEE J Solid-State Circuits 40(12):2609–2619
9. Enz C, Vittoz E (1996) CMOS low-power analog circuit design. In: Designing Low power digital systems, emerging technologies, pp 79–133

MIX
Papier aus verantwortungsvollen Quellen
Paper from responsible sources
FSC® C105338

If you have any concerns about our products,
you can contact us on
ProductSafety@springernature.com

In case Publisher is established outside the EU,
the EU authorized representative is:
Springer Nature Customer Service Center GmbH
Europaplatz 3, 69115 Heidelberg, Germany

Printed by Libri Plureos GmbH
in Hamburg, Germany